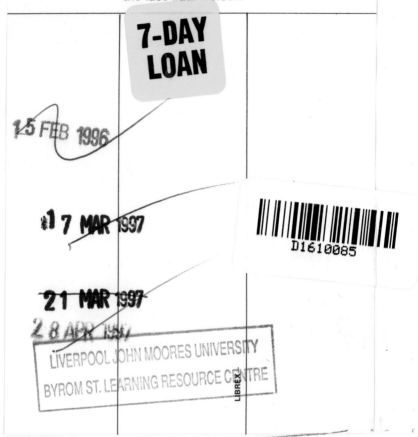

THE RAY SOCIETY
INSTITUTED 1844

This volume is No 162 of the series
LONDON
1994
Registered Charity Number: 208082

THE PHYLOGENETIC SYSTEMATICS OF FREELIVING NEMATODES

Sievert Lorenzen

Translation

With a Foreword by H. M. Platt

© The Ray Society 1994
Registered Charity Number: 208082

ISBN 0 903874 22 9

Sold by The Ray Society

c/o Intercept Ltd., P.O. Box 716, Andover, Hants SP10 1YG.

Printed and bound by Unwin Brothers Ltd., The Gresham Press, Old Woking, Surrey GU22 9LH.
A Member of the Martins Printing Group

FOREWORD

In 1981 Dr Lorenzen published in German his thesis containing a phylogenetic analysis of freeliving nematodes. It proved to be the most comprehensive review of this important group ever published. Its significance lay not simply in its scope but also in its introduction of new characters based on an extensive investigation of actual specimens. Furthermore, Dr Lorenzen used the cladistic methodology and drew attention to, and attempted to rectify, what he saw as weaknesses in cladistic theory.

A brief survey of people in the field indicated that a translation into English would be of great benefit to not only nematologists but those interested in the theory and practice of phylogenetics. Dr Lorenzen has also taken the opportunity to revise certain parts of the original text and make additions to bring the work more up to date. This book, therefore, can be considered a second edition rather than a simple translation.

Perhaps as part of this foreword, for those readers not intimately familiar with nematodes generally and the freeliving ones in particular, I should introduce them.

Nematodes are the most abundant animal group on earth: of every five animals, four are nematodes. They are of singular zoological, economic, medical and environmental importance.

Many people know them as parasites: of ourselves (e.g. elephantiasis), our crops and our domestic animals. Of course they parasitise most other plants and animals as well. However, vast numbers live non-parasitically in terrestrial soils, the sediments of lakes and rivers and in the sea bed.

In marine sediments and terrestrial soils, there are several million nematodes per square metre – which is an order of magnitude larger than all the other animals added together. Not only are they numerically abundant, they are also very diverse in terms of species numbers and life styles. In a small patch of grassland The Natural History Museum has been investigating recently, we have recorded over 100 different species. They also feed in a variety of different ways – some feed on root hairs of plants, some ingest bacteria whilst others are predators.

A survey of nematode assemblages thus provides a reflection of environmental condition or habitat quality. This approach to the analysis of marine environmental quality has been the focus of much of my recent work at The Natural History Museum. The study of nematode assemblages is now accepted as one of the core techniques recommended by the Intergovernmental Oceanographic Committee's Group of Experts on Environmental Pollution. Several commercial companies now routinely include a nematode survey as part of their environmental impact assessment.

Despite the advantages of using nematodes to evaluate environmental quality, there has been a traditional reluctance to devote resources to this work: primarily because they are thought to be difficult to identify. I overcame this in the marine field through two major advances. We provided pictorial guides to the identification of freeliving marine nematodes and also demonstrated that with the latest techniques of data analysis, it is not always necessary to identify nematodes to named species. This has made nematode surveys much more cost-effective.

This work of Dr Lorenzen's can thus be seen as filling in the pure systematic background from which much of my more applied approach has emanated. A general introduction to nematode morphology and methods for their collection, investigation and identification can be found in the three volumes of the Synopses of the British Fauna devoted to freeliving marine nematodes (Platt and Warwick, 1983, 1988 and in preparation).

Reference

Platt H. M. and R. M. Warwick (1983, 1988, in prep.) Freeliving marine nematodes. Part I: British enoplids. Part II: British chromadorids. Part III: British monhysterids. Pictorial key to world genera and notes for the identification of British species. *Synopses of the British fauna (New Series)* Nos 28, 38, in prep. Cambridge University Press; E J Brill, Leiden; in prep.

I would like to take the opportunity here to thank the translator, Ms J Greenwood, for the care she took over translating such a complex work. The accuracy of her original translation made the work of editing it more of a pleasure than a problem.

Howard M Platt
Department of Zoology
The Natural History Museum
London

CONTENTS

Foreword	5
1. Introduction	13
1.1. The current dilemma in nematode systematics	13
1.2. Solution to the current dilemma in nematode systematics and aims of the present study	15
1.3. Acknowledgements	16
2. Revised basis for a theory of phylogenetic systematics	19
3. Material	25
4. Phylogenetic assessment of known and new characters	29
4.1. Cephalic sensilla and dereids	29
4.1.1. Number, arrangement and structure of the labial and cephalic sensilla	29
4.1.2. Postembryonic data on labial and cephalic sensilla	35
4.1.3. Dereids	38
4.1.4. Discussion	39
4.2. Form, position and postembryonic development of the amphids	47
4.2.1. General structure and description of the main types	47
4.2.2. Structure and position of different taxa	51
4.2.3. Postembryonic data	59
4.2.4. Discussion	60
4.3. Metanemes	68
4.3.1. General structure and position	68
4.3.2. Occurrence of metanemes within the Nematoda	70
4.3.3. Postembryonic data	74
4.3.4. Discussion	74
4.4. Gonads	79
4.4.1. Structure of the ovaries	80
4.4.2. Number and orientation of the ovaries	84
4.4.3. Number and orientation of the testes	86
4.4.4. Position of the gonads relative to the intestine	89
4.4.5. Discussion	94
4.5. Supplementary copulatory organs	102
4.5.1. Supplementary copulatory organs in free-living Adenophorea	103

4.5.2.	Supplementary copulatory organs in freeliving Secrenentea	105
4.5.3.	Discussion	106
4.6.	Caudal glands, phasmids and phasmata	107
4.6.1.	Caudal glands	107
4.6.2.	Phasmids and phasmata	110
4.6.3.	Discussion	111
4.7.	The cervical gland	114
4.8.	Excretory ducts of epidermal glands in "Enoplidae" and Phanodermatidae	116
5.	Outline of a phylogenetic system for the freeliving nematodes	119
5.1.	Classification of the freeliving nematodes into subclasses	120
5.1.1.	Assessment of currently conflicting concepts	120
5.1.2.	Criteria for a gross systematic classification of the freeliving nematodes	123
5.2.	Outline of a phylogenetic system for the Chromadoria	126
5.2.1.	Classification of the Chromadoria into orders	127
5.2.2.	Classification of the Chromadorida	132
	CHROMADORIDA	132
	CHROMADORINA	133
	Chromadoroidea	135
	Chromadoridae	136
	Ethmolaimidae/Neotonchidae	140
	Ethmolaimidae	140
	Neotonchidae	141
	Achromadoridae	142
	Cyatholaimidae	143
	Selachinematidae	146
	Desmodoroidea	148
	Desmodoridae	149
	Epsilonematidae	153
	Draconematidae	154
	Microlaimoidea	156
	Microlaimidae	156
	Aponchiidae	158
	Monoposthiidae	159
	DESMOSCOLECINA	160
	Desmoscolecoidea	161
	Desmoscolecidae	163
	Meyliidae	164
	LEPTOLAIMINA	166

Leptolaimidae	167
Chronogasteridae	170
Plectidae	171
Teratocephalidae	173
Peresianidae	175
Haliplectidae	176
Aulolaimidae	177
Rhadinematidae	178
Tarvaiidae	179
Aegialoalaimidae	179
Ceramonematidae	181
Tubolaimoididae	184
Paramicrolaimidae	186
Ohridiidae	187
Bastianiidae	189
Prismatolaimidae	191
Odontolaimidae	192
Rhabdolaimidae	194
5.2.3. Classification of the Monhysterida	196
MONHYSTERIDA	196
Monhysteroidea	197
Monhysteridae	198
Xyalidae	200
Sphaerolaimidae	202
Siphonolaimoidea	203
Siphonolaimidae	203
Linhomoeidae	204
Axonolaimoidea	207
Axonolaimidae	208
Comesomatidae	209
Diplopeltidae	213
Coninckiidae	215
5.3. Outline of the phylogenetic system for the Enoplia	216
5.3.1. Classification of the freeliving Enoplia into orders	216
5.3.2. Classification of the Enoplida	220
ENOPLIDA	220
ENOPLINA	221
ENOPLACEA	222
Enoploidea	223
Enoplidae	224
Thoracostomopsidae	226
Anoplostomatidae	231
Phanodermatidae	232
Anticomidae	234

	Ironoidea	236
	Ironidae	236
	Leptosomatidae	239
	Oxystominidae	241
	ONCHOLAIMACEA	245
	Oncholaimoidea	245
	Oncholaimidae	249
	Enchelidiidae	254
	TRIPYLOIDINA	259
	Tripyloididae	259
	Tobrilidae	261
	Tripylidae	264
	Triodontolaimidae	266
	Rhabdodemaniidae	267
	Pandolaimidae	269
5.3.3.	Classification of the Trefusiida	270
	TREFUSIIDA	270
	Simpliconematidae	271
	Trefusiidae	271
	Onchulidae	272
	Lauratonematidae	274
	Xenellidae	275
5.3.4.	Remarks on the Dorylaimidae	276
	BATHYODONTINA	278
	Bathyodontidae	278
	Cryptonchidae	278
	Monochulidae	278
	Diptherophoridae	279
	Trichodoridae	279
	Isolaimiidae	279
	Alaimidae	279
	MONONCHINA	280
5.4.	Genera *incertae sedis* of the freeliving Adenophorea (except the Dorylaimida)	280
5.5	Remarks on the gross systematic classification of the freeliving Secrenentea	281
	RHABDITIDA	284
	TYLENCHIDA	285
5.6.	Summary of the outline for a phylogenetic system for freeliving nematodes	285
5.7.	Concluding remarks: the attraction of an incomplete system	285
6.	Summary	299
7.	Bibliography	307

FOREWORD

8. Appendix ... 317
   8.1. Summary in tabulated form of the characters of the labial and cephalic sensilla in freeliving nematodes ... 317
   8.2. Summary in tabulated form of the gonad characters in freeliving nematodes ... 324
      8.2.1. Personal results ... 324
      8.2.2. Concluding summary of data ... 346
9. Index ... 360

1. Introduction

1.1 The current dilemma in nematode systematics

Any given theory of systematics has conflicting aims and functions. On the one hand, it is expected to provide a scientific theory on the inter-relationships of groups of organisms; on the other, it is expected to bring some kind of order to species diversity, so that the latter can be dealt with from a practical viewpoint. A theory of systematics should, then, stand up equally well both to scientific and to pragmatic demands. However, this is not entirely possible because a scientific theory should, ideally, be subjected to repeated enquiry, whereas the most stable system possible is preferred from a pragmatic point of view.

It is not this conflict, however, which causes a dilemma in the theory of systematics of a group of organisms. Solutions can always be found which, although they may not be perfect, are justifiable from both scientific and pragmatic viewpoints. The dilemma comes about when it is unclear from systematic statements whether they are intended as an advancement of a scientific theory or simply as a more complete survey of current knowledge. Such systematic statements defy criticism; they do not, as a rule, present any clear problems and thus fail to stimulate new hypotheses to advance the development of the scientific discipline. The result is serious: a theory of systematics, which remains without a clear aim, can hardly be regarded as a science, but is nothing more than a subsidiary science, and thus easily lapses into a state of stagnation. The systematics of a large number of organisms, and nematodes in particular, have reached such a point.

The dilemma of nematode systematics began, as did that of other groups of organisms, because too little attention was paid to the basic principles involved in the investigation of phylogenetic relationships. Research was concentrated on creating some order amongst nematode characteristics and correspondingly little attention was paid to their phylogenetic significance. As a result, most features and systems of nematodes were, hitherto, only of diagnostic value and had little to offer from a scientific point of view, and the solution of systematic controversies was seen not so much as a scientific problem, but more as a matter of convention, where scientific arguments were largely superfluous. The growth of nematode systematics is in a state of stagnation: since 1937 no new systematic criterion has been produced which could be used to demonstrate the inter-relationships of higher ranking nematode taxa and which, as a result, would provide a new basis for the assignment of taxa from the lower to the higher ranks.

The most recent criterion was developed in 1937 – 1942 by Chitwood & Chitwood and concerns the location of the openings of the pharyngeal glands (summarized in Chitwood & Chitwood, 1950; here the pharyngeal glands are called oesophageal glands). The majority of other systematic criteria for freeliving nematodes were developed between 1918 and 1937, mostly by Filipjev (1918/1921). The systematic criteria for freeliving nematodes are made up of the following features, which currently govern the diagnosis of taxa of different ranks: length and thickness of the body; structure of the cuticle and presence of setae on it; number, length and arrangement of the labial and cephalic sensilla; occurrence and position of the dereids; form and position of the amphids; structure of the buccal cavity; structure of the pharynx (often called the oesophagus); structure of the spicular apparatus; structure and arrangement of male supplementary copulatory organs; number of ovaries (also occasionally their structure); shape of the tail; occurrence and position of the phasmids; occurrence of caudal glands. Filipjev's criterion of ovarian structure and structure of the vestibulum in the buccal cavity (section 5.2.1. and discussion point 66 of this study) were later considered erroneously to be of no significance.

The dilemma of nematode systematics is illustrated by an example of a systematic controversy: do the Tripyloididae (a family of freeliving marine nematodes) belong to the "Enoplida"[1] because of similarities to *Tobrilus* in the structure of the anterior end and despite the presence of spiral amphids (the view held by Filipjev, 1918/1921, 1934; Gerlach, 1966; Gerlach & Riemann, 1974) or, precisely because of the presence of spiral amphids, do they belong to the "Araeolaimidae" (the view held by de Coninick and Stekhoven, 1933; de Coninck, 1965; Andrassy, 1976) or to the "Chromadorina" (the view held by Chitwood & Chitwood, 1950)? So far, the question has always been settled one way or the other with arguments which have been common knowledge since 1918, and arguments supporting opposite views have not been disproved, just ignored.

---

[1] If the name of a taxon appears in inverted commas (e.g. "Enoplida"), this means that the range of this taxa is altered in the course of the present study.

INTRODUCTION

1.2. Solution to the current dilemma in nematode systematics and the aim of the present study

The only way to extricate nematode systematics from the current dilemma is to make a clear distinction between scientifically based statements and those of a purely pragmatic nature. Scientifically based statements on the inter-relationships of species groups are elaborated on the basis of phylogenetic systematics. In the present study, the accompanying theory is realised in a new way (see Chapter 2).

For a long time it has been recognised that the understanding of inter-relationships of a species group can be deepened not so much by the discovery of new species but by the discovery of new features and by the phylogenetic interpretation of these features. In the last 40 years, no newly discovered features have been developed as systematic criteria for the gross systematics of the nematodes. So one might get the impression that the number of nematode features, which can be detected using a light microscope and which could then be interpreted phylogenetically, has already been exhausted. This, however, is simply not the case, as I first came to realise when studying the Monhysteroidea from a systematic viewpoint: this sub-family, with its many species, which are found in marine, freshwater and terrestrial habitats, could be divided in a completely new way into three families on the basis of a new feature (Lorenzen, 1978 c). This new feature is the position of the gonads relative to the intestine. In the period following this discovery, it emerged that, using this feature, the monophyly (*sensu* Hennig) of other nematode groups can also be demonstrated. These findings stimulated the search for further new features, and these were forthcoming. In particular, a hitherto unknown system of organs was discovered. This consists of serially arranged structures, called metanemes, which are thought to be stretch receptors (Lorenzen, 1978 d; section 4.3 of this book).

In the present study, known, neglected and new features are described, compared and then assessed from a phylogenetic point of view (Chapter 4). The results obtained form the basis on which a phylogenetic system for the freeliving nematodes is outlined (Chapter 5). The new system is regarded as no more than an outline, mainly because the freeliving nematodes only represent a non-monophyletic part of all nematodes, and because not all the features available have been analysed in detail.

Three resources have considerably facilitated the work: 1) a collection of permanent preparations which contains approximately 700 species of nematodes from marine, freshwater and terrestrial habitats. 2) a card-index containing filing cards (size DIN A4), onto which have been stuck, in a systematic order, copies of almost all the taxonomic descriptions of freeliving species of the Adenophorea (excluding the Dorylaimida). This card-index allows one to obtain rapidly a summary of every taxon recorded. 3) The Bremerhaven Checklist by Gerlach & Riemann (1973/1974), which played an important part in the organisation of the many thousands of species descriptions.

This book represents a revised version of the thesis written by the author on his appointment to a university lectureship and submitted in May 1979 to the Science and Mathematics Faculty of Kiel University (Lorenzen, 1979 b). A major change in the present version is the use of the terms "holophyly" and "holapomorphy" instead of the terms "monophyly" (sensu Hennig) and "synapomorphy". The reason for this change is given in the newly included Chapter 2.

Throughout the text a number of "discussion points" are identified, numbered 1–117. These relate to the holapomorphies used to justify the cladograms depicted in Figs 37–42.

1.3. Acknowledgements

The author was able to collect and study many specimens of freeliving marine nematodes thanks to the support of the Deutsche Forschungsgemeinschaft (Bonn) and Orplan Los Canales (Puerto Montt, Chile). The Deutsche Forschungsgemeinschaft supported the author's research on nematodes from the sublittoral zone of the German Bay (in particular from the titanium waste disposal area near Helgoland, April 1969 – June 1970) and contributed towards the financing of a five-week research expedition from Valdivia (Chile) to the coast of Colombia (South America, February/March 1974). Orplan Los Canales supported the research on marine nematodes from the Island of Chiloé and from Patagonia.

For several years I have worked desk to desk with Dr Schminke (now a professor in Oldenburg) in the Institute of Zoology at the University of Kiel. I wish to thank him for the stimulating discussions on problems, large and small, of systematics; and on the concept of the present work Professor Gerlach and Dr Riemann (both from Bremerhaven), as well as other nematologists who have given their critical appraisal of this work in its earlier form. I owe them many

INTRODUCTION 17

thanks for their encouragement in the re-writing of the text. Lastly, I wish to thank Frau Fedderson (Kiel) for the care she took in producing the camera-ready copy for the German text; the appendix (Chapter 8) was the only section taken in printed form straight from the original thesis.

2. REVISED BASIS FOR A THEORY OF PHYLOGENETIC SYSTEMATICS

The theoretical basis for phylogenetic systematics, as it is known today, was developed by Henning in 1966 (summary). Since that time nobody has remarked on certain problems with the logic of the theory. In this chapter, these logical deficiencies are outlined, and suggestions are made as to how they might be dealt with. A detailed description of the deficiencies is given elsewhere.

1) *Holophyly.* The term "monophyly" *sensu* Hennig is replaced by the equivalent term "holophyly" (Ashlock, 1971 : 65). The reason for this is that, by using the concept of synapomorphy, only monophyly in its old sense ("descent from a common ancestor") can be demonstrated, but not holophyly. Holophyly is defined in the following way: "a species group, which in particular cases may consist of a single species, is referred to as holophyletic if it is probable that all its species are descended from a common ancestral species, and if, outside of the known species, no other species can be discovered which are also probably descended from this ancestral species". This definition corresponds to that of Hennig (1966 : 73) in the strict relationship between ancestral species and holophyletic species groups. The new aspects are the extension of the definition to include species groups which contain only one known species, and the emphasis on the probability of the presence of a character, an emphasis which any concept of holophyly must surely have from the start. Using Hennig's terminology it was not possible to define monotypic taxa (taxa which consist of only one species) as holophyletic or as paraphyletic or polyphyletic. The probability of a character was included in the definition because, without it, one could, at the best, only describe a taxon as "probably holophyletic" but never as "holophyletic".

2) *Homology.* The term homology is understood in a number of different ways. Bock (1974) played a major role in producing a more precise definition of the concept (1974 : 386f): "A feature (or condition of a feature) in one organism is homologous to a feature (or condition

of a feature) in another organism if the two features (or conditions) can be traced phylogenetically to the same feature or condition in the immediate common ancestor of both organisms". With reference to this definition, homology is now defined as follows: "The presence of a feature in two or more organisms is referred to as a homology if it seems probable that the immediate common ancestor of the organisms concerned also possessed the feature". This definition differs from Bock's definition in two respects: 1) "Feature (or condition of a feature)" is replaced by "feature"; this is to emphasise the view that naming a feature is always the result of abstraction. Features, and not organisms, are homologized. Many features can be recognised in a single organism, but the organism itself is not a feature. Greater emphasis is placed on the probable presence of a feature than in Bock's definition; in his definition this is contained in the word "phylogenetically": every phylogenetic statement is a statement involving the probable presence of a character.

Every time the term homology is used, it constitutes a statement which is tested with the help of the criteria of a homology. It is stressed that, in the case of a homology, it must be possible to identify the feature concerned in all the organisms observed.

3) *Synapomorphy and symplesiomorphy.* The meaning of the terms synapomorphy and symplesiomorphy has hitherto been largely misunderstood. Hennig (1953) introduced the terms and only defined them thirteen years later (Hennig, 1966 : 89): "We will call the presence of plesiomorphous characters in different species *symplesiomorphy*, the presence of apomorphous characters *synapomorphy*, always with the assumption that the compared characters belong to one and the same transformation series". In the same place Hennig made the following statement: "We will call the characters or character conditions from which transformation started (a, b) in a monophyletic group *plesiomorphous*, and the derived conditions (a', a", b', b") *apomorphous*".

It was probably a mistake on Hennig's part that, in the first definition quoted, he refers to "characters" in the plural and not to "a character" in the singular; in any case, only a single feature and never a group of features is generally referred to as synapomorphic or symplesiomorphic. The word "character" was probably used by Hennig in the sense of a concrete structure and not an abstract concept. In the following, as was the case with homology, it is understood in the second sense.

Synapomorphies and symplesiomorphies differ from each other in two respects: 1) Establishment of a synapomorphy always rests on a decision that a homologous feature is probably absent at a *primary* level in certain species, which therefore do not belong to the holophyletic species group, whose immediate common ancestor probably had the homologous feature. Establishment of a symplesiomorphy, on the other hand, always rests on the decision that a homologous feature is probably absent at a *secondary* level in certain species (i.e. it was probably lost in the course of phylogenesis); these species belong to the holophyletic species group, whose immediate common ancestor probably possessed the homologous feature concerned. 2) A synapomorphy is always formulated with regard to organisms which probably do not possess the homologous feature concerned at a primary level, whereas a symplesiomorphy is always formulated with regard to organisms which do possess the homologous feature concerned.

Up until now, whenever the terms synapomorphy and symplesiomorphy were used, two serious mistakes, which were not recognised logically, have always been made: 1) A synapomorphy was always regarded as necessary and sufficient condition for the establishment of the holophyl of a species group; as soon as a synapomorphy was demonstrated, the relevant species group was automatically referred to as holophyletic. 2) Symplesiomorphies were regarded as useless in the process of establishing the holophyly of a species group.

The first objection: the definition quoted for synapomorphy does not contain either an explicit or an implicit rule to the effect that *all* known species of a holophyletic species group should be identified, i.e. all known species which possess the relevant homologous feature or which probably do not possess it at a secondary level. Since this rule is not contained in the definition, a synapomorphy can, at best, be a necessary condition for holophyly. In the same way this is also true of homologies and symplesiomorphies. One could argue that this rule is superfluous to the definition of synapomorphy, because it is already contained in the definition of holophyly. This argument itself suggests evidence for the objection which has been raised.

The second objection: from the first objection it emerges that, hitherto, the identification of all known species of a holophyletic species group was not regarded as a scientific problem. Decisions on synapomorphy and decisions on symplesiomorphy likewise play an important part in the solution of this problem. In particular, all decisions as to whether a feature is probably absent at a secondary level in certain species of a holophyletic species group, are based on decisions of symplesiomorphy.

4) *Holapomorphy*. In the course of the present study the new term *holapomorphy* is introduced, the use of which is necessary and suf-

ficient for the establishment of the holophyly of species groups. Definition of the term (according to Lorenzen, 1982): "The conformity which two or more organisms show in a feature is referred to as holapomorphy, provided that the following two conditions are satisfied:
1) The conformity in the feature is considered a homology.
2) All known species, which have the homologous feature or which probably do not have it at a secondary level, are identified and joined together to form a single species group".

The term homology is understood in the sense defined above. As a rule, every decision on holapomorphy entails several decisions on synapomorphy and symplesiomorphy.

There are several criteria (see Remane 1956, Hennig 1966), the use of which leads to the decision whether a feature in a certain species is probably absent at a primary or a secondary level. However, no criterion has as yet been given which permits such a decision purely technically on the basis of homologies and the hierarchical structure of phylogenetic systems alone. Such a criterion is developed in the course of the present work. It is as follows:

"The conformity in a feature α is considered a homology of a species group A. α is then probably absent at a secondary level in species outside A, which form a group B', if 1) on the basis of the homologous feature β, the species in B' form a group B with some of the species in A, and if 2) β is probably absent at a primary level in at least one species in A."

If the absence of a feature cannot be interpreted as a secondary condition on the basis of any criterion, then primary absence can be assumed as the probable condition.

Reasons for the criterion: If, according to the specifications of the criterion, α was probably absent at a primary level in species of B', then the following argument would be conclusive, that α evolved independently in A' (subgroup of A, whose species probably do not possess β at a primary level) and B, which would contradict the homology of α. Such a condition could come into being if β was to exist in none or all species of A or if it was probably absent at a secondary level. The establishment of holophyly in the Enoplida (discussion point 32) in the present work provides a detailed example of the use of the new criterion.

5) *Paraphyly and polyphyly.* Paraphyletic and polyphyletic taxa are not holophyletic and thus do not contribute to the knowledge of

probable phylogenetic relationships of a group of organisms. Both terms can therefore be defined according to purely pragmatic criteria. The following definition is considered useful in the case of paraphyly (according to Lorenzen, 1976 : 229, modified): "If one or more holophyletic sub-groups are separated off from a holophyletic group G, and the holophyly of the remainder of the group G cannot be demonstrated, then the remainder forms a paraphyletic species group". The definition is useful, because a holophyletic taxon can often not be divided completely into holophyletic sub-taxa, with the result that a paraphyletic group is left over. A definition of polyphyly is not needed in the present work.

6) *Phylogenetic system.* Hennig (1965 : 105) made rigid demands that a phylogenetic system of a group of organisms should only contain holophyletic taxa (monophyletic *sensu* Hennig). Such a requirement cannot be met and therefore is pointless. It seems much more realistic to create an appropriate place in a phylogenetic system for an incomplete understanding of phylogeny. Such a system is defined as follows:

"A phylogenetic system is a system of a holophyletic species group, which is developed on the basis of phylogenetic systematics, and which contains, as its constituent parts, the systematic arrangement of taxa together with the arguments on which this arrangement is based, and which clearly allows one to recognise the possibility that it may be incomplete".

There are two criteria by which the incompleteness of a phylogenetic system may be recognised: 1) non-holophyletic taxa are present; 2) the relevant phylogenetic tree has branches which are not dichotomous. In the latter instance, it is not possible to decide in what order the different branches have probably become differentiated. One must always reckon with the possibility that during phylogenesis lines of phyla have branched off in clusters.

A system, which reflects a completely accurate picture of the phylogenesis of organisms, is referred to as *the ideal phylogenetic system*. It represents an unobtainable goal that can only be approximated, whereby each approximation is seen as *a phylogenetic system*.

3. MATERIAL

The specimens studied consisted exclusively of glycerine preparations of freeliving nematodes. These originated from the biotopes listed below; the various headings used to describe the biotopes are the same as those used in section 8.2.1.

Marine biotope or biotope with marine bias:

Flats: mud flats or sandy mud flats along the North Sea coast of Schleswig-Holstein in the Meldorf Bay and the Hamburg Hallig; samples from 1965 to 1968; the material forms the basis of studies by the author (Lorenzen, 1966, 1969 a and b).

Salt marshes: salt marshes (*Spartina, Puccinellia, Festuca*) in the same locality as above; this material was also used in the studies cited above.

Sewage '*priel*': a '*priel*' which has been contaminated by waste water from a petroleum refining station in the Meldorf Bay; samples from 1968.

Helgoland: sublittoral coarse sand, fine sand, muddy sand and mud in Helgoland; samples from April 1969 to June 1970; the material provides in part the basis of studies by Lorenzen (1974 a) and was collected, in part, by Dr Rachor (Bremerhaven) and Dr Apelt (formerly of Bremerhaven).

Helgoland, on *Ligia*: *Gammarinema ligiae* (Monhysteridae) from the sea-slater *Ligia oceanica*, supralittoral region of the Island of Helgoland, 29th September 1977.

Kattegatt: sublittoral coarse sand, fine sand and muddy sand from the Kattegatt and from the southern Skagerrak, 29th–30th June 1976.

Baltic: sublittoral sand from Kiel Bay, January and July 1969; samples were collected by Dr Scheibel from Kiel.

Baltic on *Gammarus*: *Gammarinema gammari* (Monhysteridae) from the amphipod *Gammarus* sp., 'Aufwuchs' from Kieler Innenförde, 15th March 1978.

Brackish water: sand from shallow water in Tvärminne (south Finland), July 1968.

Salt marshes, Finland: salt marshes from the same region as above, July 1968.

Sublittoral, Chile: sublittoral sand and mud from south Chile (the Island of Chiloé and the adjacent, southern archipelago, 42°–46.5° S), December 1971 to March 1972.

Beach, Chile: coarse sand beach on the Island of Chiloé (43° 20' S), December 1971 to September 1973.

'Aufwuchs', Chile: Byssus threads of the mytilid *Aulacomya ater* from water a few metres deep on the Island of Chiloé, January 1972.

Beach, Colombia: clean beach in Santa Marta and on the Island of San Andrés (Carribean coast), February/March 1974.

Red Sea: sublittoral region along the Sudanese shores of the Red Sea, summer 1978, collected by D. Betz (Hamburg).

Freshwater biotope:

Freshwater: Greater Lake Plöner and small fresh waters in Kiel, 1975 and 1976.

Lake Plöner: shallow water with sandy substrate on the Greater Lake Plöner in Gut Ruhleben, October 1975.

Lake Seleter: sandy shore of Lake Selenter, October 1976.

Stream in a wood: sandy stream bed in Schierenseebach, between Lake Schieren and Westen Lake, Kiel, August 1976.

Banks of the Rhine: coarse sand and gravel 90 cm below the surface of the substrate on the banks of Rhine at St Goarshausen, 11th October 1976, collected by Prof. Schminke (Oldenburg).

River bank, U.S.A.: 40–100 cm deep in the sandy banks of various rivers in the south west U.S.A. (Ocanolluftee River, River Establishment, Frazer River), August 1973, collected by Prof. Noodt (Kiel).

River bank, Africa: underground water in the banks of various rivers in South Africa, August 1973, collected by Prof. Schminke (Oldenburg).

Stream, Malaysia: underground water in the banks of streams in Malaysia, summer 1978. Collected by Prof. Schminke (Oldenburg).

River bank, Brazil: limnotic underground water, Brazil, 1968, collected by Prof. Noodt (Kiel).

Terrestrial biotopes:

Woods: mixed woods, Kiel, layer of chaff, autumn 1975.

Meadows: non-agricultural meadows on the edge of a wood, Kiel, dead blades of grass and top layer of soil, autumn 1975.

Moorlands: damp *Sphagnum* bank from the Kaltenhofer Moors, Kiel, October 1975.

Terrestrial, south Germany and the Azores: various terrestrial biotopes from south Germany (early summer 1974 and 1975) and the Azores (early summer 1969), collected by Dr Sturhan (Münster).

4. PHYLOGENETIC ASSESSMENT OF KNOWN AND NEW CHARACTERS

4.1 Cephalic sensilla and dereids

The description of the number, arrangement and length of the labial and cephalic sensilla in adult nematodes is a standard inclusion in every species description. Until recently, however, it has relatively seldom been noted whether the labial and/or cephalic sensilla are jointed or not, and whether any features of the sensilla change at all during postembryonic development.

Dereids are known only in the Secernentea and in a few Adenophorea (they are found in the "Plectidae"). They are described in conjunction with the labial and cephalic sensilla because recent evidence suggests that it is possible that dereids are displaced cephalic sensilla from the third sensilla circle.

Information on the length and form of the labial and cephalic sensilla in adult freeliving nematodes has been compiled from literature sources and from personal findings, and is presented in the appendix (Secion 9.1).

4.1.1. Number, arrangement and structure of the labial and cephalic sensilla

The setiform or papilliform sense organs at the anterior end of the nematode's body are referred to as labial and cephalic sensilla. In a very large number of nematodes from the most diverse orders the number of sensilla is found to be constant: 6 labial and 6+4 cephalic sensilla (Figs 1–4). They are often arranged in a series of three circles, with the 6 labial sensilla in the first (anterior) circle, the anterior 6 cephalic sensilla in the second (middle) circle, and the remaining 4 cephalic sensilla in the third (posterior) circle. With the exception of *Kinonchulus* (Onchulinae, "Enoplida"), the first and second circles are never found at the same level, in contrast to the second and third circles which often are. In the following text the term "6+4 cephalic sensilla" will be used, regardless of whether the latter are arranged in two separate circles or are found at the same level.

The 6 labial sensilla and the 6 foremost cephalic sensilla are lateral and submedial, the rear 4 cephalic sensilla are only submedial and are always slightly closer to the lateral line of the body than are the submedial cephalic sensilla of the other two sensilla circles. This is particularly clear when the second and third sensilla circles are at the same level and the sensilla of one circle are longer than those of the other circle.

The labial sensilla are usually papilliform and are often hardly visible; personal findings and literature research show them to be setous only in the following taxa:

"Desmodorida": "Richtersiidae".
"Chromadorida": "Cyatholaimidae" partim, "Choniolaimidae" partim, "Selachinematidae" partim.
"Enoplida": Onchulinae partim, *Ironella* ("Ironidae"), Tripyloididae partim, Trefusiidae partim (Fig. 2), "Oxystominidae" partim (Fig. 3c-d), *Barbonema* (Leptosomatidae), "Enoplidae" partim, *Pontonema ardens* (Oncholaimidae), Enchelidiidae partim.
Dorylaimida: Cephalodorylaiminae.

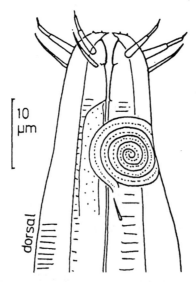

Fig. 1: *Cervonema allometrica* ("Chromadorida", "Comesomatidae"). The cephalic setae in the second sensilla circle are jointed. The 6+6+4 sensilla are in three separate circles. The dotted line in the amphids has a double outline and either represents a slit-shaped aperture, or corresponds to 2 or more sensory cilia. The opening of the dorsal pharyngeal gland can be seen clearly; the openings of the subventral pharyngeal glands could not be detected. Adult male from sublittoral mud, Island of Chiloé (south Chile), 15th December 1971.

PHYLOGENETIC ASSESSMENT 31

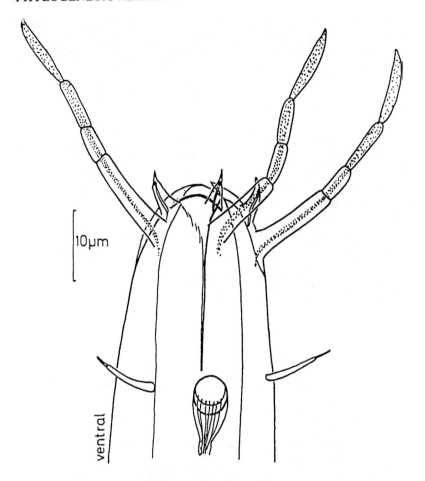

Fig. 2: *Rhabdocoma* sp. ("Enoplida", Trefusiidae). The 6+6+4 sensilla are all jointed and are in 3 separate circles; the dark marks in the labial sensilla are of an unknown nature. The amphids are non-spiral and pocket-shaped. Behind the amphids there are no setae which could be thought to be homologous with dereids. Juvenile from a sandy beach, the Island of Chiloé (south Chile), 12th December 1971.

Within the freeliving Adenophorea (with the exception of the Dorylaimida) the cephalic sensilla are usually setiform in at least one, and often in both circles. Where the 6 cephalic sensilla of the second circle are longer than the sensilla of the first and third circles, the second and third circles of sensilla are usually found at the same level. This is the case in Xyalidae ("Monhysterida"), in many "Comesomatidae",

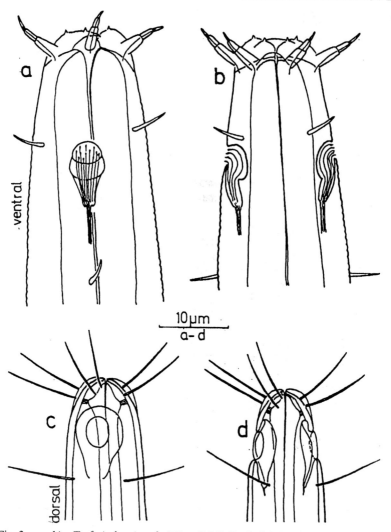

Fig. 3: a – b) *Trefusia longicauda* ("Enoplida", Trefusiidae);
c – d) *Thalassoalaimus pirum* ("Enoplida", "Oxystominidae").
The 6+6+4 sensilla are in 3 separate circles; the second and third circles are particularly far apart. In *T. longicauda* the 6 cephalic setae of the second sensilla circle are jointed. In *T. pirum* the sensilla of all 3 circles are more or less equal in length, which is uncommon in nematodes with setiform sensilla. The amphids in both species are non-spiral and pocket-shaped; this is particularly easy to see in the dorsal views (b and d). In *T. longicauda* the two bristles (one left and right) behind the amphids are the only two somatic setae. It is possible that they are homologous with the dereids and are absent in juvenile animals. a – b: 2 adult females from sublittoral "schillsand" in the Kattegatts, 30th June 1976; c – d: 2 adult males from salt marshes on the North Sea coast of Schleswig-Holstein, 28th October 1968.

Euchromadorinae and "Cyatholaimidae" (formally "Chromadorida"), in most families of "Enoplida" and in a few small taxa from different orders. Where the 4 cephalic sensilla of the third circle are longer than the remaining sensilla, the second and third sensilla circles are usually separate. This is the case in almost all families of the "Araeoelaimida", in the "Desmoscolecida", in the majority of the "Chromadorida" and in a few taxa of the "Monhysterida". Other combinations of sensilla length and position of the second and the third sensilla circles are scattered throughout different orders and are listed in the appendix (Section 8.1).

Within most Dorylaimida and Secernentea all sensilla are insignificant or only slightly setiform; the few exceptions are listed in the appendix (Section 8.1).

Jointed labial and/or cephalic sensilla have, in fact, been described in quite a number of nematode species. Personal observations show that they are however, more common than previously reported. A joint might well have been overlooked, because in many cases it can only be detected using phase contrast. The use of phase contrast was, however, not common practice until recently in the preparation of species descriptions, because most characteristics, can be analysed sufficiently well using normal brightfield illumination. Literature research and personal observations show jointing in labial and cephalic sensilla in the following taxa, all of which belong to the free-living Adenophorea:

- "Araeolaimida", Bastianiidae: personal observations show that the 6 foremost cephalic setae in *Bastiania gracilis* have three joints (Fig. 9a-c); as far as the author is aware, jointing in cephalic setae was, until now, unknown in the Bastianiidae.
- "Desmoscolecida", "Desmoscolecidae": the posterior cephalic setae have two joints in some species of the Desmoscolecinae and Tricominae (examples in Timm 1970, Lorenzen 1972a, Decraemer 1974b and 1975).
- 'Monhysterida", Xyalidae: the 6 cephalic setae of the second sensilla circle are very often divided into 2-4 sections and the 4 cephalic setae of the third sensilla circle into 2 sections (numerous examples in Lorenzen, 1977a); articulation in the cephalic setae was hitherto considered rare in the Xyalidae.
- "Desmodorida", "Richtersiidae": the sensilla in all three circles can be divided into 2 segments.
- "Chromadorida", "Comesomatidae": personal observations show that the 6 cephalic setae of the second sensilla circle in *Cervonema allometrica* have 3 sections (Fig. 1); joints in labial

or cephalic sensilla were hitherto unknown in the "Comesomatidae".
- "Chromadorida", "Cyatholaimidae": personal research shows that within this family jointing in the 6 cephalic setae of the second sensilla circle is much more widespread than was hitherto thought. The presence of 2 segments in the 6 cephalic setae was found in nine out of 16 cases, where the cephalic setae were at least 8–9μm long, but not in the remaining seven cases where the cephalic setae measured no more than 5–6μm. The nine species with the longer cephalic setae belong to the genera *Longicyatholaimus*, *Paracanthonchus*, *Paracyatholaimus* and *Pomponema*, the species with the shorter cephalic bristles belong to the genera *Achromadora*, *Neotonchus*, *Paralongicyatholaimus* and *Paracyatholaimus* (*P. intermedius*).
- "Chromadorida", "Selachinematidae" and "Choniolaimidae": the sensilla in all three circles may be jointed, with 2 to 3 segments.
- "Enoplida": reports in the literature and personal research indicate that the cephalic setae of the second sensilla circle have 2 to 3 segments in Onchulinae partim, Tripyloididae partim, Trefusiidae partim (Figs 2 and 3a-b), and "Tobrilinae" partim (Fig. 35), and 2 segments in the Prismatolaiminae (Fig. 9d), "Tripylinae" partim (Fig. 36) and in *Pandolaimus* ("Anoplostomatidae"). In addition to jointing in the sensilla of the second circle, the sensilla of the first and/or third circle may also be jointed, as personal research showed to be the case in some *Bathylaimus* species (Tripyloididae) and in *Rhabdocoma* sp. (Trefusiidae, Fig. 2).

One may conclude then that, in the case of jointing in the sensilla, the 6 cephalic setae of the second sensilla circle are usually jointed, whereas in the first and/or third sensilla circle joints are found less frequently and that the 6 cephalic setae of the second sensilla circle are usually longer than the rest.

In adult animals of many species of Xyalidae and some species of the "Linhomoeidae" (both "Monhysterida") 2 lateral and, less often, 4 submedial cephalic setae occur in addition to the 6+4 cephalic setae; the 2 additional lateral setae are almost always situated ventrally to the lateral cephalic setae of the second sensilla circle. The additional cephalic setae are at the same level as the 6+4 cephalic setae.

Coomans (1979) and Wright (1980) have collected together and discussed the research literature on the ultrastructure and function of the labial and cephalic sensilla (the sensilla of the first, second and third circles are referred to as inner labial sensilla, outer labial sensilla and cephalic sensilla respectively). In the body cavity a socket cell and a pocket cell are attached to every sensilla. These cells are non-neural and probably represent modified epidermal cells. Every sensillum is also attached to one or more sensory cells. In *Rhabditis elegans* (Rhabditida) each sensory cell has exactly one sensory cilium. It is unknown whether the labial and cephalic sensilla of other nematodes have more than one sensory cilia per sensory cell. All that is known is that in many cases labial and cephalic sensilla have several sensory cilia. The Dorylaimida (Adenophorea) have more sensory cilia per sensillum than do the Secernentea, namely 4–5 in the labial sensilla, 3–4 in the 6 cephalic sensilla of the second circle and 2–3 in the 4 cephalic sensilla of the third circle, whereas the Secernentea only have one (species of the Criconematoidea) or two sensory cilia per sensillum in the labial sensilla, and only one in the cephalic sensilla of the second and third circles. So far, hardly any electron microscope studies have been carried out on the labial and cephalic sensilla of freeliving namatodes (with the exception of the Dorylaimida); there are only a few, incomplete data on cephalic setae in *Chromadorina* sp. ("Chromadorida") and *Enoplus communis* ("Enoplida") (Croll and Smith, 1974).

In many cases labial and cephalic sensilla act as combined mechano- and chemoreceptors. In the first case, at least two sensory nerve cells are probably always present per sensillum (this has only been demonstrated for *Caenorhabditis elegans* by Ward et al., 1975).

4.1.2. Postembryonic data on labial and cephalic sensilla

Data on the postembryonic development of the labial and cephalic sensilla from the first or second juvenile stage to adulthood are known to the author in the following: 4 species of the "Araeolaimida", 12 species of the "Monhysterida", 2 species of the "Desmodorida", 2 species of the "Chromadorida", 14 species of the "Enoplida" and one species of the Rhabditida. The names of the species are listed under point 4 below. The data are as follows:

1) Where 6+4 sensilla occur in adult animals, they are recognizable in the same number from the earliest known juvenile stage. An exception is found in the Secernentea: in the Diplogasteroidea (Rhabditida) there are 6+6+4 sensilla in adult males, but only 6+6 sensilla in adult females and juveniles. *Sphaerocephalum chabaudi* ("Linhomoeidae") may also present an exception among the freeliving Adenophorea, for Inglis (1962) describes only 4 cephalic setae for the first juvenile stage and 6+4 cephalic setae for all following stages.

2) The length of the sensilla in one circle relative to that of the sensilla in the other two circles remains more or less the same during the entire period of postembryonic development.

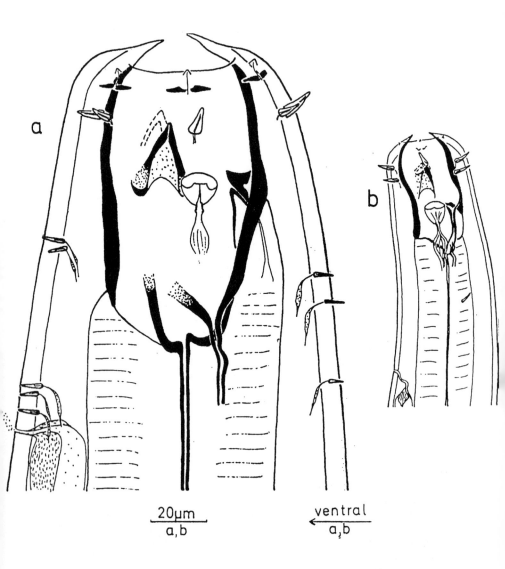

Fig. 4: *Pontonema ditlevsensi* ("Enoplida", Oncholaimidae). a) Adult female (body length 17 mm); b) Juvenile at stage I of development (body length 2.1 mm). The second and third sensilla circles are at the same level in adult animals (Fig. 4a), but behind one another in early juvenile stages (Fig. 4b). In addition, the dorsal onchium, the amphids and the cervical pore lie relatively further forward in adult animals than in early juvenile stages. The animals drawn here are from sublittoral coarse sand near Helgoland, 14th April 1970.

3) Research so far suggests that, unlike the 6+6+4 sensilla, additional cephalic sensilla only appear during postembryonic development (in Xyalidae, Lorenzen 1978c, in "Linhomoeidae", Hendelberg 1979).

4) In the course of postembryonic development, either the second and third sensilla circles draw nearer to one another, or both circles remain more or less the same distance apart; no record is known of the two circles drawing away from one another in the course of postembryonic development.

In individual species the following is the case:

- In *Bastiania gracilis* the second and third sensilla rings are clearly separate from one another in early juvenile stages (approximately stages I and II), but by the adult stages, at the latest, they are directly one behind the other ("Araeolaimida", Bastianiidae, personal observations, Fig. 9a-c), or even at the same level in the following 5 species, all of which belong to the "Enoplida": *Tobrilus longus* ("Tobrilinae", after Riemann, 1966c), *Adoncholaimus thalassophygas* (Oncholaimidae, after von Thun, 1968 and from personal observations), *Pontonema parocellatum* and *P. ditlevsensi* (Oncholaimidae, after Wieser, 1954 and from personal observations, Fig. 4a-b), *Lauratonema* aff. *spiculifer* (Lauratonematidae, from personal observations) and *Enoplus brevis* ("Enoplidae", personal observations).
- In the remaining species, listed below, the distance between the second and third sensilla circles shows practically no change during the entire period of postembryonic development:

"Araeolaimida": *Plectus cirratus* ("Plectidae", personal observations), *Axonolaimus helgolandicus* and *Odontophoroides monhystera* (= *Synodontium m.*) (both "Axonolaimidae", after Lorenzen, 1972b and 1973a).

"Desmodorida": *Epsilonema byssicola* and *Perepsilonema papulosum* (both Epsilonematidae, after Lorenzen, 1973b).

"Monhysterida": *Siphonolaimus cobbi* ("Siphonolaimidae", personal observations), *Sphaerocephalum chabaudi* ("Linhomoeidae", after Inglis, 1962), *Tripylium carcinicola* (Monhysterida, after Riemann, 1970b), *Gammarinema gammari* and *G. ligiae* (Monhysteridae, personal observations), *Steineria ericia* and *S. pilosa* (Xyalidae, after Lorenzen, 1978e), *Amphimonhystera anechma* and *Valvaelaimus maior* (both Xyalidae, after Lorenzen, 1977a), *Sphaerolaimus gracilis*, *Subsphaerolaimus litoralis* and *Parasphaerolaimus paradoxus* (all Sphaerolaimidae, after Lorenzen, 1978e).

"Chromadorida": *Sabatieria celtica* ("Comesomatidae"), *Paracyatholaimus pentodon* ("Cyatholaimidae") (both from personal observations).

"Enoplida": *Chaetonema riemanni, Enoploides labrostriatus* and *Epacanthion buetschlii* (all "Enoplidae", personal observations), *Anoplostoma vivipara* ("Anoplostomatidae", personal observations), *Parironus bicuspis* ("Ironidae", personal observation), *Phanodermopsis necta* (Phanodermatidae, personal observations) and *Rhabdodemania minor* (Rhabdodemaniidae, personal observations).
Rhabditida: *Rhabditis (Caenorhabditis) elegans* (Rhabditidae, after Ward *et al.*, 1975).

4.1.3. Dereids

The dereids are lateral sense organs, which, as far as is known, always take the form of papillae or short bristles. They are located at about the same level as, or slightly behind the nerve rings and are, in fact, not exactly lateral, but are slightly displaced towards the dorsal side. Dereids have been described in many Secernentea and within the Adenophorea of the "Plectidae" (for "Plectidae", see Maggenti, 1961 and Allen & Noffsinger, 1968). Ward *et al.*, (1975) present arguments to suggest that dereids are probably lateral cephalic sensilla of the third sensilla circle which have been displaced posteriorly (see point 2 of the discussion; recent nematodes have only 4 submedial and no lateral sensilla in the third sensilla circle).

Personal research on freeliving nematodes from all orders of the Adenophorea showed that the presence of dereids could likewise only be established with any certainty in species of the "Plectidae" ("Plectinae" and Wilsonematinae). The following result is, however, noteworthy: within the Trefusiidae, "Oxystominidae" and "Ironidae" (all "Enoplida") there are species where the adult animals have only 2 additional setae on their entire bodies besides the 6+6+4 sensilla. Each of these two additional setae is situated laterally between the amphids and the nerve ring, i.e. in a position similar to that of the dereids. The following species were involved: *Trefusia longicauda* (after Riemann, 1966a and personal observations), *T. litoralis* (after Riemann, 1966a). *Rhabdocoma americana* (from personal observations); "Oxystominidae": *Thalassoalaimus pirum*, *T. tardus* and *Oxystomina alpha* (all personal observations); "Ironidae": *Ironella prismatolaima* (after Cobb, 1920 and Riemann, 1966). However, there are also species, at least within the named genera of Trefusiidae and "Oxystominidae", which, besides labial and cephalic sensilla, either have no additional setae on their bodies (*Trefusia* sp. from Chile) or have 4–6 lateral bristles behind the amphids (i.e. 2–3 setae on both sides of the body (*Trefusia helgolandica*, after Riemann, 1966a *Rhabdocoma* sp. from the North Sea, personal observations). Personal research on *Onchulus nolli* and *Stenonchulus troglodytes* failed to establish the presence of any lateral setae behind the amphids in the Onchulinae.

Personal studies of *Plectus cirratus* revealed dereids close behind the nerve ring in juveniles as early as stage I of development. No setae were observed behind the amphids of the few available juvenile *Trefusia*, but they were seen to be present in the few examples available of advanced juvenile stages of *Thalassoalaimus pirum*.

4.1.4. Discussion

1. On the terminology of the labial and cephalic sensilla:

The division, terminologically, of the 6+6+4 sensilla into 6 labial and 6+4 cephalic sensilla is widespread within the literature on nematodes, and is used in particular by Filipjev (1918/1921) and by Chitwood & Chitwood (1950). This division was based on the finding that the first and second sensilla circles are almost never on the same level as one another, whereas the second and third circles frequently are. According to the less widespread terminology used by de Coninck (1942a and 1965a) and by Coomans (1979), but not in the present study, the sensilla of the first and second circles are referred to as inner and outer labial sensilla on account of their grouping in 6 radiating parts, and the sensilla of the third circle are referred to as cephalic sensilla on account of their arrangement in 4 radiating parts. Since it is also common to find groups of 4 or 8 radiating body setae, the 4 cephalic sensilla, unlike the 6+6 remaining sensilla are, according to de Coninck (1942a : 61) specialized body setae.

This is unlikely for the following reasons:

- All 6+6+4 sensilla are, in contrast to the body setae, already present from the first juvenile stage of development; in this respect, then, the 4 sensilla of the third circle are like the 6+6 sensilla of the first and second circles and not like body setae.
- The process from each neural and non-neural cell to its accompaning sensillum in the third circle follows the same course as do the corresponding cell processes to the sensilla of the first and second circles (Chitwood & Chitwood, 1950: 165, Ward et al., 1975).
- It is possible that dereids are sensilla displaced from the third circle (see discussion point 2). Therefore sensilla of the third circle might also have originally formed a group of 6 radiating parts.

2. Phytogenetic assessment of the number of labial and cephalic sensilla and of the existence of the dereids:

It was possible to detect 6 labial and 6+4 cephalic sensilla, or at least the rudiments thereof, in nematodes from the most diverse orders of the Adenophorea and Secernentea; at the same time it was found that the sensilla were more or less at the same level within each circle. De Coninck (1942a) and Gerlach (1966) assume, therefore, that the existence of 6+6+4 sensilla is plesiomorphic (original) within the nematodes. Chitwood & Wehr (1934) and Chitwood & Chitwood (1950), on the other hand, consider the existence of 6+6+6 sensilla (1+1+1 laterally and submedially respectively) to be plesiomorphic, because the lips, the pharyngeal glands, the rectal glands and the caudal glands are arranged in radiating groups of three.

Ward *et al.* (1975) was able to show that in *Rhabditis elegans* (Rhabditidae) the 4 sensilla of the third circle are much more similar in their ultrastructure to the dereids than they are to the remaining 6+6 sensilla; Sulston (in Ward *et al.*, 1975) was able to demonstrate that in *C. elegans* catecholamine was only present in those 6 neurons which innervate the 4 sensilla of the third circle and the 2 dereids. From these results Ward *et al.* concluded that dereids are probably posteriorly displaced lateral sensilla of the third circle and these authors are therefore in agreement with Chitwood & Wehr (1934) in suggesting that the occurrence of 6+6+6 sensilla should be considered as plesiomorphic within the nematodes. However, since not a single recent nematode has been found with 6 cephalic sensilla at the same level in the third circle (see discussion point 3 regarding additional cephalic sensilla), it must be assumed that in primitive nematodes the lateral sensilla of the third circle had already been displaced posteriorly as dereids, and were not situated at the same level as the remaining 4 sensilla, as was assumed by Chitwood & Wehr. The above can be summed up thus: within the nematodes the following are judged to be plesiomorphic:

a) the occurrence of 6 labial and 6+4 cephalic sensilla which are found at more or less the same level within each circle, and

b) the occurrence of 2 dereids, which are probably posteriorly displaced lateral sensilla from the third circle. According to Coomans (verbal communication) a dereid and also a phasmid are, in both cases, a combination of a modified secretory cell from a lateral epidermal ridge with a sensillum; see also point 6 of the discussion.

The number 6+6+4 sensilla is considered to be holapomorphic for the nematodes, because it does not occur outside the nematodes. Conclusion (b) above is only correct if the Adenophorea and Secernentea are holophyletic groups and/or the dereids are shown to be homologous with the lateral setae posterior to the amphids which are present in a few species of the Trefusiidae, "Oxystominidae" and "Ironidae" (see p. 56). The homology of these setae with dereids is yet uncertain, because i) the setae in question are not present in all species of the genera concerned, ii) unlike the other 6+6+4 sensilla, they only appear during postembryonic development in some species and iii) their innervation is as yet unknown.

The amphids cannot be interpreted as transformed lateral sensilla of the third circle as has been done occasionally in the past and recently by Andrassy (1976 : 56). The amphids are innervated quite independently from the labial and cephalic sensilla (Chitwood & Wehr, 1934 : 280, Ward *et al.*, 1975), and their sensory cilia, unlike those of the labial and cephalic sensilla, project from the cuticle (see e.g. Storch & Riemann, 1973; Ward *et al.*, 1975; McLaren, 1976).

3. Phylogenetic assessment of additional cephalic sensilla:

In contrast to Chitwood & Chitwood (1950 : 57), the author considers the existence of additional cephalic setae which are at the same level as the 6+4 cephalic setae in adult animals of many species of Xyalidae and of some species of "Linhomoeidae", to be a case of apomorphy and not plesiomorphy in the two families. The reason, as far as we know (Lorenzen, 1978c; Hendelberg, 1979), is that the additional cephalic setae first appear in the course of postembryonic development, whereas the real 6+6+4 sensilla are already present from the first juvenile stage.

4. Phylogenetic assessment of the position of the sensilla circles and the dereids:

Postembryonic research on labial and cephalic sensilla have shown that in species from widely different orders, the second and third sensilla circles either move closer together or maintain the same distance apart during postembryonic development, but never move away from one another. From this it is concluded, in agreement with de Coninck (1942a) and Gerlach (1966) and in contrast to Chitwood & Wehr (1934) and Chitwood & Chitwood (1950 : 56) that the separation of the two sensilla circles is plesiomorphic within the nematodes. Gerlach's arguments in favour of this include, among other things, the postembryonic data for *Tobrilus longus* quoted on p. 37 and the observation that body setae of the cervical region are also displaced towards the cephalic sensilla independently in several families. There is a further argument in favour: since dereids are probably posteriorly displaced sensilla from the third circle (disscusion point 2), this would be considered to be a case of plesiomorphy within the nematodes, if the sensilla of the third circle and the dereids are located very close to one another. Since the dereids are always located posterior to the amphids, it follows that it must be seen as plesiomorphic within the nematodes that the sensilla of the third circle are located well posterior to those of the second circle and indeed behind the amphids. Furthermore, it must be considered plesiomorphic within the nematodes if the dereids are situated as close as is possible behind the amphids. If the lateral seta behind each amphid in several species of Trefusiidae, "Oxystominidae" and "Ironidae" is really homologous to a dereid in each case, then it would be considered a particularly clear example of plesiomorphy.

Chitwood & Wehr (1934) and Chitwood & Chitwood (1950) based their view, that the 6+6+4 cephalic sensilla were originally together in a single circle, solely upon the finding that the cephalic structures in *Rhabditis* are particularly plesiomorphic within the nematodes. This argument arises from the incorrect interpretation of similarities which are said to exist between the Plectidae and the Rhabditidae:

- According to Chitwood & Chitwood (1933: 130; 1950: 57, 196), since the Plectidae belong to the Adenophorea and the Rhabditidae to the Secernentea, the members of both families should be, on account of the existing similarity of several characteristics, particularly primitively related. However, such a conclusion is only correct if the Adenophorea and the Secernentea are both holophyletic, which has not been shown to be the case by the authors.
- In the Rhabditidae the 6+4 cephalic sensilla are arranged in a single circle in the labial region, whereas in the Plectidae they are in two distinct circles, with the 4 posterior cephalic sensilla clearly posterior to the labial region. In this point, then, there is no similarity between the two families.

5. Phylogenetic assessment of the jointed nature of the labial and cephalic sensilla:

Jointed sensilla (usually in the second circle) have been found in various species of "Araeolaimida", "Desmoscolecida", "Monhysterida", "Desmodorida", "Chromadorida" and "Enoplida" (see p. 33). Joints in a few and, in some cases, all 6+6+4 sensilla are also common within the Adenophorea; it thus seems probable that their presence is homologous and not analogous. Since joints in sensilla occur in particular in many species of "Chromadoria" and "Enoplia", which together form the Adenophorea, it seems probable that, at least within the Adenophorea, the presence of joints in sensilla is plesiomorphic and the absence of joints is apomorphic. This interpretation is supported by the fact that jointed somatic setae are almost never found on the anterior body region in Adenophorea (they are jointed only in a few species of Desmoscolecinae and Epsilonematidae). From this argument one can specifically conclude that, contrary to the view of Chitwood & Wehr (1934) and Chitwood & Chitwood (1950:56), the setiform development of the labial and cephalic sensilla should be regarded as plesiomorphic, at least within the Adenophorea, because the jointed sensilla are always setiform.

If, within the Adenophorea, the setiform development of the labial and cephalic sensilla is seen as plesiomorphic and the papilliform development as apomorphic, then it seems probable that this might also be the case in the Secernentea. In the latter the sensilla are almost always papilliform or short and setiform, with the result that this condition is considered apomorphic within the nematodes. As the Adenophorea and Secernentea together represent all nematodes, it can be concluded, in summary, that the setiform appearance of the labial and cephalic sensilla is considered plesiomorphic and the papilliform appearance apomorphic within the nematodes.

6. Phylogenetic assessment of the difference in length between the sensilla of the three circles:

It has been shown that only the sensilla of the second circle may have 3 or even 4 segments. This was observed in nematodes of the species "Monhysterida", "Chromadorida" and "Enoplida", and is thus common within the Adenophorea. It is thus considered plesiomorphic, at least within the Adenophorea, that the sensilla of the second circle have 3 or 4 segments and are longer than the sensilla of the other 2 circles, where, according to current data, the sensilla have no more than 2 segments. Where the sensilla of the second circle are smaller than those of the first and/or third circle, then this is considered to be a case of apomorphy and not plesiomorphy within the Adenophorea, contrary to the view put forward by de Coninck & Stekhoven (1933) and by de Coninck (1942: Fig 27). De Coninck and Stekhoven's argument is based on their, as yet, unsubstantiated statement, that within the Chromadoria the "Axonolaimidae" are particularly primitive with regard to their head structures. In "Axonolaimidae' the 4 sensilla of the third circle are almost always longer than the 6 sensilla of the second circle.

7. Phylogenetic assessment of the similarity in length of the sensilla within a single sensilla circle:

Within each circle the sensilla are almost always the same length. There are only a few exceptions: in some "Comesomatidae" (e.g. *Comesoma heterosetum, Paracomesoma hexasetosum, P. inaequale* Jensen & Gerlach, 1977[1], *P. sipho*), in some "Siphonolaimidae" (e.g.

---

[1] Authorities for species are only given when the species concerned are not listed in Gerlach & Riemann's checklist (1973/1974).

Siphonolaimus cobbi, S. ewensis Warwick & Platt, 1973) and in *Linhomoeus hirsutus* ("Linhomoeidae") the lateral sensilla of the second circle are longer than the submedial sensilla of the same circle, whereas in *Halinema varicans* ("Linhomoeidae") they are shorter; in *Theristus (Penzancia) aculeatus* (Xyalidae) the subventral sensilla of the second and third circles are longer than the remaining sensilla of the circles concerned. From the very rare occurence of dissimilarity in sensilla length within a circle, it is concluded that similarity in length of the sensilla within a circle is a case of plesiomorphy within the nematodes.

8. Phylogenetic assessment of dereid form:

Since dereids are probably displaced sensilla from the third circle (discussion point 2), it follows from discussion point 7 that originally the dereids were probably similar in form to the cephalic sensilla of the third circle.

9. Phylogenetic assessment of the fact that the labial sensilla are usually smaller than the cephalic sensilla:

Where the sensilla length is different from circle to circle, the sensilla of the first circle are almost always the shortest, though very seldom those of the second circle ("Selachinematidae": *Gammanema conicauda*; "Oxystominidae": *Wieseria hispida, W. inaequalis, W. pica*) or those of the third circle are the shortest ("Cyatholaimidae": *Pomponema*, partim; "Richtersiidae"; *Richtersia*, partim; Tripyloididae: *Bathylaimus*, partim: Leptosomatidae: *Barbonema horridum*). From this it follows that:

- It is considered plesiomorphic within the nematodes that the shortest sensilla are in the first circle.
- In cases where only 6+4 sensilla can be recognized within a nematode species, in all probability they belong to the second and third circle and not to the first and third circle.

10. Particularly plesiomorphic and particularly apomorphic expression of characteristics in the labial and cephalic sensilla of the Adenophorea:

According to points 4, 5 and 6 of the discussion, within the Adenophorea the Trefusiidae (Figs 2 and 3a-b) and the Onchulinae (both "Enoplida") have particularly plesiomorphic characteristics in their labial and cephalic sensilla, because:

- the second and third sensilla circles are far apart from one another,

PHYLOGENETIC ASSESSMENT

- the sensilla of the second circle are longer than those of the first and third circle,
- the sensilla of the second circle often have 3 segments and those of the first and third circles have no more than 2 segments.

According to points 4, 5 and 6 of the discussion, the Dorylaimida and isolated taxa among the remaining orders within the Adenophorea have particularly apomorphic characteristics in their labial and cephalic sensilla, because:

- the second and third circles of sensilla are located at the same level, and
- the sensilla of all three circles are papilliform and unsegmented.

11. Particularly common combinations of apomorphies and plesiomorphies with regard to the labial and cephalic sensilla in the Adenophorea:

Two apomorphies with regard to the labial and cephalic sensilla are particularly common within the Adenophorea:

- The second and third sensilla circles are at the same level (apomorphic), and the 6 sensilla of the second circle are longer than the 4 of the third circle (plesiomorphic). This combination of an apomorphy and a plesiomorphy is found in one section of the Bastianiidae ("Araeolaimida"), in the Xyalidae and in some smaller taxa of the "Monhysterida", in most "Cyatholaimidae", some smaller taxa of the "Chromadorida" and most families of the "Enoplida" (see table in Section 8.1).
- The second and third sensilla circles are separate from one another (plesiomorphic), and the sensilla of the third circle are longer than those of the second circle (apomorphic). This combination of an apomorphy and a plesiomorphy which is widespread within the "Araeolaimida", "Desmoscolecida", "Desmodorida" and "Chromadorida" occurs only occasionally in the "Monhysterida", and is absent in the "Enoplida" (see table in Section 8.1).

12. The insignificant value of labial and cephalic sensilla characteristics as criteria for the establishment of holophylies:

Although, according to points 10 and 11 of the discussion, apomorphies regarding the form and position of the 6+6+4 sensilla occur within the Adenophorea, they can be used only in a few cases to demonstrate holophyly in taxa. The reason for this is the following: as can be seen from the table in section 8.1 of the appendix, sensilla characteristics are distributed in such a heterogenous way within the Chromadoria (which consists of the "Araeolaimida", "Desmoscolecida", "Monhysterida", "Desmodorida" and "Chromadorida") and within the Enoplia (which consists of the "Enoplida" and Dorylaimida, as well as some of the parasitic nematodes not dealt with here), that it is hardly possible to determine with any degree of certainty the range of taxa for which sensilla characteristics are circumstantial evidence for holophyly. For example if, by way of experiment, one was to ascribe the apomorphy mentioned second in point 11 of the discussion to holapomorphy and then, also by way of trial experiment, one were to compile the corresponding taxon, the holapomorphy would then contradict other holapomorphies which are revealed by the study of ovarian characteristics (discussion point 66), the study of gonad position (discussion point 52) and the study of the number of testes (points 44 and 45 of the discussion).

13. Phylogenetic assessment of the 4 additional cephalic sensilla in adult males of the Diplogasteroidea:

In all Diplogasteroidea 6+6+4 sensilla have been found only in the adult males; adult females and juveniles have only 6+6 sensilla. This sexual dimorphism is unique among the freeliving nematodes and is thus considered a holapomorphic feature of the Diplogasteroidea. The 4 posterior cephalic sensilla in the adult males are probably not homologous with the 4 cephalic sensilla of the third sensilla circle, but represent 4 additional sensilla, which are innervated by 4 additional neurons. The argument for this is that 4 additional neurons were also found behind the 4 cephalic sensilla of the third sensilla circle in adult males of *Rhabditis*, but were not present in adult females and juveniles (Ward et al., 1975 : 327). The additional neurons in *Rhabditis* each have one sensory filament, but none of the neurons project into a papilla or seta.

14. Phylogenetic assessment of the relative length of the cephalic sensilla in the Sphaerolaimidae:

In all Sphaerolaimidae the 6+4 cephalic sensilla are situated at the same level, and the 6 cephalic sensilla of the second sensilla circle are always shorter than the 4 of the third sensilla circle. This characteristic is considered as holapomorphous for the Sphaerolaimidae because it is unique within the holophyletic group Sphaerolaimidae plus Xyalidae (see Lorenzen, 1978c).

15. Phylogenetic assessment of the arrangement of the 6+4 cephalic sensilla in one circle in the Cyatholaimidae:

In the Cyatholaimidae (excluding Neotonchinae and Achromadorinae) the 6+4 cephalic sensilla are almost always arranged in a circle. This characterstic is considered a holapomorphic feature of the family,

because it is unique within the holophyletic group of Chromadorina and occurs in combination with the following features:
— The 6 cephalic sensilla of the second sensilla circle are always longer than the 4 of the third sensilla circle.
— The amphids are always spiral and have several turns.
— The buccal cavity usually contains one distinct, dorsal tooth, whereas the subventral teeth are very small or absent altogether.
— The anterior and posterior gonads are always located on different sides of the intestine.

The strength of this judgement is lessened by the fact that, in some examples of the "Choniolaimidae", "Selachinematidae" and Achromadorinae, the second and third sensilla circles are located at the same height (table in Section 8.1).

4.2 Form, position and postembryonic development of the amphids

Nematodes possess two amphids which are located laterally directly posterior to the anterior end of the body. The description of the amphids forms a standard part of all species descriptions of freeliving Adenophorea and some freeliving Secernentea. Since Filipjev (1918/1921), the form of the amphids has played an important part as a systematic criterion in distinguishing between the Enoplia and the Chromadoria, in identifying the Secernentea and in differentiating between many taxa over the complete range within the Enoplia and Chromadoria. Despite their great systematic significance, the structure of the amphids and the inter-relationship between the various types of amphid have until recently remained relatively unresearched and have only within the last 10 years been basically clarified with the use of light microscopes and electron microscopes in research (see review article by Coomans, 1979).

4.2.1. General structure and description of the main types

Studies by various authors using light microscopes and electron-microscopic techniques (see review article by Coomans, 1979 and Fig. 5) have shown the structure of an amphid to be as follows: there is a pocket or depression sunk into the upper surface of the body; this is known as the fovea. It is lined with a thin cuticle and is connected to the outside through the aperture. In the fovea there is the corpus gelatum which, according to Riemann et al. (1970) sometimes projects

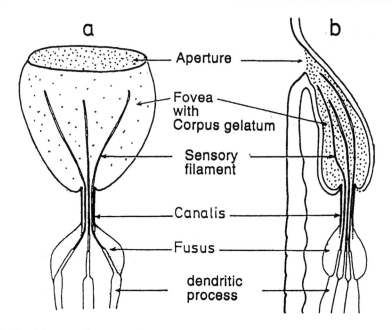

Fig. 5: Diagram of a nematode amphid. a) Plan view; b) Side view (in longitudinal section). Based on light and electron microscopic studies of the amphids of *Tobrilus* ("Enoplida"). Only 3 of the 6 dendritic processes and 3 of the sensory filaments which are present in *Tobrilus* are shown here. No study was made of the existence and position of the amphid gland and socket cell in *Tobrilus*. a: combined from Storch & Riemann (1973: Fig. 4, structural characteristics) and from the author's light microscopic observations on *Tobrilus grandipapillatus* (outlines); b: from Storch & Riemann (1973: Fig. 4).

out of the aperture. There are several sensory filaments imbedded into the corpus gelatum; these filaments travel through the narrow, cuticularized canalis into the fusus, which lies inside the body. Dendritic neural processes stretch from within the body into the fusus; the sensory filiaments emerge from these processes. Further details are known from other studies: the neurons of the amphids have 0, 1 or 2 sensory filaments in *Rhabditis* (Rhabditida; according to Ward et al., 1975), whereas in *Oncholaimus vesicularis* ("Enoplida"; according to Burr & Burr, 1975) from 3 neurons 28–36 sensory filiaments travel to the fovea, and from a further neuron roughly 10 sensory filiaments connect with the pigment spots. According to Coomans (1979), the Adenophorea have, in general, more sensory filiaments per amphid than the Secernentea. According to McLaren (1976) the corpus gelatum is isolated from the unicellular amphid gland, which surrounds the fusus like a ring and which is called a "sheath cell" by Ward et al. (1975). According to Ward et al. (1975) and Coomans (1979), the canalis is surrounded by the socket cell. The amphid gland and the socket cell correspond in their layout to the pocket and socket cell of the labial and cephalic sensilla, and are likewise not neural and thus are probably of epidermal origin.

The main types of amphids have hitherto been described as spiral, round, pocket-shaped and pore-shaped. This characterization of types arises from the shape of the fovea (spiral, round, symmetrically extended) and from the relative size of the fovea to the aperture (fovea roughly equal in size to the aperture in spiral and round amphids, and larger in pocket-shaped and pore-shaped amphids). This nomenclature is neither clear-cut nor does it encompass all types of amphid:

— Using the criteria above the amphids of all Trypyloididae and many Trefusiidae (both "Enoplida", Fig 7j, k and o), as well as many species of Chromadoria (e.g. *Metachromadora* sp., Fig. 7g) could be called both pocket-shaped and spiral.

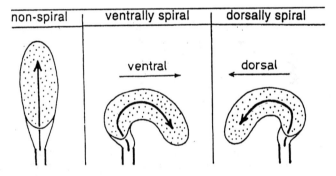

Fig. 6: Diagram of the 3 main types of amphids in nematodes: a) non-spiral; b) ventrally spiral; c) dorsally spiral. The dotted area indicates the aperture and the thick contours indicate the canalis.

— The distinction between pocket-shaped and pore-shaped amphids is not a clear one, because it is determined solely by the size of the aperture.
— The extended amphids of the Halalaiminae (Fig. 7e) and of the adult male *Chaetonema* (Fig. 11) (both "Enoplida") cannot be grouped with any main type.
— The term "spiral amphid" fails to give any indication of the direction in which the spiral turns.

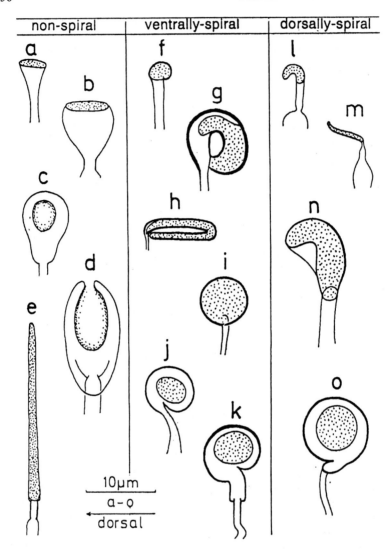

Fig. 7: Examples of non-spiral (a – e), ventrally spiral (f – k), and dorsally spiral (l – o) amphids in freeliving nematodes. All the amphids were drawn from adult animals. Sensory filaments have not been included. The dorsal side is to the left in the picture.

As a result of the deficiencies indicated above, the current nomenclature for amphids is not satisfactory. Therefore a new nomenclature is suggested, where the following three main types are distinguished:

Type 1: *Non-spiral amphids* (Fig. 6 and 7a–e). From the dorsal view, non-spiral amphids look roughly bilaterally symmetrical, whereby the axis of symmetry is parallel to the lengthways axis of the body and runs through the canalis. The non-spiral amphids correspond extensively to the pocket-shaped and pore-shaped amphids of the original nomenclature.

Type 2: *Ventrally-spiral amphids* (Figs 6 and 7f–k). If one follows the turns of a ventrally-spiral amphid from the canalis, one comes to the anterior region of the amphid on the ventral side of the body. If one draws a straight line through the canalis parallel to the lengthways axis of the body, the large part of the amphids will usually (but not always — see Fig. 7k) be found ventral to this straight line.

Type 3: *Dorsally-spiral amphids* (Figs 6 and 7l–o). The direction of the turns in a dorsally-spiral amphid is exactly opposite to that of the ventrally-spiral amphid. If one draws a straight line through the canalis parallel to the body axis, the larger part of the amphid usually (but not always — see Fig. 7o) lies dorsal to the straight line.

The left and right amphid of a nematode are mirror images of one another and thus always belong to the same amphid type.

4.2.2. Structure and position in different taxa
a) Amphids in the Chromadoria.

The amphids in the Chromadoria are non-spiral in a number of instances, ventrally-spiral in most cases, and dorsally-spiral in a very few cases.

Non-spiral amphids, in which the canalis lies posterior to the aperture, occur within the Chromadoria in the following taxa: within the "Araeolaimida" in *Anaplectus* (Fig. 7a, "Plectidae", the structure of the amphid was first described by Allen & Noffsinger, 1968), "Rhabdolaimidae", various genera of the "Leptolaimidae" (*Anomonema, Assia, Chronogaster, Stephanolaimus* partim, *Leptolaimoides, Leptolaimus* partim), Aulolaimidae, Isolaimiidae and within the "Desmodorida" in the Monoposthiidae and Xenellidae (personal observations). Non-spiral amphids, in which the canalis lies in the region of the aperture, occur in the "Desmoscolecidae" ("Desmoscolecida") and in *Paramonohystera elliptica* ("Monhysterida", Xyalidae, according to Riemann, 1966a : 24); the amphids of these animals appear blistered, with the outer skin of the blisters probably forming as a result of hardened material from the corpus gelatum. This theory needs verifying using electron microscopic techniques, for it is possible that the outer skin is, in fact, a thin cuticle (see also discussion point 16).

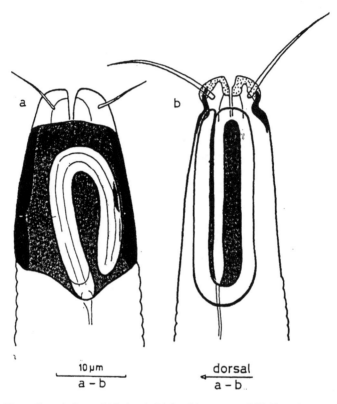

Fig. 8: Ventrally spiral amphids in a) *Diplopeltis* sp. ♀ and b) *Tarvaia* sp. ♀ (both "Axonolaimidae"). In *Diplopeltis* sp. the amphids are like an inverted "U" in shape, which is very common within the Chromadoria. In *Tarvaia* sp., the loops of the amphids, starting at the canalis (thin double line), become wider and not narrower which is very unusual in nematodes. Both animals were from the sub-littoral zone of Kiel Bay.

In most taxa of the Chromadoria the amphids are ventrally-spiral. They are often a rounded spiral (Fig. 7g) or like an upside-down 'U' in shape (Fig. 8a). In cases where the amphid is multi-spiral, the turns almost always run from the canalis round towards the centre of the spiral and only seldom away from it (Fig. 8b: *Tarvaia*; indications of this also in *Cricolaimus, Rhadinema* and *Pseudonchus*). In inverted U-shaped amphids the canalis always lies at the beginning of the dorsal leg; the ventral leg is often equal in length to the dorsal leg, but is considerably longer in *Parodontophora, Psuedolella* and *Campylaimus* partim (all "Axonolaimidae"). Occasionally the inverted U-shape closes posteriorly to form an 'O' (e.g. in *Ascolaimus*). In the following taxa the fovea and aperture of the amphid are round: *Aegialoalaimus* ("Axonolaimidae"), Haliplectidae, "Meyliidae", "Siphonolaimidae", many "Linhomoeidae" (Fig. 7j), Monhysteroidea, "Microlaiminae" and isolated species of other families. In round amphids the canalis lies either on the edge of, or inside, the aperture; usually the canalis lies in such a way that it is clear in which direction the connecting loops are twisted and thus that the amphid belongs to the ventrally-spiral group.

Personal observation has shown that, within the Chromadoria, dorsally-spiral amphids only occur in *Bastiania gracilis* (Bastianiidae, "Araeolaimida"); in this species the amphids are either ventrally or dorsally-spiral (Fig. 9a–c).

Within each of the following families the shape of the amphid is extremely constant: "Rhabdolaimidae", "Isolaimiidae", Haliplectidae, "Desmoscolecidae", "Siphonolaimidae", Monoposthiidae, "Comesomatidae" and "Cyatholaimidae". In the remaining families of the Chromadoria the amphids vary in shape with the family, particularly in "Leptolaimidae", "Axonolaimidae", "Desmodoridae" and "Chromadoridae" which have many species.

Within the Chromadoria the canalis of the amphids lies more or less on the lateral line of the body in all species with non-spiral amphids, dorsal to the lateral line in all species with ventrally-spiral amphids and ventral to the lateral line in those animals of the species *Bastiania*, whose amphids are dorsally-spiral.

Sexual dimorphism in the formation of the amphids only occurs very rarely in the Chromadoria. It is usually limited to differences in size: the males have larger amphids than the females. In addition to differences in size, differences in shape also occur in *Richtersia* ("Richtersiidae"; there are 1 to 1½ loops in the amphids of the females and 3 to 3½ loops in those of the males), and in *Metepsilonema laterale* Lorenzen, 1973b (Epsilonematidae; the amphids of the females have one loop, whereas the amphids of the males are no more than an inconspicuous depression in the body). Differences in shape are most pronounced in *Paraeolaimus nudus* ("Axonolaimidae"; according to Lorenzen, 1973a, the amphids of the females form a small spiral with one loop, whereas those of the males are large and curve into an 'O' shape).

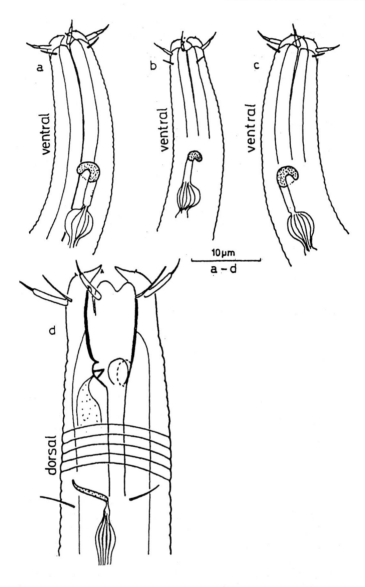

Fig. 9: Different orientations in amphid loops. In *Bastiania gracilis* ("Araeolaimidae") the amphids are either ventrally (a, b) or dorsally spiral (c); in *Prismatolaimus dolichurus* ("Enoplida") the amphids are always dorsally spiral. The 6 cephalic sensilla of the second sensilla circle are jointed in both species. In *B. gracilis* the second and third sensilla circles are situated further apart in juvenile animals (b) than in adult animals (a, c). a, c and d are drawn from adult females; b is from a juvenile. All animals come from the Kaltenhof Moor near Kiel, 17th October 1975.

In a very few cases amphids may be absent within the Chromadoria, e.g. in *Retrotheristus* Lorenzen, 1977a and in *Gnomoxyala* Lorenzen, 1977a (both Xyalidae).

b) Amphids in the Enoplia

The amphids in the Enoplia are usually non-spiral and only seldom ventrally or dorsally-spiral.

The non-spiral amphids in the Enoplia are, in general, bilaterally symmetrical: the fovea is usually significantly larger than the aperture, so that the amphids are pocket-shaped. Only in the Halalaiminae (Fig. 7e) and in the adult male *Chaetonema* (Fig. 11a, "Enoplidae") are the fovea and the aperture of the non-spiral amphids roughly the same size. In cases where non-spiral amphids occur in the Trefusiidae they often consist of two chambers which together form the shape of a question mark (Fig. 3a–b). This type of amphid is not known in other nematodes and could be called laterally-spiral.

Ventrally-spiral amphids occur within the Enoplia only in all species of Tripyloididae and in *Rhabdocoma* sp. (Fig. 7k, Trefusiidae).

From personal observation, dorsally-spiral amphids occur within the Enoplia in species from 4 families, i.e. in species of Prismatolaiminae, within the Trefusiidae in *Rhabdocoma americana, Rhabdocoma*. sp. (Fig. 7o) and *Cytolaimium exile* (sensu Cobb, 1920 nec Gerlach, 1962), within the "Enoplidae" in *Gairleanema angremilae* (Warwick & Platt, 1973) (Fig. 34, re-examination of the syntypes) and within the Enchelidiidae in *Belbolla, Ditlevsenella* (Fig. 7n), *Eurystomina, Megeurystomina* and *Pareurystomina*. The amphids in species of *Eurystomina* and *Pareurystomina* have already been described as dorsally-spiral by other authors.

Amphids have as yet not been observed in *Rhabdodemania* (Rhabdodemaniidae). In their place I have now found a sinusoidal subcuticular structure which can be traced as far as the nerve ring (Fig. 10); the posterior half of this structure is straight. The structure is present from the first juvenile stage. Amphids have not been observed in many species of "Enoplidae", so it appears that they are absent. The Rhaptothyreidae, which have very large, non-spiral and completely irregularly shaped amphids, have recently been considered to be related to the Mermithoidea (Hope, 1977).

In all species of Enoplia the amphids lie posterior to the labial region and behind the 6 cephalic sensilla of the second sensilla circle. Within

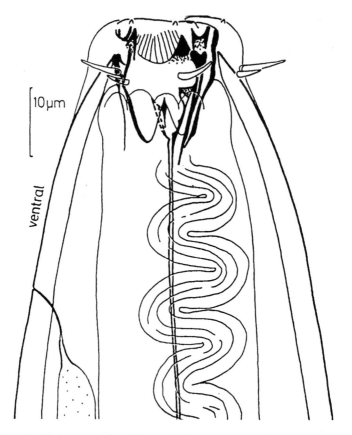

Fig. 10: *Rhabdodemania minor* ("Enoplida"). In the place of an amphid there is a hitherto unknown sinusoidal subcuticular structure posterior to the lateral cephalic seta. Adult female from sublittoral sand in the Kattegatt, 29th June 1976.

the "Oxystomininae" and Trefusiidae, where the 4 cephalic sensilla of the third circle lie well posterior to the 6 sensilla of the second circle, the amphids may be situated anterior to the 4 posterior cephalic sensilla.

The position of the canalis in relation to the lateral line of the body is not uniform within the Enoplia and is in no way correlated with belonging to one of the three amphid types. The canalis is situated on the lateral line in the "Tripylidae", Onchulinae, Odontolaiminae, "Ironidae", Cryptonchidae, Tripyloididae partim, Trefusiidae (also in species with ventrally or dorsally-spiral amphids), "Alaimidae", "Oxystominidae", Lauratonematidae, Anticomidae, "Anoplostomatidae", Chaetonematinae, Oncholaimidae partim and Dorylaimida; it lies dorsal to the lateral line and thus even more dorsally than the lateral cephalic sensilla in Tripyloididae partim, Leptosomatidae, Triodontolaimidae, Oncholaimidae partim and Enchelidiidae; the canalis lies ventral to the lateral line and thus further ventral than the lateral cephalic sensilla in the Prismatolaiminae (dorsally-spiral amphids), Phanodermatidae, "Enoplidae" (excluding the Chaetonematinae) and *Ingenia* (Trypyloididae: in *Ingenia* the amphids are well displaced towards the ventral side). Special mention should be made of the "Eurystomininae" and "Thoonchinae", where not only are the canales displaced towards the dorsal side, but the amphids are also dorsally-spiral (Fig. 33), something which is unique within the nematodes.

Sexual dimorphism in the development of the amphids is rare in the Enoplia. It is known in *Cytolaimium exile* (*sensu* Gerlach, see Gerlach, 1962, Trefusiidae), possibly in a few other species of Trefusiidae, where males and juveniles have different shaped amphids but females are as yet unknown (e.g. *Trefusia varians*), in *Platycoma* and *Platycompsis dimorphus* (Leptosomatidae) and in the Enchelidiinae, where sexual dimorphism is also marked in the rest of the body structure. In all cases listed, the amphids of the males differ in appearance from those of the females and juveniles. Personal observations have revealed a further case of sexual dimorphism in the development of the amphids which far exceeds all hitherto known. Intra-species differences between male and female amphids (Fig. 11): in *Chaetonema* ("Enoplidae") the adult males have extended amphids without a pocket, similar to those found in the Halalaiminae, whereas the adult females and juveniles have very small, non-spiral amphids which have an aperture that is much smaller than the fovea. The amphids of the males had in fact been described earlier, but were mistaken for Steiner's organs, which do not occur outside the genus *Chaetonema*; the amphids of the females and juveniles were hitherto unknown. Within the Nematoda, *C. riemanni* is the only species in which the amphids extend posteriorly beyond the nerve ring.

c) The amphids of the freeliving Secernentea.

Generally speaking, the amphids of the freeliving Secernentea only have a pore-shaped aperture; they lie in the labial region or directly posterior to it and are displaced slightly towards the dorsal side (Chitwood & Chitwood, 1950). According to Goodey (1963), species where the amphids clearly lie posterior to the labial region are only known in the Diplogasteridae, Myolaimidae and Chambersiellidae;

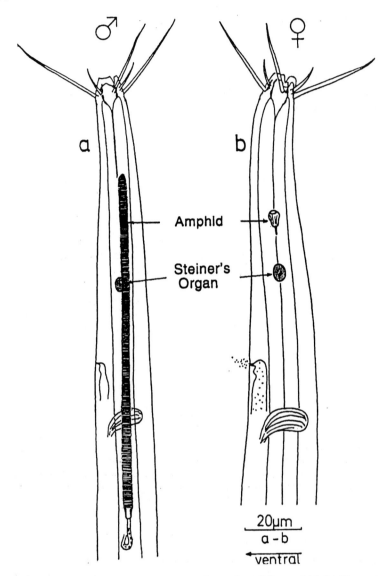

Fig. 11: Extreme sexual dimorphism in the formation of the amphids in *Chaetonema riemanni* ("Enoplidae"). The extended amphids of adult males in *Chaetonema* have until now been mistaken for the so-called Steiner's organ, while the amphids of adult females and juvenile animals were hitherto unrecognised. Steiner's organs are unknown outside the genus *Chaetonema*. On only one occasion have the amphids of adult males in *Chaetonema riemanni* been found to extend behind the nerve ring. The examples drawn are adults from sublittoral fine sand in Helgoland, 18th April 1969 (♂) and 9th May 1970 (♀).

in these species the amphids are situated on the lateral line of the body.

In *Diplogaster rivalis* (Diplogasteridae) sexual dimorphism is evident in the development of the amphids (Fig. 12). The amphids are ventrally-spiral in adult males and non-spiral in adult females and juveniles. No further cases are known to the author where adult animals of the Secernentea have spiral or sexually dimorphic amphids.

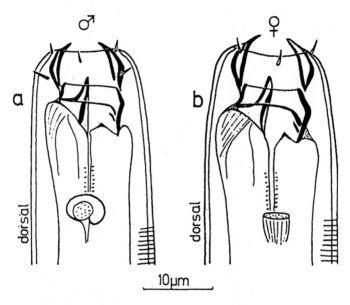

Fig. 12: Sexual dimorphism in the formation of the cephalic sensilla and the amphids in *Diplogaster rivalis* (Rhabditida, Diplogasteridae). The adult males have 6 plus an additional 4 cephalic sensilla and ventrally spiral amphids, while the adult females and juveniles have 6 cephalic sensilla and non-spiral amphids. Adult male (a) and adult female (b) from a small, freshwater pool near Kiel, 27th April. 1976.

4.2.3. *Postembryonic data*

The species in which the postembryonic development of the amphids has been studied from the first or second juvenile stage up to adulthood are the same as those in which the postembryonic development of the labial and cephalic sensilla were studied (p.35 onwards).

In cases where the shape of the amphids (within a family) is highly constant, the amphids of all juvenile stages are similar to the amphids of the adults. Thus, from the first juvenile stage onwards the amphids are round within the Monhysteroidea, spiral within the "Comesomatidae" and "Cyatholaimidae", and non-spiral and pocket-shaped in the "Tobrilinae", Leptosomatidae, "Ironidae", Lauratonematidae, Phanodermatidae, "Enoplidae" (with the exception of *Chaetonema*), Oncholaimidae and Dorylaimida. In the freeliving Secernentea the amphids in the juveniles are also similar in structure to those of the adults.

However, in cases where the shape of the amphids is variable within a family, it also tends to change during the course of postembryonic development. In the species of "Axonolaimidae" and Epsilonematidae studied, the amphids form a spiral with one loop in the first juvenile stage and in the second juvenile stage onwards are bent into an inverted 'U'-shape ("Axonolaimidae": *Axonolaimus* and *Odontophoroides*, Fig. 13) or are spiral, with more than one loop (Epsilonematidae). In *Sphaerocephalum chabaudi* ("Linhomoeidae") the amphids are round at an early stage and curved into an 'O'-shape in adults (an inverted, 'U'-shaped curve closing to form an 'O' posteriorly).

According to Chitwood & Chitwood (1950 : 57), the amphids of embryos in *Rhabditis* (Secernentea) are similar to those in *Plectus* (Fig. 7f), whereas in adult animals they have a pore-shaped aperture.

In cases where sexual dimorphism occurs in the form and/or size of the amphids, the amphids of all known juvenile stages are similar to those of the adult females.

Postembryonic data show that the relative distance between the labial region and the front edge of the amphids remains more or less the same or becomes slightly less during postembryonic development (see e.g. Fig. 4: *Pontonema ditlevseni*); in embryos of *Rhabditis* the amphids are situated posterior to the labial region (Chitwood & Chitwood, 1950 : 57). Very occasionally the distance between the labial region and the canalis increases during the course of postembryonic development, e.g. in *Odontophoroides* (Fig. 13), *Axonolaimus* (both "Axonolaimidae") and in males of *Chaetonema* (Fig. 11, "Enoplidae").

4.2.4. Discussion

16. Homology of the amphids of nematodes with the lateral sensory organs of the gastrotrichs:

According to Remane (1936) and Teuchert (1977 : 234), the amphids of nematodes and the lateral sensory organs of gastrotrichs are homologous for the following reasons (Teucherty): they are present in pairs at the anterior end of the body and contain approximately 12 to 14 sensory nerve cells, the axons of which travel directly to the central

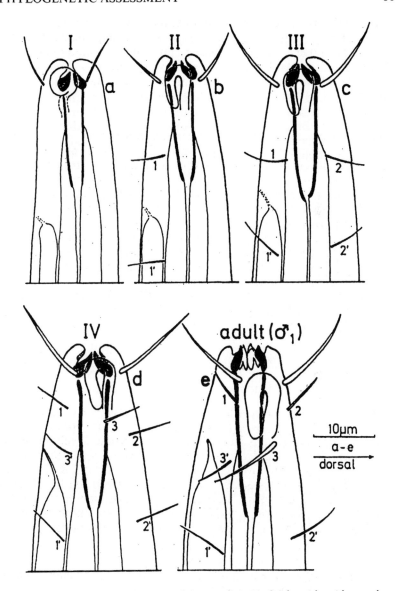

Fig. 13: Postembryonic development of the anterior end of *Odontophoroides monhystera* (= *Synodontium m.*, "Araeolaimida", "Axonolaimida"). a – d, Juvenile stages I – IV; e, adult male. The amphids in the first juvenile stage (I) are similar to those in *Plectus*, but are an inverted U-shape in the stages that follow. Changes also occur in the number of subcephalic setae (numbered) and in the position of the cervical pores. From Lorenzen (1973a); drawing of the first juvenile stage (a) has been altered slightly following re-study.

commissure of the brain and the dentrites of which have a ciliary region, whereby the microtubules pass through the outer segments of the dendritic processes. The main distinguishing feature between the amphids and the lateral sensory organs is that the former open to the exterior and the latter are cut off from the exterior by a thin, culicular wall. It should be stressed that the bubble-shaped amphids of the Desmoscolecoidea (Chromadoria) are also cut off from the exterior by a thin membrane, but it is as yet unknown whether this membrane consists of a hardened outer layer of the corpus gelatum or whether it is made of cuticle.

17. Existing phylogenetic assessment of the form and position of the amphids:

Filipjev (1918/1921) gave up an attempt at a phylogenetic assessment of the form and position of amphids. Since then there have been 4 different and contradictory views on the possible plesiomorphic form of the amphids within the nematodes. Phylogenetically, the position of the amphids has been assessed less frequently. The separate views are as follows:
— Chitwood & Chitwood (1950 : 57) considered the plectoid amphids (Fig. 7f, a spiral with one loop), which lie posterior to the labial region, to be plesiomorphic within the nematodes, because this combination of characteristics occurs in *Plectus* (Adenophorea) and in embryos of *Rhabditis* (Secernentea) and because the Adenophorea and Secernentea are seen as sister groups.
— Gerlach (1966 : 30 and Figs 4 and 5) and Riemann (1966a : 21, 1972 : 71), both of whom deal with the evolution of the amphids in the Adenophorea, consider the non-spiral type to be the primary form, because it occurs both within the Chromadoria and the Enoplia (Gerlach). According to Riemann's interpretation (1972 : 71), the primary form can in no way be considered plesiomorphic within the nematodes, because there is no corresponding reference point in the ontogeny.
— Stekhoven & de Coninck (1933 : Fig. 2) and de Coninck (1965 : 601) consider an O-shaped amphid (as in Fig. 7g) to be plesiomorphic within the Chromadoria, but offer no further explanation.
— Andrassy (1976 : 57), in accordance with views that were disproved long ago, considers the papilliform amphids situated in the labial region to be plesiomorphic within the nematodes, because he considers the amphids to be modified lateral cephalic sensilla of the third sensilla circle. This theory of homologies is not possible on account of the difference in innervation of the

6+6+4 sensilla on the one hand, and of the amphids on the other (Chitwood & Wehr, 1934: 280; Chitwood & Chitwood, 1950: 163; Ward et al., 1975).

18. Phylogenetic assessment of the position of the amphids:

In agreement with Chitwood & Chitwood (1950), the position of the amphids posterior to the labial region is considered to be a plesiomorphic feature and their position within the labial region an apomorphic feature in the nematodes for the following reasons:

— According to current postembryonic data, the relative distance between the amphids and the labial region decreases, or remains more or less the same, during postembryonic development in species of Chromadoria, Enoplia and Secernentea. The amphids never move posteriorly in their entirety, though the accompanying canales may occasionally do so (Fig. 13: *Odontophoroides*: male *Chaetonema*).
— In all nematodes where the 6 cephalic sensilla of the second sensilla circle are longer than the 4 of the third sensilla circle, the amphids are situated posterior to the labial region. Since it was considered a plesiomorphy within the nematodes, if the 6 cephalic sensilla of the second sensilla circle are longer than the 4 of the third circle (discussion point 6), the position of the amphids posterior to the labial region should therefore also be considered a plesiomorphic feature within the nematodes.
— The lateral sensory organs, which are homologous with the amphids in nematodes (see discussion point 16), are situated posterior to the labial region in the gastrotrichs.

19. Phylogenetic assessment of the ventrally-spiral amphids of the Chromadoria:

In those cases where the amphids are looped within the Chromadoria (with the exception of the Bastianiidae), they are always of the ventrally-spiral type. Moreover, in the ventrally-spiral aphids of the Chromadoria each accompanying canalis is always situated dorsal to the lateral line of the body. It follows from the wide distribution and absolute constancy of these two characteristics within the Chromadoria, that with this group the ventral orientation of the loops of the amphids is probably attributable to homology and not analogy, and thus that ventrally-spiral amphids with a dorsally displaced canalis are probably plesiomorphic within the Chromadoria. In the present work, the Bastianiidae and Prismatolaiminae are classified as Chromadoria despite their dorsally-spiral amphids (discussion point 23).

20. Phylogenetic assessment of the non-spiral amphids of the Enoplia:

Within the Enoplia non-spiral amphids with a distinct aperture are the most frequent, though ventrally-spiral (Tripyloididae, Trefusiidae partim) and dorsally-spiral amphids ("Eurystomininae", "Thoonchinae", Trefussidae partim) also occur. Despite the occurence of both types of spiral amphids, it is likely that the non-spiral type is due to homology and not analogy within the Enoplia (except Prismatolaiminae), and therefore that non-spiral amphids are plesiomorphic within the Enoplia. Reasons:
— Despite the dorsal direction of the loops of the amphids in the "Eurystomininae" and "Thoonchinae", the canalis is situated dorsal to the lateral line of the body, whereas in the Chromadoria the dorsally displaced position of the canalis is always accompanied by a ventral orientation in the loops of the associated amphid.
— The aperture is always considerably smaller than the accompanying fovea in the spiral amphids of the Tripyloididae and Trefusiidae, unlike that of the spiral amphids of the Chromadoria, and the lateral line of the body always (Trefusiidae) or often (Tripyloididae) runs through the canales and always runs through the respective amphids.

According to discussion point 22, the existence of non-spiral amphids can even be considered a holapomorphic feature of the Enoplia.

21. Phylogenetic assessment of the amphids in the Secernentea.

It was concluded (discussion points 19 and 20) that within the Chromadoria ventrally-spiral amphids with a dorsally-displaced canalis, and within the Enoplia (except Prismatolaiminae) non-spiral amphids, are probably plesiomorphic. Since dorsally-spiral amphids are probably derived from non-spiral amphids (in the case of the "Eurystomininae", "Thoonchinae", Trefusiidae) or from ventrally-spiral amphids (in the case of the Bastianiidae and Prismatolaiminae, see discussion point 23), the ventrally-spiral amphids with a dorsally displaced canalis and the non-spiral amphids represent the two basic types within the Adenophorea. In addition, the small size of the labial and cephalic sensilla and the arrangement of the cephalic sensilla in two closely situated circles in the Secernentea were considered apomorphic features within the nematodes (discussion point 6), so that it seems likely that the smallness of the amphids in the Secernentea is apomorphic within the nematodes. Consequently, the small amphids must be derived from larger amphids. Where larger amphids are concerned, only the ventrally-spiral amphids with a dorsally displaced canalis, or non-spiral amphids occur, so that one of the two amphid types must also be plesiomorphic within the Secernentea. Since, within the Adenophorea, the "Plectidae" (Chromadoria) are similar to the Rhabditidae (Secernentea) in a particularly large number of features (the structure of the pharynx, the position of the gonads, the position of the dereids posterior to the nerve ring), and since the Secernentea are considered to be holophyletic due to the presence of a single testis (discussion point 44), the Secernentea are probably more closely related to representatives of the Chromadoria than to representatives of the Enoplia. Thus the presence of ventrally-spiral amphids with a dorsally displaced canalis is considered to be a plesiomorphy within the Chromadoria plus Secernentea, which is in agreement with the view put forward by Chitwood & Chitwood (1950) (discussion point 17). This argument is supported by the fact that the canalis of the amphids lies on or dorsal to the lateral line of the body in both the Secernentea and the Chromadoria.

22. The probable nature of the plesiomorphic amphid within the Nematoda.

It still remains to be decided whether the ventrally-spiral amphids with a dorsally displaced canalis, or the non-spiral amphids, should be considered as plesiomorphic within the whole of the Nematoda. Within the Enoplia, the ventrally-spiral amphids probably developed from the non-spiral ones, which however, — and this is important — do not resemble the amphids of the Chromadoria (discussion point 20). Conversely, within the Chromadoria, probably independently, at least within the "Plectidae", "Leptolaimidae", Adulolaimidae and Monoposthiidae, the non-spiral amphids have developed from ventrally-spiral amphids with a dorsally displaced canalis. These non-spiral amphids — and this again is important — resemble those of the Enoplia. It is thus considered likely that ventrally-spiral amphids with a dorsally displaced canalis are plesiomorphic vis-a-vis non-spiral amphids and thus, in agreement with Chitwood & Chitwood (1950) and contrary to Gerlach (1966) and Riemann (1966a, 1972) (see discussion point 17), represent the plesiomorphic amphid type within the entire nematode class. From this it follows that the existence of non-spiral amphids probably represents a holopomorphy in the Enoplia.

23. Phylogenetic assessment of the amphids in the Bastianiidae and Prismatolaiminae:

The Bastianiidae and Prismatolaiminae contain only *Bastiania* and *Prismatolaimus* respectively as well established genera. Current knowledge shows that only dorsally-spiral amphids occur in *Prismato-*

laimus and both ventrally and dorsally-spiral amphids occur within species in *Bastiania*. In ventrally-spiral amphids the canalis always lies dorsal to the lateral line of the body and in dorsally-spiral amphids it always lies ventral to the lateral line. Therefore the relative position of the canalis depends alone on the direction of the loops in the amphids — thus differing from the situation in the Trefusiidae, "Eurystomininae" and "Thoonchinae". It is thus considered probable that the dorsally-spiral amphids in *Bastiania* and *Prismatolaimus* are directly descended from ventrally-spiral amphids with dorsally displaced canalis. For this reason, the Bastianiidae and the Prismatolaiminae are classified as Chromadoria.

The Prismatolaiminae and Bastianiidae probably do not constitute a holophyletic taxon because of the differences in structure of their buccal cavities and in the shape of the amphids. As a result, the constant dorsal direction in the loops of their amphids is considered to be holapomorphic in the Prismatolaiminae.

24. Phylogenetic assessment of the plectoid, inverted U-shaped and spiral amphids of members of the Chromadoria:

It has been established that, in the juvenile stage I, plectoid amphids and, in the following stages, inverted U-shaped (Fig. 13: "Axonolaimidae") or spiral amphids with several turns (Epsilonematidae) occur in species of the "Axonolaimidae" and Epsilonematidae. From this finding it follows that the plectoid formation of the amphids vis-a-vis the inverted U-shape or spiral form is probably plesiomorphic, at least within the "Axonolaimidae" and Epsilonematidae. Since, on the basis of gonad characteristics, the species of "Axonolaimidae" and Epsilonematidae studied are divided into two different sub-groups, which together form the Chromadoria (Fig. 22), plectoid amphids are considered plesiomorphic within the Chromadoria, contrary to the view of Stekhoven & de Coninck (1933) and de Coninck (1965) (see discussion point 17).

25. Phylogenetic assessment of round amphids as frequently found within the Chromadoria:

Round amphids, where the fovea and aperture are, as a result, equal in size because the canalis lies at times on the edge of, or in, the region of the aperture, occur within nematodes almost exclusively in various families of the Chromadoria. Since plectoid amphids are considered plesiomorphic within the Chromadoria (discussion point 24), and since the aperture is always smaller than the fovea in the remaining nematodes, round amphids are considered an apomorphic feature within the Chromadoria.

26. Phylogenetic assessment of the amphids in *Tarvaia* (Chromadoria):

In *Tarvaia* alone the loops of the ventrally-spiral amphids run from the canalis outwards to the outer edge of the spiral (Fig. 8b), a condition which is at the utmost only faintly evident within the Chromadoria (p.66). This characteristic is thus considered a holapomorphy in *Tarvaia*.

27. Phylogenetic assessment of the spiral amphids of the Comesomatidae:

In the Comesomatidae (with the exception of the Acantholaiminae) the amphids always form a spiral with at least 2½ loops. This feature is considered a holapomorphy of the Comesomatidae. Reason: the feature is unique within nematodes with extended ovaries, most species of which (among them the Comesomatidae except "Acantholaiminae") are united in one holophyletic order (discussion point 51).

28. Phylogenetic assessment of the bubble-shaped amphids of the Desmoscolecoidea:

Amphids of the Desmoscolecoidea are oval or round in outline and almost always have a bubble-shaped corpus gelatum with a very tough outer skin (p.51). Since this feature only occurs at the most in an indistinct form within the remaining nematodes, it is considered a holapomorphy of the Desmoscolecoidea.

29. Phylogenetic assessment of the sinusoidal "amphid" in the Rhabdodemaniidae ("Enoplida"):

In the Rhabdodemaniidae the amphids are replaced by a sinusoidal structure, which can be traced as far as the nerve ring (p.55 and Fig. 10). As far as is known this structure is unique within the nematodes and is thus considered a holapomorphy of the Rhabdodemaniidae.

30. Phylogenetic assessment of extreme sexual dimorphism in the amphids of the Chaetonematinae ("Enoplida"):

The amphids of the Chaetonematinae show extreme sexual dimorphism (Fig. 11). This sexual dimorphism does not occur even in an indistinct form in other nematodes and is thus considered a holapomorphic feature of the Chaetonematinae.

31. Phylogenetic assessment of the extended amphids of the Halalaiminae ("Enoplida"):

Within the Enoplia the aperture of the amphids forms a longitudinal groove (Fig. 7e) both in the juvenile and in the adult animals of the

Halalaiminae. Due to the uniqueness of this feature it is considered a holapomorphy of the Halalaiminae. Outside the Halalaiminae this feature only occurs in adult male Chaetonematinae (discussion point 30), and outside the Enoplia the amphids only have an extended aperture in species of the genus *Leptolaimoides* (Chromadoria, "Leptolaimidae").

4.3 Metanemes

4.3.1. General structure and position

Metanemes are serially arranged, subcuticular, thread-like organs which lie in the region of the lateral epidermis borders (Greek: meta = one after another; nema = thread). They have only recently been discovered and are thought to be stretch receptors (Lorenzen, 1978d).

Fig. 14a shows the structure of a metaneme based on studies using a light microscope. The main part is the scapulus (diminutive of the Latin word *scapus* meaning *shaft*), which has a diameter of 1–3 μm and measures from 5–15 μm, thus taking up about 2.5–6.5% of the total length of the metaneme. The wall of the scapulus is cuticularized in appearance. Inside the scapulus there is a clear zone, which probably contains liquid. A fine thread projects from the anterior end of the scapulus into the clear zone and ends at the caudal end of the zone. At the frontal end the fine thread continues into the frontal filament, which is between 20 μm and 400 μm long and is cuticularized in appearance. At the frontal end the frontal filament tapers out into a fine point. In many species a caudal filament attaches itself to the scapulus; this filament is also cuticularized in appearance and is 30–170 μm long. In many species it is absent. In many instances the scapulus is clearly associated with a cell-like structure which may possibly represent the sensory cell of the metaneme.

Metanemes can, to some extent, be recognized at a magnification of 10×40 (e.g. in *Enoplus* and *Deontostoma*, both "Enoplida"). It is therefore surprising that they have not been found earlier by other authors. In a number of species metanemes can only be detected at a magnification of 10×100 and only in very bright light (e.g. in *Ironus* and *Tripyla*, both "Enoplida").

Metanemes are thought to be stretch receptors, firstly on account of the mechanically weak connection between scapulus and frontal filament and secondly on account of their positioning on the inside

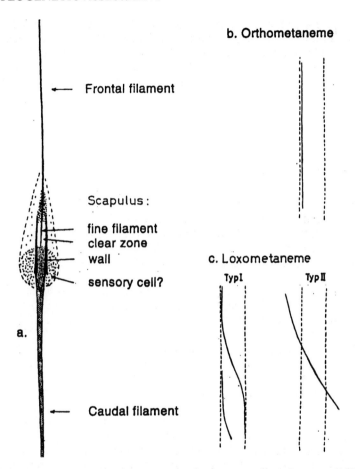

Fig. 14: Schematic drawing of the structure of a metaneme (from Lorenzen (1978d)). a, b, Schematic representation of a) an orthometaneme, b) two loxometanemes of type I and one loxometaneme of type II. The dotted lines represent the outer limits of the epidermal borders. Only in type II loxometanemes do the filiaments penetrate into the region between the cuticle and the musculature.

of the body in places where stretching occurs both as a result of increase in volume and through twisting movements of the body (Lorenzen, 1978d). These suggestions need to be substantiated using electron-microscopic techniques.

In cases where metanemes actually occur, adult animals have, depending on the species, between 6 and 115 of them on either side of the body (Table 1). The anteriormost are found in the pharyngeal region, the posteriormost in the caudal region.

Metanemes run either parallel or obliquely to the longitudinal axis of the body (Fig. 14b, c). The first sort are referrred to as orthometanemes (Greek: *orthos* = straight), and the second sort as loxometanemes (Greek: *lox* = oblique). Two types can be distinguished within the loxometanemes: in type I the frontal and caudal filaments are strictly confined to the region of the lateral epidermal border, whereas in type II the frontal filament and often also the caudal filament penetrate into the area between the cuticle and the musculature.

Orthometanemes always occur on the dorsal and often also on the ventral edge of the lateral epidermal border. Those on the dorsal edge are referred to as dorsolateral and those on the ventral edge as ventrolateral orthometanemes. The frontal filament of an orthometaneme often passes close by the posterior section of the nearest metaneme in front of it. In all these cases the frontal filament lies closer to the adjoining edge of the lateral epidermal border than does the posterior section of the nearest metaneme in front of it. On account of this constant relationship in the position, it is possible that the dorsolateral orthometanemes should be considered homologous with those loxometanemes which run subdorsally from the anterior and subventrally to the posterior; these loxometanemes are therefore known as dorsolateral loxometanemes. Correspondingly, the ventrolateral loxometanemes run subventrally from the anterior and subdorsally towards the posterior, and are homologous with the ventrolateral orthometanemes. Both dorsolateral and/or ventrolateral metanemes can occur within a species.

4.3.2. Occurrence of metanemes within the Nematoda

Metanemes have hitherto been found exclusively within the order "Enoplida", in 13 of the 19 families listed by Gerlach & Riemann (1974) (see Table 1). Detailed descriptions of metanemes in species of the "Enoplida" are to be found in Lorenzen (1981a). In comparison with freeliving nematodes of other orders, the species of "Enoplida" are rather large, i.e. often 3mm long. In addition, their cuticle has a smooth appearance when seen through a light microscope.

Metanemes have not been found outside the "Enoplida", neither in species over 3mm in length nor in smaller species, neither in smooth-skinned animals nor in those with striated cuticles, neither in freeliving nor in parasitic species. Metanemes have been looked for in the following taxa outside the "Enoplida", but with negative results.

"*Araeolaimida*" : *Anaplectus, Aphanolaimus, Ascolaimus* (*A. elongatus* 1♂ : L=6500μm), *Axonolaimus, Deontolaimus, Leptolaimus, Plectus, Procamacolaimus, Stephanolaimus* (*S. elegans*, 1♀ : L=2900μm).

PHYLOGENETIC ASSESSMENT

Table 1: Occurrence, position, structure and number of metanemes in adult nematodes of the order "Enoplida". The range of the order and the sequence of the families are in accordance with Gerlach & Riemann (1974).

Metaneme type: Ortho : orthmetaneme
Loxo : type I loxometaneme
Loxo II : type II loxometaneme

Caudal filament: + : caudal filament present
                         − : caudal filament absent

n : the number of individuals in which the metanemes on one side of the body were counted.

Taxon	Metaneme type	Caudal filament	Number of metanemes on each side of the body in adults		n
			dorso-lateral	ventro-lateral	
"Tripylidae" Tripylinae:					
Tripyla glomerans	Ortho	+	5–6	5–6	1
"Tobrilinae"					
Tobrilus gracilis	Loxo II	+	13–15	−	4
Monochromadorinae	?	?	?	?	−
"Prismatolaimidae" Onchulinae:					
Onchulus nolli	−	−	−	−	+
Stenonchulus troglodytes	−	−	−	−	+
Prismatolaiminae:					
Prismatolaimus dolichurus	−	−	−	−	+
Odontolaiminae:					
Odontolaimus aquaticus	−	−	−	−	+
"Ironidae"					
Ironus ignavus	Ortho	−	c.10	c.10	1
Parironus bicuspis	Loxo II	+	16–18	−	6
Trissonchulus acutus	−	−	−	−	1
"Cryptonchidae"					
Cryptonchus tristis	−	−	−	−	1
Tripyloididae					
Bathylaimus australis	Loxo II	+	1–2	12–15	3
Tripyloides marinus	Loxo II	+	−	11	1
Trefusiidae Trefusiinae:					
Trefusia cf. longicauda	−	−	−	−	2
Trefusia helgolandica	−	−	−	−	+
Rhabdocoma sp.	−	−	−	−	1
Halanonchinae:					
Halanonchus sp.	−	−	−	−	+

Table 1 continued

Taxon	Metaneme type	Caudal filament	Number of metanemes on each side of the body in adults		n
			dorso-lateral	ventro-lateral	
"Alaimidae"					
Alaimus primitivus	–	–	–	–	+
Amphidelus sp.	–	–	–	–	+
"Oxystominidae"					
"Oxystomininae"					
Thalassoalaimus septentrionalis	Ortho	+	c.7	–	1
Thalassoalaimus tardus	Ortho	+	+	–	1
Oxystomina elongata	Ortho	+	+	–	1
Halalaiminae					
Halalaimus aff. longicaudatus	Ortho	+	+	+	1
Halalaimus gracilis	–	–	–	–	+
Paroxystomininae	?	?	?	?	–
Lauratonematidae					
Lauratonema aff. spiculifer	–	–	–	–	+
Lauratonema hospitum	–	–	–	–	+
Leptosomatidae					
Leptosomatinae:					
Platycoma sudafricana	Loxo I	+	c.56	c.11	1
Cylicolaiminae	?	?	?	?	–
Synonchinae:					
Synonchus longisetosus	Loxo I	+	c.71	–	1
Thoracostomatinae:					
Deontostoma arcticum	Ortho, seldom Loxo I	+	c.55	c.60	1
Triodontolaimidae					
Triodontolaimus acutus	Loxo II	+	19	–	1
Anticomidae					
Anticoma acuminata	Loxo I, seldom, Loxo I	+	6	–	1
Anticoma trichura	Loxo I	+	2–3	–	4
	Loxo II	+	15–18	3–10	
Paranticoma tubuliphora	Loxo I	+	10	–	1
	Ortho	+	2	–	
Phanodermatidae					
Crenopharynx marioni	Ortho	–	20	31	1
Phanoderma campbelli	Ortho, seldom Loxo I	–	c.20	c.4	1
Phanodermopsis necta	Ortho	–	1–2	–	+

PHYLOGENETIC ASSESSMENT

Table 1 *continued*

Taxon	Metaneme type	Caudal filament	Number of metanemes on each side of the body in adults		n
			dorso-lateral	ventro-lateral	
"Enoplidae"					
Thoracostomopsinae:					
Thoracostomopsis barbata	Ortho	–	c.27	–	1
Chaetonematinae:					
Chaetonema riemanni	Loxo II	+	*14*	–	6
Trileptiinae:					
Trileptium sp.	Ortho	–	c.4	–	2
"Oxyonchinae"					
Fenestrolaimus sp.	Ortho	–	13–14	–	2
Oxyonchus dentatus	Ortho	–	18–21	–	3
Saveljevia sp.	Ortho	–	16	–	1
"Enoplolaiminae"					
Enoplolaimus connexus	Ortho	–	c.27	–	1
Enoplolaimus propinquus	Ortho	–	10–12	–	2
Mesacanthion diplechma	Ortho	–	22–26	–	5
"Enoploidinae"					
Enoploides labrostriatus	Ortho	–	17–19	–	3
Epacanthion buetschlii	Ortho	–	26–27	–	3
Enoplinae:					
Enoplus brevis	Loxo II	+	16	14	1
Rhabdodemaniidae					
Rhabdodemania minor	Ortho	+	c.19	c.19	1
"Anoplostomatidae"	⎧Loxo I	+	6–7	–	6
Anoplostoma vivipara	⎨seldom				
	⎩Loxo II				
Pandolaimus latilaimus	Ortho	+	5	5	1
Oncholaimidae					
Pelagonematinae	?	?	?	?	–
Krampiinae	?	?	?	?	–
Oncholaimellinae:					
Oncholaimelloides vonhaffneri	Ortho	+	+	+	1
Viscosia rustica	Ortho	+	+	+	1
Adoncholaiminae:					
Adoncholaimus thalassophygas	Ortho	+	c.22	c.28	1
Oncholaiminae					
Oncholaimus brachycercus	Ortho	+	+	+	+
Pontonematinae:					
Pontonema ditlevseni	Ortho	+	+	+	1
Octonchinae	?	?	?	?	–
"Dioncholaiminae"	?	?	?	?	–
Enchelidiidae					
"Thoonchinae":					
Ditlevsenella danica	Ortho	+	+	+	1
"Eurystomininae":					
Belbolla asupplementata	Ortho	+	+	+	+
Enchelidiinae:					
Calyptronema maxweberi	Ortho	+	+	+	–
Rhaptothyreidae	?	?	?	?	–

"Desmoscolecida" : *Desmoscolex, Greeffiella, Haptotricoma, Meylia, Tricoma.*
"Monhysterida" : *Anticyclus* (*A.* sp, 1♂ : L=4470μm), *Daptonema, Desmolaimus, Eleutherolaimus, Gammarinema, Gonionchus, Monhystera, Parasphaerolaimus, Siphonolaimus* (*S. cobbi,* 1♀ : L=4500μm), *Sphaerolaimus, Theristus, Tubolaimoides, Valvaelaimus, Xyala.*
"Desmodorida" : *Desmodora, Epsilonema, Ixonema, Leptonemella, Metachromadora, Microlaimus, Monoposthia, Onyx, Richtersia, Spirinia* (*S. laevis,* 1♂ : L=3180μm), *Synonema, Xenella.*
"Chromadorida" : *Chromadora, Comescoma, Dorylaimopsis, Ethmolaimus, Halichoanolaimus, Latronema, Neochromadora, Paracanthonchus, Paracyatholaimus, Sabatieria* (*S. celtica,* 1♀ : L=3470μm), *Synonchiella.*
Dorylaimina : *Actinolaimus, Aporcelaimus, Dorylaimus, Eudorylaimus, Longidorus, Paractinolaimus*; several of the species studied are over 3mm in length.
Mononchina : *Judonchulus, Mononchus, Mylonchulus, Prionchulus.*
Mermithoidea : adult animals of two species. 1♂ : L=5800μm.
Rhabditida : *Cephalobus, Diplogaster, Mononchoides, Panagrolaimus, Rhabditis.*
Tylenchida : *Aphelenchoides, Helicotylenchus, Hemicycliophora, Pratylenchus, Tylenchorhynchus.*
Strongylida : *Chabertia* sp. and *Dictyocaulus* sp. from the intestinal tract of wild deer.

4.3.3. Postembryonic data

The postembryonic development of metanemes has been recorded in 9 species of the "Enoplia" (Table 2). During the course of this development within a species, the early juvenile stages always have fewer metanemes than the adult animals, but with two exceptions, the structure and arrangements of the metanemes do not change. The two exceptions are as follows: 1) In *Deontostoma* the metanemes lie nearer to the edge of the lateral epidermal border in early juvenile stages than they do in adult animals. 2) In *Enoplus* all juvenile stages have only type I dorsolateral loxometanemes, whereas the adult animals have type II dorsolateral and ventrolateral loxometanmes.

4.3.4. Discussion

32. Phylogenetic assessment of the existence of metanemes:

As far as is known, metanemes occur only within the "Enoplida". They are always built on the same principle, and the anterior end of a metaneme always lies further away from the lateral line of the body than does the posterior end of the nearest metaneme in front. It is concluded from their uniqueness and their fundamental uniformity

PHYLOGENETIC ASSESSMENT

Table 2: Number of metanemes on each side of the body in juvenile stages and in adults in 9 species from 5 families of the "Enoplida".

Taxon	Number of metanemes on each side of the body:				
	in juvenile stages				in adults
	I	II	III	IV	
"Ironidae"					
Parironus bicuspis	5		17–19		16–18
Tripyloididae					
Bathylaimus australis			7		13–16
Leptosomatidae					
Deontostoma arcticum	c.22				c.115
"Enoplidae"					
Oxyonchus dentatus	0	1	12	20	18–21
Mesacanthion diplechma		5	26	26	22–26
Epacanthion buetschlii	4	27	26	27	26–27
Enoplus brevis	3	3	4	0	30
Chaetonema riemanni	1	4	13	14	14
"Anoplostomatidae"					
Anoplostoma vivipara	0	0	2	6	6–7

in structure and position that metanemes can only have evolved once. Thus the existence of metanemes represents a criterion by which the holophyly can be established of a taxon 'A', which contains at least all studied species which have metanemes. The extent of Taxon A might be more exactly described as follows.

1) Table 1 shows that metanemes were found in species from 13 of the 19 families of the "Enoplida" listed by Gerlach & Riemann (1974). With the exception of *Tobrilia*, the representatives of the Monochromadorinae (all "Tripylidae") and *Porocoma* ("Oxystominidae"), all species of the 13 families concerned can be combined on the basis of other criteria (see chapter 5 of this work) to form holophyletic subtaxa within Taxon A. From the single origin of the metanemes and the holophyly of the subtaxa it must be concluded that within these subtaxa metanemes occur at a primary level from time to time, and thus also occur in those species which it was not possible to study. Consequently Taxon A encompasses at least the "Tripylinae", "Tobrilinae" (except *Tobrilia*), "Ironidae", Tripyloididae, "Oxystominidae" (except *Porocoma*), Leptosomatidae, Triodontolaimidae, Anticomidae, Phanodermatidae, "Enoplidae", Rhabdodemaniidae, "Anoplostomatidae", Oncholaimidae, Enchelidiidae. In

particular then the Tripyloididae also belong to Taxon A; their systematic position was hitherto uncontested (see Introduction). In instances where metanemes are absent in species from the families and subfamilies listed, this absence must represent, according to the above argument, a secondary condition.

2) With the exception of *Adorus* ("Alaimidae") and *Syringolaimus* ("Araeolaimida"), no genus of the Monochromadorinae, "Prismatolaimidae", Cryptonchidae, Trefusiidae, "Alaimidae", Lauratonematidae, Rhaptothyreidae (all "Enoplida") or of any other nematode order can be specified which forms a holophyletic group with a subtaxon of Taxon A on the basis of any criteria.

3) Within Taxon A there are many large species (longer than 3mm), all of which possess metanemes; metanemes can only be absent in smaller species (e.g. in *Halalaimus gracilis*; L=790μm). Outside Taxon A most freeliving species are smaller than 2mm, so when dealing with small species it is impossible to decide whether the absence of metanemes outside Taxon A is primary or secondary. However, metanemes are also absent in all the larger species outside Taxon A which have been studied, so it seems that the absence of metanemes, at least in these species, is probably primary.

From points 1 to 3 it follows that in addition to those taxa listed under point 1 only *Adorus* and *Syringolaimus* can be considered as belonging to Taxon A and that the absence of metanemes is probably primary in all species outside Taxon A. In this way the extent of Taxon A has been established using current knowledge. In chapter 5 of this work, Taxon A is defined as the order Enoplida.

33. Phylogenetic assessment of the existence of dorsolateral and ventrolateral metanemes:

According to Table 1, only in three families (Rhabdodemaniidae, Oncholaimidae, Enchelidiidae) do all the species studied have both dorsolateral and ventrolateral metanemes, whereas species with dorsoventral and ventrolateral metanemes, as well as species with only dorsolateral or with only ventrolateral metanemes occur in 9 families ("Tripylidae", "Ironidae", Tripyloididae, "Oxystominidae", Leptosomatidae, Phanodermatidae, Anticomidae, "Enoplidae", "Anoplostomatidae"); the one species of Triodontolaimidae only has dorsolateral metanemes. From these data it emerges that, within Taxon A (discussion point 32), the combined occurrence of dorsolateral and ventrolateral metanemes in one species is very probably plesiomorphic and the single occurrence of dorsolateral *or* ventrolateral metanemes is apomorphic. Reason: if the opposite were the case, then either the dorsolateral or the ventrolateral metanemes must have evolved repeatedly and independently from the remaining metanemes. This would, in general, contradict the idea that metanemes only evolved once, and in particular, it would be at variance with the structural similarity between dorsolateral and ventrolateral metanemes.

34. Phylogenetic assessment of the difference between loxometanemes of types I and II and orthometanemes:

Most of the species studies in which metanemes were found had, both as adults and juveniles, either only loxometanemes of type I, only loxometanemes of type II or only orthometanemes. The following species were the only exceptions: in *Deontostoma arcticum* (Leptosomatidae), *Paranticoma tubuliphora* (Anticomidae) and *Phanoderma campbelli* (Phanodermatidae) both type I loxometanemes and orthometanames occur; in *Anticoma trichura* (Anticomidae) loxometanemes of both types I and II are found; and in *Enoplus brevis* ("Enoplidae") adult animals have dorsolateral and ventrolateral loxometanemes of type II, and juveniles have only dorsolateral loxometanemes of type I. These exceptions allow one to draw the following conclusions:

1) Since the difference between loxometanemes of type I and orthometanemes is to some extent only poorly defined in *D. arcticum, P. tubuliphora* and *P. campbelli*, it is considered a plesiomorphic feature within Taxon A (see discussion point 32) that a species has both loxometanemes of type I and orthometanemes. Consequently it is considered apomorphic within Taxon A that either of the two types occurs on its own within a species.

2) The postembryonic development of metanemes in *Enoplus brevis* offers circumstantial evidence that loxometanemes of type I should be considered apomorphic vis-à-vis loxometanemes of type II. This argument is supported by the observation that the dorsolateral or the ventrolateral loxometanemes of type II are always dominant within a species or may even occur on their own, something which, according to discussion point 33, is considered to be an apomorphy within the taxon.

35. Phylogenetic assessment of the existence of a caudal filiament:

In instances where metanemes occur within a species, they usually have a caudal filament. This filament is absent only in *Ironus*, the Phanodermatidae and in 5 of the 7 subfamilies of the "Enoplidae". Since the species, where metanemes do not have caudal filaments, do not, on the basis of any other criteria, form a sister group with those species where metanemes do have caudal filaments, the existence of a caudal filament is considered plesiomorphic and the absence apomorphic within Taxon A (see discussion point 32).

36. Indication of possible polyphylies within the "Enoplida":

With regard to metaneme features, some families of the "Enoplida" are at times homogenous so that there is no indication of possible polyphylies (e.g. Tripyloididae, "Oxystominidae", Phanodermatidae, Oncholaimidae, Enchelidiidae). However, the five families "Tripylidae", "Ironidae", Anticomidae, "Enoplidae", and "Anoplostomatidae" are very heterogenous as regards metaneme features and are therefore possibly polyphyletic in their present extent, for sometimes their metanemes show marked differences in structure and arrangement (Table 1) and sometimes transitional forms are absent.

37. Phylogenetic assessment of the metanemes of the Enoplinae:

According to current knowledge, the Enoplinae are the only taxon with the following combination of features: loxometanemes of type II occur dorsolaterally and ventrolaterally in adult animals in more or less equal numbers, and loxometanemes of type II develop from loxometanemes of type I during the course of postembryonic development. From this it is concluded that loxometanemes of type II have evolved independently from loxometanemes of other taxa. Consequently, the combination of features is considered to be holapomorphic.

38. Phylogenetic assessment of the metanemes of the Oncholaimoidea:

All species of Oncholaimoidea studied so far (Oncholaimidae and Enchelidiidae) have numerous dorsolateral and ventrolateral orthometanemes, which show such conformity in structure, delicateness and position that this conformity is considered a holapomorphic feature of the Oncholaimoidea.

39. Phylogenetic assessment of the metanemes of the Tripyloididae:

The Tripyloididae are the only taxon in which exclusively, or almost exclusively, ventrolateral loxometanemes of type II occur. On account of the uniqueness of this characteristic, it is considered a holapomorphic feature of the Tripyloididae.

40. Phylogenetic assessment of the metanemes of the "Enoplidae" excluding the Enoplinae and Chaetonematinae:

Dorsolateral orthometanemes without caudal filaments occur in isolation within the "Enoplida" only in the "Enoplidae" (excluding

Enoplinae and Chaetonematinae), and are a constant characteristic of this species group. It is thus considered a holapomorphy of the species group concerned.

41. Phylogenetic assessment of the metanemes of the Triodontolaimidae and "Tobrilinae" (except *Tobrilia*):

The presence of dorsolateral loxometanemes of type II is considered a holapomorphic feature in Triodontolaimidae on the one hand and in "Tobrilinae" (except *Tobrilia*) on the other, because, as far as is known, the feature is constant in these taxa and because it is highly probable that neither taxon is either related to the other or to any other taxa which possess this characteristic; the characteristic has therefore probably evolved independently in the two taxa.

42. Phylogenetic assessment of the metanemes of *Ironus* ("Ironidae") and the Tripylinae:

In *Ironus* and in the Tripylinae alone, dorsolateral and ventrolateral orthometanemes occur in strongly alternating series. Since neither taxon is closely related to the other, this characteristic is considered holapomorphic for each of them respectively.

4.4 Gonads

Nematodes are normally unisexual. Only occasionally do hermaphrodites and intersexes occur. The female sexual opening (vulva) is situated ventrally and is separate from the anus; within the freeliving nematodes the two openings are joined only in *Lauratonema* ("Enoplida"). The male sexual opening is always joined with the anus to form a ventrally located cloaca. The spicule apparatus, which plays an important role in copulation, develops in this region in adult males.

Gonad characteristics are popular subjects for phylogenetic discussion because all adult nematodes possess gonads. Although differences do occur, gonad characteristics have hitherto played a surprisingly insignificant role as systematic criteria: For example:

- In species descriptions only the number of ovaries is almost always given and it has a recognized and clear-cut phylogenetic significance.
- As was pointed out in the Introduction, a structural characteristic of the ovaries – whether the ovaries are reflexed or outstretched – was accorded a systematic role by Filipjev (1918, 1934). Since

then the systematic significance of this characteristic has either been ignored or has been strongly disputed on the grounds of insufficiently detailed reasoning (Chitwood & Chitwood, 1950). It is regrettable that from about 1930 the feature has in many cases disappeared from the canon of described features.
- The number of testes has until now hardly featured as a systematic criterion: it has only been mentioned in approximately 10% of species' descriptions. It has been known for a long time now that the species of Secernentea (in the sense understood so far) have only one testis. But this feature does not appear in the diagnosis of the Secernentea in Chitwood & Chitwood (1950 : 12), Paramonov (1962 : p 340 of the 1968 translation) and de Coninck (1965 : 591), though it is used by Andrassy (1976 : 138). In addition, Clark (1961) lists the number of testes for some taxa of the Enoplia (referred to as Enoplida by him).
- Only very recently has it been shown that the position of the gonads relative to the intestine is a feature that can play a role as a systematic criterion (Lorenzen, 1978a, c). This feature has hitherto almost never been mentioned in species descriptions and consequently plays no part at all as a systematic criterion.

In order to be able to assess the phylogenetic significance of the four gonad features listed above as reliably as possible, these features were studied in 420 species of freeliving nematodes from all orders in the course of the current work; in addition, an anlaysis was made of the descriptions of almost all freeliving Adenophorea species (with the exception of many Dorylaimida species) and many freeliving Secernentea species. The results of the author's studies are presented in tabular form in section 8.2.1. of the appendix and are combined with the data from the literature in section 8.2.2. of the appendix. The most important results are given below and their phylogenetic significance is assessed in the accompanying discussion.

4.4.1. Structure of the Ovaries

Ovaries in nematodes consist of a germinal zone, a growth zone and an oviduct. The three zones are, without exception, clearly differentiated from each other in the different taxa. Until recently only two ovary types were distinguished: reflexed and outstretched. Personal observations have shown that the reflexed ovaries are represented by two distinct types, which have been termed antidromously and homodromously reflexed ovaries (Lorenzen, 1978a). The three different ovary types have the following characteristics:

a) *Antidromously reflexed ovaries* (Fig. 15a-b): the entire area of the germinal and growth zones is folded over alongside the oviduct; the

PHYLOGENETIC ASSESSMENT

Table 3: Survey of ovarian structure in freeliving nematodes. The table was constructed using data from section 8.2.2.

	Ovaries antidromously reflexed	Ovaries homodromously reflexed
Adenophorea		
"ARAEOLAIMIDA"	Teratocephalidae "Plectidae "Rhabdolaimidae" partim Isolamiidae Aulolaimidae Haliplectidae "Leptolaimidae" except Peresianinae Bastianiidae "Axonolaimidae" partim "Cylindrolaiminae" partim "Diplopeltinae" partim "Campylaiminae" partim	"Rhabdolaimidae" partim Peresianinae ("Leptolaimidae") "Axonlaimidae" partim "Cylindrolaiminae" partim "Diplopeltinae" partim "Campylaiminae" partim "Axonolaiminae"
DESMOSCOLECIDA	Meyliinae	all Taxa except Meyliinae
"MONHYSTERIDA"	*Cyartonema* ("Siphonolaimidae") *Tubolaimoides* ("Linhomoeidae")	all Taxa except *Cyartonema* and *Tubolaimoides*
"DESMODORIDA"	all Taxa except Aponchiidae and "Microlaiminae" partim	Aponchiidae "Microlaiminae" partim
"CHROMADORIDA"	all Taxa except "Comesomatidae" partim	"Comesomatidae" except "Acantholaiminae"
"ENOPLIDA"	all Taxa except *Cytolaimium*	*Cytolaimium* (Trefusiidae)
DORYLAIMIDA	all Taxa	
Secernentea		
RHABDITIDA	Diplogasteridae	
TYLENCHIDA		all Taxa

	Ovaries homodromously reflexed
RHABDITIDA	Rhabditoidea

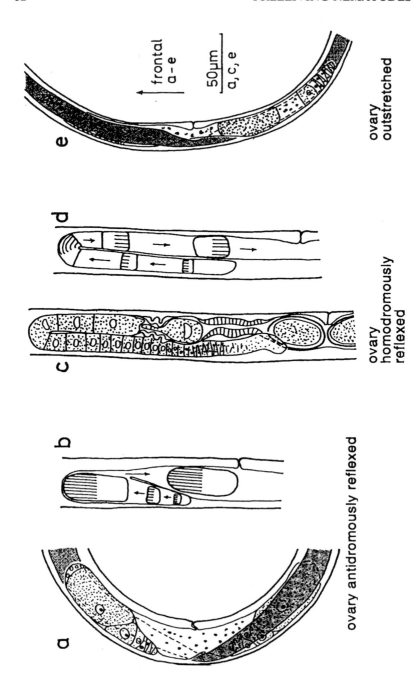

ova travel from the germinal area in one direction until they reach the fold and then continue in the opposite direction towards the vulva. The reflexed part of the genital tract is almost always shorter than the oviduct and consequently almost never reaches the region of the vulva.

b) *Homodromously reflexed ovaries* (Fig. 15c-d): the terminal zone and only a part of the growth zone are folded over the rest of the genital tract; the ova travel initially in one and the same direction to the fold and then on to the vulva. The folded part of the ovary is usually long enough for its tip to lie beyond or at the same level as the vulva.

c) *Outstretched ovaries* (Fig. 15e): the germinal zone, the growth zone and the oviduct lie in a straight line. In isolated species folds may appear at the tip of the ovaries; in these cases, unlike homodromously reflexed ovaries, the tip of the respective germinal zone generally points away from the vulva.

According to the literature and to personal observations, ovaries within freeliving nematodes are antidromously reflexed in all those Adenophorea whose ovaries were hitherto described simply as reflexed, and in the Diplogasteridae (Secernentea) too; the ovaries are, without exception, homodromously reflexed in several families of the Rhabditida (Secernentea); and, finally, they are outstretched in many Adenophorea and in the Tylenchida (Secernentea). Only 2 of the 9 existing orders of freeliving nematodes are homogenous as regards ovary structure: the Tylenchida always have outstretched ovaries and the Dorylaimida always have antidromously reflexed ovaries. Within the 7 remaining orders two differing ovary types occur: antidromously reflexed and outstretched ovaries within the "Araeolaimidae", "Desmoscolecida", "Monhysterida", "Desmodorida", "Chromadorida", and "Enoplida", and antidromously and homodromously reflexed ovaries within the Rhabditida. At the family level ovarian structure is usually very homogenous: within the 81 families of freeliving nematodes listed in section 8.2.2. of the appendix

---

Fig. 15: Ovary types in freeliving nematodes: antidromously reflexed (a – b), homodromously reflexed (c – d) and outstretched (e) ovaries. a) Ovaries from *Molgolaimus tenuispiculum* ("Desmodorida", "Microlaimidae", body length 180 μm), the anterior ovary lies to the right of the intestine and the posterior ovary to the left of it; c) anterior ovary from *Rhabditis marina* (Rhabditida, Rhabditidae, body length 1040 μm), the anterior ovary lies to the right of the intestine and the posterior ovary to the left of it; e) ovaries from *Cylindrolaimus* sp. ("Araeolaimida", "Axonolaimidae", body length 700 μm), the anterior ovary lies to the left of the intestine and the posterior ovary to the right of it. b and d are sketches. From Lorenzen (1978a).

there are only 8 families where 2 types of ovary occur; in particular, antidromously reflexed and outstretched ovaries occur within these families. The eight families are "Rhabdolaimidae", "Leptolaimidae", "Axonolaimidae" (all "Araeolaimida"); "Siphonolaimidae", "Linhomoeidae" (both "Monhysterida"); "Desmodoridae", ("Desmodorida"); Trefusiidae ("Enoplida"). In all the remaining families only a single ovary type is found. In no one family do all 3 ovary types occur.

Data in the literature, where the ovaries are classed as either reflexed or outstretched, are not always very reliable. In about 60 instances in the footnotes of section 8.2.2. of the appendix they are considered doubtful, because in each case closely related taxa invariably have a different ovarian structure according to other data in the literature and according to personal observations. In at least three cases the data in the literature can even be shown to be false, because in the text the ovaries are described as outstretched, whereas in the accompanying figures they are clearly antidromously reflexed (see footnotes 17 and 21 in section 8.2.2. of the appendix).

Very occasionally, within large taxa with antidromously reflexed ovaries, single individuals are found in which one ovary is antidromously reflexed and the other is outstretched. Examples are *Actinolaimus microdentatum* (according to Thorne, 1939 : 82) and *Crassolaibium australe* (according to Yeates, 1967b; both Dorylaimida).

4.2.2. Number and orientation of the ovaries

The occurrence of two ovaries is widespread within all nine orders of freeliving nematodes; the anterior ovary lies in front of the vulva and the posterior ovary lies behind the vulva. In *Porocoma* ("Enoplida", "Oxystominidae") alone the anterior ovary moves forward very slightly and then bends round, so that to a large extent it lies next to the posterior ovary, behind the vulva (Cobb, 1920 : 236; Gerlach, 1962 : 88). Less frequently than two ovaries, an anterior ovary may occur on its own; this feature is found in all orders except the "Desmoscolecida". A single posterior ovary occurs only seldom within the freeliving nematodes; this condition was found in a total of 18 families within the "Araeolaimida", "Enoplida", and "Dorylaimida". Table 4 shows in how many of the nine orders of freeliving nematodes two ovaries, only an anterior and/or only a posterior ovary, occur. It is noteworthy that in 49 families the number of ovaries is constant, whereas in 31 other families species with 2 ovaries occur as well as species with only an anterior and/or posterior ovary; only in one family, the Rhaptothyreidae, is the number of ovaries not known.

Table 4: Number and position of the ovaries within the individual orders of the freeliving nematodes

For each order the number of families is given in which the species always have:
- two ovaries
- only one anterior ovary
- only one posterior ovary

or have:
- two ovaries or only one anterior ovary
- two ovaries or only one posterior ovary or
- two ovaries or only one anterior ovary or only one posterior ovary

The table was constructed using information from section 8.2.2. of the appendix.

Taxon	Always 2 ovaries	Always only 1 anterior ovary	Always only 1 posterior ovary	2 ovaries or 1 anterior ovary	2 ovaries or 1 posterior ovary	2 ovaries or 1 anterior or 1 posterior ovary	Number of ovaries unknown	Total number of families
"Araeolaimida"	7	–	–	–	–	2	–	9
Desmoscolecida	2	–	–	–	–	–	–	2
"Monhysterida"	–	3	–	2	–	–	–	5
"Desmodorida"	5	1	1	1	–	–	–	8
"Chromadorida"	4	–	–	1	–	–	–	5
"Enoplida"	7	2	–	2	4	3	1	19
Dorylaimida	1	–	1	3	2	5	–	12
Total Adenophorea	26	6	2	9	10	2	1	60
Rhabditida	3	3	–	4	–	–	–	10
Tylenchida	1	8	–	2	–	–	–	11
Total Secernentea	4	11	–	6	–	–	–	21
Total freeliving nematodes	30	17	2	15	6	10	1	81

While the occurrence of two ovaries or only a single anterior ovary is hardly correlated with ovarian structure, a single posterior ovary occurs almost exclusively in nematodes with antidromously reflexed ovaries; the only exceptions are *Synodontium* and *Odontophoroides* (both "Araeolaimida", "Axonolaimidae"), which have a single, posterior, outstretched ovary.

4.4.3. Number and orientation of the testes

Since the number and orientation of the testes are cited in only approximately 10% of species descriptions, the following information is based mostly on the author's own results, which are presented in tabular form in section 8.2.1. of the appendix.

While the occurrence of two ovaries is common within all orders of freeliving nematodes, two testes only occur within all orders of freeliving Adenophorea, whereas all Secernentea, as they were hitherto understood, have only one testis, which is thought to be homologous with the anterior testis of the Adenophorea (see discussion point 44) and which is always folded over at the tip in the Rhabditida. It is still unknown to what extent the occurrence of only a single testis is common within the freeliving Adenophorea. The following picture emerges from information dispersed throughout the literature and from the author's own research (sections 8.2.1. and 8.2.2. of the appendix):

As far as is known, all species of the following 16 sub-families and families of the freeliving Adenophorea only have one single anterior testis:

"Araeolaimida":	Teratocephalidae, Wilsonematinae (testis folded), "Rhabdolaimidae".
Desmoscolecida:	Desmoscolecinae, Greeffiellinae.
"Monhysterida":	Monhysteridae (sensu Lorenzen, 1978c).
"Desmodorida":	"Desmodoridae" (excluding "Microlaiminae" with outstretched ovaries), "Xenellidae", "Richtersiidae", Draconematidae, Epsilonematidae.
"Chromadorida":	"Chromadoridae" (excluding *Ethmolaimus*), "Acantholaiminae".
"Enoplida":	Tripyloididae, "Alaimidae".
Dorylaimida:	Diphtherophoridae.

In at least 18 additional subfamilies and families of the freeliving Adenophorea a few species have only a single anterior testis, while most of the remaining species have two testes.

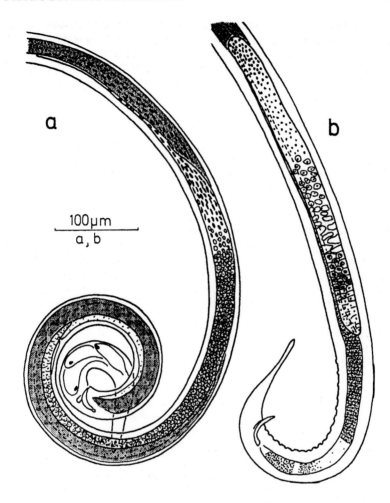

Fig. 16: Occurrence of a single posterior testis in freeliving nematodes. In both examples the posterior testis lies at least in the germinal and growth zone to the left of the intestine. a) *Lauratonema* aff. *spiculifer* ("Enoplida", Lauratonematidae) adult male from the shore of the island San Andrés (Carribean), 6th March, 1974; b) *Pomponema multipapillatum*, adult male from muddy sand 80 m deep, Skagerrak, 30th June 1976.

Until recently it was completely unknown whether there were nematodes which only had a single posterior testis. This seemed plausible as there are nematodes with only a single posterior ovary. Cobb (1920: 319) and Riemann (1976: 300) report that in *Pomponema litorium* (=*Anaxonchium l.*) and *P. multipapillatum* respectively, and in an undescribed *Pomponema* species (all "Cyatholaimidae") there is only one testis which curves backwards along its entire length. However, it remained unclear from these data whether the testis was an anterior or posterior one. Personal research has shown that nematodes with only a posterior testis do in fact exist; they were found in the following nematodes which come from three orders of the Adenophorea:

"Araeolaimida", "Leptolaimidae": *Antomicron elegans, Aphanolaimus aquaticus, Leptolaimus limicola, L. papilliger, Leptolaimoides thermastris, Manunema annulata, M. proboscidis*;
"Chromadorida", "Cyatholaimidae": *Pomponema multipapillatum* (Fig. 16b);
"Enoplida", Oncholaimidae: *Adoncholaimus thalassophygas*.

In the cases listed above it was possible to identify the single testis because 1) the entire section forming the germinal and growth zones is folded posteriorly against the remaining part of the genital tract (Fig. 16), and because 2) in the Lauratonematidae and in all those species where the anterior ovary invariably lies to the right of the intestine and the posterior ovary to the left of the intestine, the single posterior testis also lies to the left of the intestine; in the Adenophorea the testes and ovary always show agreement as regards their position in relation to the intestine (see section 4.4.4.).

There is no relationship between the occurrence of only one testis and the occurrence of only one ovary, because in a large number of families there are species where in one sex there are two gonads and in the other only one.

In species where two testes are present, they always lie one behind the other. The anterior testis usually faces forwards and the posterior testis usually faces backwards. The testes open into a common sperm sac, which is joined to the exterior by a single spermatic duct. The following six species, according to the literature are the only species where both testes face forwards: *Anticoma typica* (according to Cobb, 1891), *A. lata* (according to Cobb, 1889; both "Enoplida", Anticomidae); *Ironella prismatolaima* (according to Cobb, 1920, his formula; "Enoplida", "Ironidae"); *Alaimella truncata* (according to Cobb, 1920), *Cynura klunderi* (according to Murphy, 1965; both "Araeolaimida", "Leptolaimidae") and *Araeolaimus gajevskii* (according to Paramonov, 1929; "Araeolaimida", "Axonolaimidae"). Personal research has shown that the existence of two forward facing testes is much more common than was previously thought: numerous species with this feature were found in the families "Leptolaimidae" and "Axonolaimidae" within the order "Araeolaimidae"; in many of these cases the posterior testis has a fold, which however never extends further back than the point at which the anterior and posterior testes join. The occurrence of two forward pointing testes in *Anticoma*, as claimed by Cobb, has been confirmed by personal observations on *Anticoma trichura*; only a single, forward-pointing testis was found in *A. acuminata*, the other *Anticoma* species available to the author.

Two forward pointing testes are also occasionally observed in *Meloidogyne* (Tylenchida). This observation is of interest, because the occurrence of only one testis is otherwise constant in the Secernentea, as they were hitherto understood. Triantaphyllou has found an explanation for this exception (according to de Coninck 1965a : 202): males with two testes, unlike normal males which only have one testis, have developed from juvenlies which were originally intended as females and which had developed two potential ovarian structures, but which later turned into male animals.

4.4.4. Position of the gonads relative to the intestine

The position of the gonads relative to the intestine has hitherto been noted in only approximately 0.1% of species descriptions; the following information is therefore based almost exclusively on data from personal research, which are presented in tabular form in section 8.2.1. of the appendix.

The gonads are usually situated laterally or subventrally to the intestine. They are positioned ventrally only in a few cases, e.g. in *Manunema* ("Araeolaimida", "Leptolaimidae"), *Greeffiella* ("Desmoscolecida"), "Tripylidae", Odontolaiminae, Tripyloididae, Triodontolaimidae (all "Enoplida").

The gonads are never situated dorsally to the intestine. Where the gonads lie laterally or subventrally to the intestine there are four combinations of gonad positions (Fig. 17), all of which occur:

— Anterior gonad on the left and posterior gonad on the right of the intestine $\genfrac{(}{)}{0pt}{}{l}{r}$
— Anterior gonad on the right and posterior gonad on the left of the intestine $\genfrac{(}{)}{0pt}{}{r}{l}$
— Anterior and posterior gonads both on the left of the intestine $\genfrac{(}{)}{0pt}{}{l}{l}$
— Anterior and posterior gonads both on the right of the intestine $\genfrac{(}{)}{0pt}{}{r}{r}$.

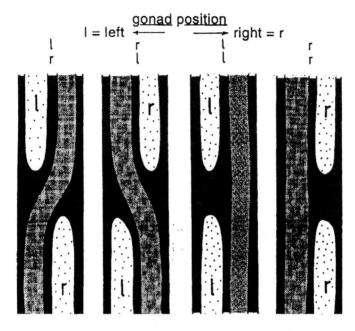

Fig. 17: Diagram to show the 4 possible positions of the anterior and the posterior gonad relative to the intestine in freeliving nematodes (viewed from the dorsal side of the body). The gonads are represented by dotted areas and the intestine by grey shaded areas.

In many cases only the germinal and/or growth zone, and not the entire gonad lies laterally or subventrally to the intestine. For the use in descriptions of "gonads on the left of the intestine" and "gonads on the right of the intestine" the criterion should be that at least the germinal zone of the gonads concerned should be situated laterally or subventrally to the right or left respectively of the intestine.

Several generalizations can be made as regards the position of the gonads relative to the intestine:

a) The males and females of a species almost always show conformity where gonad position is concerned. If, for example, the males of a given species always have the anterior testis to the left of the intestine and the posterior testis to the right of it, then the females also have the anterior ovary to the left of the intestine and the posterior ovary to the right of it. There are only a very few exceptions: according to section 8.2.1. of the appendix, the males of the Diplogasteridae, "Richtersiidae" and some species of the "Desmodoridae" and Epsilonematidae have only one testis which lies ventrally to the intestine, whereas the females have two ovaries which lie subventrally or laterally to the intestine; finally, in *Camacolaimus tardus* ("Leptolaimidae") both testes lie to the left of the intestine, whereas the anterior ovary lies to the right of the intestine and the posterior ovary lies to the left of it.

b) The position of the gonads relative to the intestine can be variable or constant within a given species. It is possible to distinguish between 3 types of intra-species variability regarding gonad position:

Type (1): ($^{lrlr}_{rllr}$ in Table 5): the position of the gonads relative to the intestine is still completely variable, i.e. within species all four combinations of lateral or, more commonly, subventral gonad position are still found.

Type (2): ($^{lr}_{rl}$ and $^{lr}_{lr}$ in Table 5): within species the position of the gonads relative to the intestine is partly fixed, in that either the anterior and posterior gonad are always situated on different sides $^{lr}_{rl}$ of the intestine, or they are always situated on the same side of it $^{lr}_{lr}$; in other words, for every gonad position within a given species there is only one other possible gonad position, that of a mirror image.

Type (3): ($^{l}_{r}$, $^{r}_{l}$, $^{l}_{l}$ and $^{r}_{r}$ in Table 5): within species the position of the gonads relative to the intestine is always constant, and is identical in both males and females of a given species. All four combinations of intraspecies gonad position are found within the freeliving nematodes (Table 5).

c) In many taxa, from genus to super family level, all species show the same pattern of gonad position (section 8.2.1. of the appendix).

d) Attention should be drawn to two correlations between ovarian structure and gonad position:

— Species in which all four combinations of gonad position occur almost all belong to the Adenophorea with antidromously reflexed ovaries; only very seldom do they belong to the Adenophorea with outstreched ovaries (*Ascolaimus*, "Axonolaimidae") or to the Secernentea with homodromously reflexed ovaries (Bunonematidae).

— The position of the anterior gonad constantly to the left and the posterior gonad constantly to the right of the intestine $^{l}_{r}$ is to be found almost always in the Chromadoria with outstretched ovaries (Xyalidae, "Comesomatidae" partim), the constant position of the anterior gonad to the right and the posterior gonad to the left $^{r}_{l}$ predominates in the Chromadoria with antidromously reflexed ovaries and in the Rhabditida, and the constant position of both gonads to the left $^{l}_{l}$ or to the right $^{r}_{r}$ of the intestine occurs almost only in the "Enoplida" (section 8.2.1. of the appendix and Table 5).

Table 5: Correlation between gonad position and ovarian structure in freeliving nematodes.

$\frac{l}{r}$: anterior gonad to the left and posterior gonad to the right of the intestine

$\frac{r}{l}$: anterior gonad to the right and posterior gonad to the left of the intestine

$\frac{l}{l}$: anterior and posterior gonad to the left of the intestine

$\frac{r}{r}$: anterior and posterior gonad to the right of the intestine

The above symbols are placed together to show the corresponding gonad positions within a given species. Where a gonad is absent in both sexes, this is shown by –.

Ar: "Araeolaimida", Ds: Desmoscolecida, Mo: "Monhysterida", De: "Desmodorida", Ch: "Chromadorida", En: "Enoplida", Do: Dorylaimida, Rh: Rhabditida, Ty: Tylenchida.

The table was constructed using data from section 8.2.1. of the appendix and is complete in comparison with Table 2 in Lorenzen (1978a).

	♀♀ with antidromously reflexed ovaries	♀♀ with outstretched ovaries

lrlr
rllr
	Ar: *Syringolaimus* ("Rhabdolaimidae")
	     Rhadinema ("Leptolaimidae")
	     Aegialoalaimus ("Axonolaimidae")
	De: *Molgolaimus* ("Microlaiminae")
	     Prodesmodora ("Microlaiminae")
	     "Desmodoridae" except "Microlaiminae" partim)
	     Monoposthiidae partim
	     "Richtersiidae"
	     Epsilonematidae partim
	     Draconematidae
	     "Choniolaimidae" partim
	     "Selachinematidae" partim
	Do: Dorylaimina partim
		Ar: *Ascolaimus* ("Axonolaimidae")

lr
rl
	Ar: Haliplectidae
	     Aulolaimidae
	Ch: "Cyatholaimidae" partim
	     "Choniolaimidae" partim
	Do: Dorylaimina partim
	     Mononchina
		Ar: "Axonolaimidae" often
		Mo: "Linhomoeidae" part.
		     Sphaerolaimidae
		De: "Microlaiminae" part.
		Ch: "Comesomatidae" (except "Acantholaiminae")

lr lr	Ar: *Euteratocephalus* (Teratocephalidae) *Rhabdolaimus* ("Rhabdolaimidae") Ch: "Achromadorinae" Do: Dorylaimina partim	Ty: Tylenchoidea

Table 5: *continued*

	♀♀ with antidromously reflexed ovaries	♀♀ with outstretched ovaries
l r		Ty: Aphelenchoidea
l r	De: Monoposthiidae partim	Mo: Xyalidae Ch: "Comesomatidae" (except "Acantholaiminae") partim

r
l
	Ar: "Plectidae"
	     "Leptolaiminae" usually
	     "Procamacolaiminae"
	     "Camacolaiminae" usually
	Ch: "Acantholaiminae"
	     "Chromadoridae" (except *Ethmolaimus*)
	     "Cyatholaimidae" partim
	En: Lauratonematidae
	Rh: Diplogasteridae ♀♀
	     (♂♂: 1 testis, ventral)
		Ds: "Desmoscolecidae" often

l
l
	En: Prismatolaiminae partim
	    Phanodermatidae
	    "Enoplidae"
	    "Anoplostomatidae"	

r
r
	En: Prismatolaiminae partim
	    Parironus ("Ironidae")
	    "Alaimidae" partim
	    Oncholaimidae
	    Enchelidiidae	

r –	Ar: *Teratocephalus* (Teratocephalidae)	Mo: Monhysteridae

ventr
ventr
	De: Epislonematidae partim
	En: "Tripylidae"
	    Tripyloididae
	    Trefusiidae partim
	    Triodontolaimidae
		Ar: Peresianinae
		    ("Leptolaimidae")

no
data
	Ar: Isolaimiidae
	    Kreisonematinae
	De: Aponchiidae
	    Pseudonchinae
	    Ceramonematidae
	    "Xenellidae"
	En: Monochromadorinae	

	♀♀ with homodromously reflexed ovaries
lrlr rllr	Rh: Bunonematidae
r l	Rh: Rhabditidae
r –	Rh: Cephalobidae Panagrolaimidae

4.4.5. Discussion

43. Phylogenetic assessment of the number of gonads:

There is still some disagreement as to whether the existence of two gonads (one anterior, one posterior) or the existence of only one anterior gonad is to be considered plesiomorphic within the nematodes. Filipjev (1921 : 79 and 90) and Chitwood & Chitwood (1950 : 136) consider the existence of two gonads to be plesiomorphic within the nematodes. Osche (1952 : 206), Sudhaus (1976 : 66) and Andrassy (1976 : 62 and 63), on the other hand, maintain that the existence of a single gonad is plesiomorphic. The author is unaware of any arguments in favour of the second view, whereas the first view is supported by the following three arguments:

Argument 1: (Chitwood & Chitwood, 1950 : 136–137): "The genital primordium is identical in that it consists of two germ cells and two epithelial cells in the first stage larva of both sexes in all nemas studied. In forms with two gonads the primordial cells are thereafter separated by somatic cells, one germ cell entering each gonad, while in forms with two gonads the primordial cells are thereafter separated by somatic cells, one germ cell entering each gonad, while in forms with one gonad the germ cells remain together. . . . it would therefore seem obvious that two gonads are the primitive nemic condition." Hirschmann (see e.g. 1971 : 59) later found that in some species of Tylenchida the genital primordium contains only one primitive sex cell nucleus, whereas in other species of the same order the two usually found were present. As the Tylenchida are not a very closely related group, Hirschmann's observation hardly reduces the strength of Chitwood & Chitwood's argument.

Argument 2: Since, according to Chitwood & Chitwood's (1950) argument cited above, the genital primordium of one sex can be considered homologous with that of the other sex, and since in most nematodes two gonads are found in at least one sex, it follows that the genetic potential for the formation of two gonads is very common within the nematodes. The occurrence of only one gonad in both the sexes is infrequent within freeliving nematodes, e.g. in the Monhysteridae (sensu Lorenzen, 1978c), in some families of Tylenchida and in some smaller taxa from different orders (see Section 8.2.2. of the appendix).

Argument 3: (Lorenzen, 1978a : 114): In many taxa from subfamily or family level where the species typically have two ovaries or two testes, there are isolated species with only one ovary or only one testis (see section 8.2.2. of the appendix). However, where the species of a taxon from the sub-family to sub-class level typically have only one testis or only one ovary, one hardly ever comes across species with two testes or two ovaries (the only exception is the occurrence of two testes in *Ethmolaimus* of the "Chromadoridae", where normally only one testis is present; *Ethmolaimus* also deviates from the other "Chromadoridae" as regards gonad position). It follows then, that a gonad can often be reduced but can never be formed anew.

On the basis of the above arguments the existence of two gonads is considered plesiomorphic and the existence of a single gonad is considered apomorphic within the nematodes.

44. Cases where the existence of a single anterior testis is considered to be holapomorphic:

The existence of a single, anterior testis is considered to be holapomorphic in the following taxa:

— "Chromadoridae" excluding *Ethmolaimus* and including *Acantholaimus* ("Chromadorida"))
— The combination of the "Desmodoridae" (excluding *Bolbolaimus* and most "Microlaiminae", but including *Molgolaimus* and *Prodesmodora*), Draconematidae and Epsilonematidae (all "Desmodorida") to form the superfamily Desmodoroidea in its new sense (Chapter 5);
— The combination of the Desmoscolecinae and Greeffiellinae (Desmoscolecida) to form the family Desmoscolecidae in its new sense (Chapter 5);
— Monhysteridae (sensu Lorenzen, 1978c; "Monhysterida");
— Tripyloididae ("Enoplida");
— Diphtherophoridae (Dorylaimida).

Reasons: a) According to discussion point 43, the existence of a single gonad is considered to be apomorphic within the nematodes. b) All species within the taxa listed have only a single, anterior testis, whereas in the related taxa two testes are normally present. c) This feature, the existence of a single, anterior testis, is linked in all cases with other features, which are characteristic of the taxa concerned and which are even considered in part to be holapomorphies.

The single testis, which is found in all Secernentea in the hitherto understood sense and in the parasitic nematodes of the super-families Mermithoidea, Trichuroidea and Dioctophymatoidea, is considered to be homologous with the anterior testis of the Adenophorea, because in the most diverse families of the Secernentea there are species where the anterior ovary and the testis are both situated on the same side of the intestine (Section 8.2.1. of the appendix). As was pointed out on page 90, anterior and posterior gonads in both sexes are positioned in the same way relative to the intestine in almost all nematodes. Since the three named superfamilies of parasitic nematodes have recently been classified as Secernentea and since some species of Mermithoidea have two testes, the existence of a single, anterior testis can no longer be considered a holapomorphic feature of the Secernentea.

45. Cases where the existence of a single posterior testis is considered to be holapomorphic:

The existence of a single, posterior testis is considered to be a holapomorphic feature of the following taxa:

— Peresianinae ("Araeolaimida");
— Lauratonematidae ("Enoplida").

The argument is the same as in discussion point 44.

46. Cases where the existence of a single, anterior ovary is considered to be holapomorphic:

The existence of a single, anterior ovary is considered a holapomorphic feature of the following taxa:

— Monhysteroidea (Monhysteridae, Sphaerolaimidae and Xyalidae) ("Monhysterida");
— "Siphonolaimidae" excluding *Cyartonema* ("Monhysterida");
— Aponchiidae ("Desmodoridae");
— Aphelenchoidea (Tylenchida).

The argument is the same as in discussion point 44.

47. One case where the existence of a single, posterior ovary is considered to be holapomorphic:

The existence of a single, posterior ovary is considered a holapomorphic feature of the following taxon:

— "Oxystomininae" excluding *Porocoma* ("Enoplida").

The argument is the same as in discussion point 44.

48. Phylogenetic assessment of the orientation of the gonads and the existence of a vulva:

In the gastrotrichs, rotifers and kinorhynchs, both gonads — where two are present — lie at more or less the same level and are similar in orientation, i.e. both gonads point anteriorly or both point posteriorly.

This similarity in orientation of both gonads is thus seen as plesiomorphic within the Nemathelminthes. Within the freeliving nematodes, both testes point forwards only in a few "Leptolaimidae", "Axonolaimidae" ("Araeolaimida") and Anticomidae ("Enoplida"), whereas in the remaining nematodes the two testes point in opposite directions (the only exception: in *Porocoma* both ovaries point posteriorly). The following interpretations arise from this:

1) The orientation of the two testes in opposite directions is an apomorphy within the nematodes and cannot be considered a holapomorphy of the nematodes because in a few species both testes face forwards.

2) The orientation of the ovaries in opposite directions is, in all probability, a holapomorphy of the nematodes.

3) Since, with the exception of a very few divergent species, all female nematodes have a vulva which has an opening separate from the anus, and since the existence of two ovaries of opposite orientations is considered to be holapomorphic for the nematodes, the formation of the two characteristics seems in all probability to be connected. The existence of a vulva is thus considered to be a further holapomorphic feature of the nematodes.

The three conclusions just drawn are in agreement with the views of Filipjev (1921 : 79 and 90) and Steiner (1921 : 75 and 78), who also consider the opposing orientation of the nematode gonads to be apomorphic regarding the remaining nemathelminthes. According to Steiner, the nematode ancestors (i.e. not the primitive nematodes) must have had two forward pointing gonads in both sexes, and "the vulva in the nematode ancestors must have had a joint opening with the anus or must have opened near by." Steiner came to this conclusion after comparing the structure of the nematodes with that of the rotifers. In the meantime, the gastrotrichs have been seen as the closest relatives of the nematodes on the basis of similarities in embryology and morphology (Teuchert, 1968 and 1977). In the gastrotrichs however, the male sexual orifice may lie directly anterior to the anus in the central or in the anterior body region (Remane, 1936), and the eggs are not laid through a sexual orifice, but through a rupture in the dorsal body wall (Teuchert 1968). Thus, in contrast to Steiner's view, there is no longer any evidence for the existence of a cloaca in the ancestors of the nematodes.

49. Phylogenetic assessment of the antidromous reflex in ovaries:

Within the nematodes the existence of antidromously reflexed ovaries, in contrast to the existence of outstretched or homodromously reflexed ovaries, is considered plesiomorphic for the following reasons:

— Antidromously reflexed ovaries are much more common within the nematodes than are outstretched and homodromously reflexed ovaries (Table 3).
— The variable position of the gonads relative to the intestine — a plesiomorphic feature (discussion point 52) — is almost exclusively limited to nematodes with antidromously reflexed ovaries; the only known exceptions are *Ascolaimus* ("Araeolaimida") which has outstretched ovaries and *Bunonema* (Rhabditida), which has homodromously reflexed ovaries.
— With the exception of *Cytolaimium* (outstretched ovaries, Trefusiidae), all nematodes, which show particularly plesiomorphic traits in the structure and arrangement of the 6 labial and 6+4 cephalic sensilla, also have antidromously reflexed ovaries (Onchulinae, Trefusiidae, see discussion point 10).

50. Phylogenetic assessment of the homodromous reflex in ovaries:

Within the freeliving nematodes homodromously reflexed ovaries are restricted to the Rhabditoidea alone (sensu Goodey, 1963). On account of its uniqueness the feature is considered a holapomorphy of the Rhabditoidea. Since only species with antidromously reflexed ovaries occur otherwise within the Rhabditida, homodromously reflexed ovaries evolved in all probability from antidromously reflexed ovaries and not from outstretched ovaries.

51. Phylogenetic assessment of the outstretched condition in ovaries:

According to discussion point 49 the existence of outstretched ovaries, in contrast to the existence of antidromously reflexed ovaries, is considered apomorphic. The outstretched condition in ovaries is considered a holapomorphy in the following taxa:

a) Tylenchida (Secernentea).
b) *Cytolaimium* (Adenophorea: "Enoplida").
c) Peresianinae (Adenophorea: "Araeolaimida").
d) Taxon B, which contains all species with outstretched ovaries from the "Araeolaimida", "Monhysterida", "Desmodorida" and "Chromadorida", with the exception of Peresianinae, "Microlaiminae", Aponchiidae and "Rhabdolaimidae". Taxon B is defined in Chapter 5 as the order Monhysterida and corresponds largely to the "Monhysterata" of Filipjev (1934).

Reasons:

For a): Within the Secernentea outstretched ovaries occur only in the

Tylenchida; this order is always characterized as a holophyletic taxon on account of the stylet in its buccal cavity.

For b): Within the "Enoplida", outstretched ovaries occur only in *Cytolaimium*; this genus belongs to the Trefusiidae ("Enoplida") on account of the structure of the amphids and the structure and arrangement of the labial and cephalic sensilla.

For c): On account of the structure of the pharynx, the existence of pre-cloacal tubules in adult males and the existence of a single, posterior testis, the Peresianinae (sole genus: *Manunema*) are classed in Chapter 5 in the sub-order of the Leptolaimina, within which only antidromously reflexed ovaries otherwise occur. Moreover, in contrast to Riemann et al. (1971), *Manunema* cannot be seen as a connecting link between the "Araeolaimida" and "Desmoscolecida", because stalked somatic spines do not occur in the "Meyliidae", and in many species of Tricominae.

For d): In the case of Taxon B there is no indication that, on the basis of other features, any subtaxon of Taxon B forms a holophyletic group with other nematodes in which antidromously reflexed ovaries occur. If there were such a holophyletic group, outstretched ovaries must have developed in this group from antidromously reflexed ovaries independently from Taxon B.

Despite the presence of outstretched ovaries, the Aponchiidae and — in contrast to Filipjev (1934) and in agreement with Chitwood & Chitwood (1950: 143) — the "Microlaiminae", are not placed in Taxon B, because in these groups at least some species have a twelvefold pleated vestibulum. In Chapter 5 of the present work this feature, in agreement with Filipjev (1934), is considered to be a holapomorphy of the suborder Chromadorina, in which all species, with the exception of most "Microlaiminae", and the Aponchiidae, have antidromously reflexed ovaries. The "Microlaiminae" and Aponchiidae cannot at present be assigned to any holophyletic group because it is not yet possible to ascertain in these groups whether outstretched ovaries evolved from antidromously reflexed ovaries independently from one another or not.

Within the "Desmoscolecida", most species have outstretched ovaries with the exception of species of the Meyliinae which have antidromously reflexed ovaries. Since the order is considered holophyletic on account of its cephalic setae, outstretched ovaries must have developed from antidromously reflexed ovaries independently within Taxon B and the "Desmoscolecida".

Within the "Rhabdolaimidae", some species have antidromously reflexed ovaries and other species have outstretched ovaries; this is a marked feature in the genus *Rhabdolaimus*. The outstretched condition of the ovary must then have developed within the "Rhabdolaimidae" independently from the same condition in Taxon B, so that the family is not placed in Taxon B.

According to Chitwood & Chitwood (1950 : 143), within the "Comesomatidae", (excluding "Acantholaiminae"), both ovaries in *Mesonchium* and at least the anterior ovaries in *Comesoma* should be reflexed, whereas the ovaries of the remaining "Comesomatidae" (excluding "Acantholaiminae") and at least the posterior ovary of *Comesoma* should be outstretched. Personal observations (Section 9.2.1. of the appendix) have produced nothing to suggest that reflexed ovaries ever occur within the "Comesomatidae" (excluding "Acantholaiminae"); the germinal zone of an ovary may, however, occasionally be folded, with the result that Chitwood & Chitwood probably mistook the folds of the germinal zone for a reflex in the ovary. Thus, on the basis of ovarian structure, there is no reason to believe that the "Comesomatidae" (excluding "Acantholaiminae") belong to Taxon B.

Since the outstretched condition of the ovaries have probably occurred repeatedly within the nematodes, its value as proof of holophly for Taxon B is not very great.

52. Phylogenetic assessment of the position of the gonads relative to the intestine:

Within the freeliving nematodes a variable position for the gonads relative to the intestine ($^{lrlr}_{rl|lr}$, see Table 5) is very probably plesiomorphic in contrast to a semi-fixed ($^{lr}_{rl}$ or $^{lr}_{lr}$) or rigidly fixed position ($^{l}_{r}$, $^{r}_{l}$, $^{l}_{l}$ or $^{r}_{r}$); Reason: where, within a nematode taxon from the subfamily level to the super-family level, the gonad position is variable, other significant features are also generally plesiomorphic within larger taxa (e.g. the structure and arrangement of the labial and cephalic sensilla) or are variable in their expression (e.g. the form of the amphids, in the "Desmodorida" with antidromously reflexed ovaries besides the pharynx), with the result that in part the taxa concerned can in no way be shown to be holophyletic; on the other hand, taxa with a fixed gonad position also generally show little variation as regards other features and can easily be shown to be holophyletic (Chapter 5 of this work).

It must also be considered possible that a fixed gonad position may have developed not only from an originally variable gonad position which gradually became fixed, but may also have developed spontaneously from another, equally constant gonad position. Examples of the second possibility are probably the taxa *Dagda* ("Leptolaimidae"), *Calligyrus* ("Desmoscolecida"), *Hofmaenneria, Steineria pilosa* (both Xyalidae) and *Prismatolaimus* aff. *verrucosus* ("Prismatolaimidae"), where the fixed gonad position is a mirror image of that of the remaining species of the families concerned (see section 9.2.1. of the appendix).

53. Cases in which the constant nature of the gonad position is considered to be holapomorphic:

The semi-fixed position of the gonads ($^{lr}_{rl}$, $^{rl}_{lr}$) or the rigidly fixed gonad position ($^{l}_{r}$, $^{r}_{l}$, $^{l}_{l}$, $^{r}_{r}$) is considered to be holapomorphic for a number of taxa. The following reasons form the basis of this interpretation:

— According to discussion point 52 the semi-fixed and the rigidly fixed gonad positions are considered apomorphic relative to the variable gonad position.
— All or almost all species of the taxa concerned have developed the characteristic of the gonad position each in the same way (the exceptions are listed in discussion point 52).
— The particular gonad position does not normally occur in the species of the closely related taxa.
— The feature of the particular gonad position is at times linked with other characteristic features, which can even in part be considered as holapomorphies of the taxa concerned.

The taxa for which the particular gonad position suggests holophyly, are listed in the following section.

a) The constant position of the gonads on the same side as the intestine ($^{lr}_{lr}$) indicates holophyly in at least the following taxa:
— "Achromadorinae" ("Chromadorida");
— Tylenchida.

b) The constant position of the anterior gonad to the left and the posterior gonad to the right of the intestine ($^{l}_{r}$) indicates holophyly in at least the following taxa:
— Xyalidae ("Monhysterida");

c) The constant position of the anterior gonad to the right and the posterior gonad to the left of the intestine ($^{r}_{l}$) indicates holophyly in at least the following taxa:
— "Chromadoridae" excluding *Ethmolaimus* and including *Acantholaimus* ("Chromadorida");
— Lauratonematidae ("Enoplida");
— From Table 5 it appears that the constant gonad

position (⌐) within the Chromadoria occurs almost only in taxa of the "Araeolaimida" and "Chromadorida", which both have antidromously reflexed ovaries. The scattered and largely isolated appearance of this constant gonad position is seen as a homoiology (analogous development of given feature of a homologous structure) and is taken to be an indication of a possible holophyly of a Taxon C, which among other things contains all species with antidromous reflexed ovaries from the "Araeolaimida" (excluding Isolaimiidae), "Desmodorida" (excluding Xenellidae), "Monhysterida", "Chromadorida" and the Prismatolaiminae ("Enoplida"); Taxon C is referred to as the order Chromadorida in Chapter 5 of this work.

— The constant gonad position (⌐) is also common within the freeliving Rhabditida. This finding is interpreted in the same way as for Taxon C (Chromadorida).

d) The constant position of the anterior gonad to the right of the intestine, where the posterior gonad is absent (⌐) in both the sexes, indicates holophyly in the following taxon at least:

— Monhysteridae (sensu Lorenzen, 1978c).

e) The constant position of both gonads to the left of the intestine (⌐) indicates holophyly in at least the following taxa:

— "Enoplidae" together with "Anoplostomatidae", Anticomidae and Phanodermatidae; this taxon is referred to as the super-family Enoploidea in Chapter 5.

f) The constant position of both gonads to the right of the intestine (⌐) indicates holophyly in the following taxon at least:

— Oncholaimidae together with Enchelidiidae; this taxon is referred to as the super-family Oncholaimoidea in Chapter 5.

4.5. Supplementary copulatory organs

In adult male nematodes auxillary organs, which play a part in copulation, are common; they are known as supplementary copulatory organs. They may be in the form of papillae, tubules or copulatory bursae and are located ventrally or subventrally (though seldom ventrally and subventrally). Adult females only very occasionally have auxillary copulatory organs; they are present as ventrally located tubules in *Halaphanolaimus* sensu Boucher & Bovee (1972) and in *Stephanolaimus* (both "Araeolaimida", "Leptolaimidae"), and are present as ventrally located papillae in *Linhomoeus hirsutus* ("Monhysterida', "Linhomoeidae") and *Mesacanthion arabium* ("Enoplida", "Enoplidae").

4.5.1. Supplementary copulatory organs in freeliving Adenophorea

In those cases where males of the freeliving Adenophorea have supplementary copulatory organs, they are usually situated in a ventral row anterior to the cloaca and only seldom in two subventral rows, which extend into the pre-caudal and post-caudal region. A copulatory bursa is found only rarely in males of the freeliving Adenophorea.

In most cases the pre-cloacal, ventrally situated supplementary copulatory organs take the form of tubules or papillae; both types are found together only in some species of "Leptolaimidae" (e.g. *Leptolaimus cupulatus, L. leptaleus, L. papilliger*), where the cup-shaped papillae are always frontal to the foremost tubule.

The only other occurrence of cup-shaped, ventrally located pre-cloacal papillae (except in species of "Leptolaimidae") is in the "Chromadoroidae", and in particular in the "Chromadoridae", "Selachinematidae", "Choniolaimidae" and "Cyatholaimidae".

Ventrally situated pre-cloacal tubules are considerably more common within the Adenophorea than are the cup-shaped pre-cloacal papillae; they are found in the following taxa:

"Araeolaimida": — "Leptolaimidae": most species of "Leptolaiminae", "Procamacolaiminae" and Peresianinae.
— "Plectidae": most species of Plectinae, but not in any species of Wilsonematinae.
— Teratocephalidae: *Euteratocephalus* partim.

"Desmodorida": — "Desmodoridae": *Onyx, Polysigma, Sigmophoranema, Eubostrichus*.

"Chromadorida": — "Cyatholaimidae": *Paracanthonchus, Praeacanthonchus*.

"Enoplida": — Anticomidae, Phanodermatidae, "Enoplidae": the male has a pre-cloacal tubule in most species; some species do not possess a tubule (e.g. *Trileptium salvadoriense, Enoploides cirrhatus*), and one species (*Epacanthion multipapillatum*) has 9 tubules.

My own research on the pre-cloacal tubules of species from all the families listed produced the following data:

a) The tubules always lie with their distal end on the ventral line of the body. In larger tubules, the proximal end, which is situated inside the body, almost always lies to the right of the dorsoventral plane of the body and at the same time to the right of the intestine; only two exceptions were observed: 1) In all males and females of *Stephanolaimus elegans* ("Leptolaimidae"), the proximal end of the tubules lies to the right of the dorsoventral plane of the body and to the right of the intestine only in the posterior half of the body, whereas in the anterior half of the body it lies to the left of them. 2) Six males of the species *Rhadinema flexile* ("Leptolaimidae") were studied; in one male the proximal end of the tubules lay to the left of the intestine, but in the remaining 5 males it lay, as usual, to the right of it.

b) In most adult males with pre-cloacal tubules the posterior testis lies to the left of the intestine or it is absent altogether; only in a few cases does the posterior testis lie to the right of the intestine (*Aphanolaimus aquaticus, Rhadinema flexile* partim, *Leptolaimus elegans* partim; all "Leptolaimidae").

c) When the tubules are small, both the distal and proximal ends lie in the dorsoventral plane of the body.

d) The distal end of the tubules always communicates with the external world.

e) In many species of "Leptolaimidae" and "Plectidae" there is a clearly visible, single-celled gland situated on the proximal end of the tubules. This gland was not observed outside these two families.

The pre-cloacal, ventrally situated supplementary copulatory organs of the Adenophorea are usually restricted to the posterior half of the body; in the following taxa, however, they occur right into the pharyngeal region:

"Araeolaimida": — "Leptolaimidae": *Anonchus mirabilis, Halapthanolaimus* sensu Boucher & Bovée (1972), *Stephanolaimus elegans* (only tubules in these species); *Anonchus maculatus, A. mangrovi, Antomicron pratense, A. profundum, Leptolaimus cupulatus, L. leptaleus, L. papilliger, Procamacolaimus tubifer* (tubules in these species and cup-shaped papillae anterior to them); *Leptolaimus pumicosus, Deontolaimus papillatus* (only cup-shaped papillae in these species; the papillae are present only in the anterior body region in the last-named species).
— Bastianiidae: *Bastiania* partim (only papillae).

"Desmodorida": — "Desmodoridae": *Robbea tenax* (posteriorly,

"Chromadorida":	inconspicuous bristle papillae and in the pharyngeal region, sucker-like papillae). — "Chromadoridae", *Chromadorida pharetra* (three coniform papillae are present, all of which are located in the anterior body region).
"Enoplida":	— "Tripylidae": *Tripyla* (only inconspicuous papillae).
	— "Prismatolaimidae": *Prismatolaimus intermedius* (only papillae, see Gagrin & Kuzmin, 1972); *Onchulus longicauda* (only papillae, see Gerlach, 1966), *Kinonchulus* (pre-cloacally, small bristle papillae; cervically, very large papillae without bristles).
	— "Cryptonchidae": *Cryptonchus tristis* (only papillae).
	— Trefusiidae: *Halanonchus* (only papillae).
Dorylaimida:	— Trichodorida: *Trichodorus*.

Adult males of the Adenophorea seldom show anything other than pre-cloacal, ventrally situated, supplementary copulatory organs. A copulatory bursa is found in *Diplolaimelloides, Monhystrium, Tripylium* (all "Monhysterida", Monhysteridae sensu Lorenzen, 1978c), *Anoplostoma* ("Enoplida", "Anoplostomatidae") and *Oncholaimelloides* ("Enoplida", Oncholaimidae); two rows of subventrally situated papillae occur in the Plectinae ("Araeolaimida"), *Cytolaimium* ("Enoplida", Trefusiidae), Paroxystomininae ("Enoplida", "Oxystominidae"), in many species of the Leptosomatida ("Enoplida"), in several species of the Oncholaimidae and in *Vanderlinida* (Dorylaimida, Tylencholaimidae).

Ventrally and subventrally situated auxillary copulatory organs almost never occur together in one species. Only in the Plectinae ("Araeolaimida") do pre-cloacal, ventrally situated tubules and sub-ventral pre- and post-cloacal papillae occur together, whereas in *Paracanthonchus macrodon* ("Chromadorida", "Cyatholaimidae"), short, subventral, hooked bristles occur in addition to the pre-cloacal, ventrally situated tubules.

4.5.2. Supplementary copulatory organs in freeliving Secernentea

A copulatory bursa is commonly found in the adult males of the freeliving Secernentea; in many species, however, there may instead be two subventral rows of papillae or (e.g. Diplogasteridae) bristles, the majority of which are situated pre-cloacally and the minority of which post-cloacal. Ventrally situated, supplementary copulatory organs are absent altogether in the Secernentea, if, in accordance with

Gerlach & Riemann (1973), one considers the Teratocephalidae as belonging to the Adenophorea, and not to the Secernentea.

4.5.3. Discussion

54. Phylogenetic assessment of the existence of pre-cloacal tubules, of the continuation of the ventral supplementary copulatory organs towards the anterior right into the cervical region, and of the existence of subventrally located supplementary copulatory organs:

Pre-cloacal, ventrally located tubules are found only in the orders Enoplida and Chromadorida (both in the newly understood sense). They are thus restricted to those Adenophorea in which the females have antidromously reflexed ovaries. The only exception is *Manunema* which belongs to the Chromadorida and which has outstretched ovaries. Since the tubules are always basically similar in structure and since the larger tubules almost always lie with their proximal end to the right of the dorsoventral plane of the body and to the right of the intestine, their development probably depends on homology and not analogy. Since no comparable pre-cloacal tubules occur outside the nematodes, their existence is considered to be a holapomorphic feature of a Taxon D which comprises, at least, all species with pre-cloacal tubules and, at most, all nematodes. The extent of Taxon D cannot be determined at the moment with any degree of certainty. Reason: very probably the Enoplida and Chromadorida do not together form a holophyletic taxon (see section 5.1.2.), with the result that Taxon D probably contains other nematodes besides those of the two named orders. The holophyletic taxon of the lowest possible rank which contains both the Enoplida and the Chromadorida is, as far as is currently known, the entire class of the nematodes (see section 5.1.2.), thus making the latter identical with Taxon D. The pre-cloacal tubules in males of the Monhysterida (in the new sense) and Secernentea must then have been absent at the secondary level. In accordance with this view the development of homodromously reflexed and outstretched ovaries from antidromously reflexed ovaries (discussion points 50 and 51) would be linked with the loss of pre-cloacal tubules in males.

The considerations outlined above are valid in the same way for the continuation of the pre-cloacal papillae towards the anterior right into the cervical region and in a similar way for the existence of subventrally located, auxillary copulatory organs (the latter also occur in the Monhysterida — as understood in the new sense — and in the Secernentea), as long as each conglomeration of features is considered homologous. In both cases, the diverse occurrence of the feature concerned and the same location on the body act as favourable evidence for homology, whereas the differences in the structure of the organs are contrary evidence for homology.

55. Phylogenetic assessment of the existence of only one pre-cloacal tubule in the Anticomidae, Phanodermatidae and "Enoplidae":

Within the three named families the males of almost all the species have a single, pre-cloacal tubule (only seldom is the tubule absent: in one case 9 tubules were present). On account of the striking frequency of this feature in almost all other nematodes (which either have more than one tubule or none at all), the existence of only one pre-cloacal tubule is considered a holapomorphy of Taxon E, which contains, at the minimum, the Anticomidae, Phanodermatidae and "Enoplidae". From the remaining nematodes only *Anoplostoma* ("Anoplostomatidae") can be placed in Taxon E because in this genus, as in the other species of Taxon E, both gonads are always situated to the left of the intestine (this is considered a holapomorphy of Taxon E in discussion point 53), and because *Anoplostoma* together with *Chaetonema* ("Enoplidae") forms a holophyletic sub-taxon on account of the similarity in structure of the amphids and structure and position of the metanemes. In Chapter 5 of this work Taxon E is described as the superfamily Enoploidea.

56. The relation between pre-cloacal tubules and cup-shaped papillae:

Within the "Leptolaimidae" there are several species in which, pre-cloacally, the males have tubules as well as cup-shaped papillae. Both types of auxillary copulatory organs are thus considered homologous to one another.

4.6. Caudal glands, phasmids and phasmata

4.6.1. Caudal glands

Caudal glands are found in most Adenophorea but are absent in all Secernentea (both in the former and current sense). They are also absent in some taxa of the Adenophorea, whereby only the Dorylaimina and Dioctophymatina are named in general data in the literature (see Chitwood & Chitwood, 1950:21, de Coninck, 1965:591). According to specific data in the literature and as personal research has shown, caudal glands are also absent in most Teratocephalidae (caudal glands are only described for *Euteratocephalus crassidens* by de Coninck, 1935), in the Isolaimiidae and in the Aulolaimidae (all "Araeolaimida"); in the "Prismatolaimidae", "Alaimidae" (except *Adorus*) and Rhaptothyreidae (all "Enoplida"); in the Mermithoidea and Trichuroidea; in addition they are also absent in single species of other families. e.g. in *Leptolaimoides thermastris, Leptolaimus acicula* (both "Araeolaimida", "Leptolaimidae") and in *Pareurystomina* ("Enoplida", Enchelidiidae).

Where caudal glands are visible, they almost always occur in threes. The presence of only two caudal glands is known only in some Xyalidae (see Lorenzen, 1978c : 528), some Monhysteridae (see Riemann, 1970b), *Pandolaimus* ("Anoplostomatidae", according to Jensen, 1976 and personal observations) and in the Rhabdodemaniidae (personal observations). Finally, Hopper (1968) specifies only one functional and two rudimentary caudal glands in *Chromadorita tenuis* ("Chromadoridae").

The caudal glands are always unicellular and in all Chromadoria and many Enoplia they do not extend beyond the caudal region, and are thus limited to the end section of the body posterior to the rectum. It has been known for some time that all caudal glands extend well into the pre-anal region in the Oncholaimidae, in many Leptosomatidae and in some Phanodermatidae (formerly "Enoplida") (Fig. 18a). New data in the literature and personal observations, which are summarized in Table 6, show that this feature is more widespread than was previously thought. However, not all the data in the literature are reliable: it is reported that the caudal glands are situated completely in the tail in many species of the "Enoplidae" (excluding Enoplinae and Chaetonematinae). However, personal observations failed to reveal any evidence to show this to be the case in any species from this group; presumably the (sometimes) ampulla-shaped extension of the excretory duct was hitherto mistaken for the cellular part of the caudal gland containing the nucleus, while the genuine cellular part containing the nucleus and the narrow duct leading to it were overlooked (see Fig. 18a). My own observations showed that the caudal glands in the Enoplinae are situated completely in the tail.

Usually the caudal glands open through a common pore on the tip of the tail. Only seldom are the openings separate; according to Lorenzen (1978c : 528) this is known to be the case for species of the Xyalidae and Sphaerolaimidae (in cases where the caudal gland openings could be detected at all), for *Diplopeltula incisa, D. breviceps, Paraeolaimus nudus* (formally known as "Axonolaimidae") and *Ixonema* ("Microlaiminae"). Where the openings are separate, one always lies dorsally and two lie subventrally on the tip of the tail.

Table 6: Number and position of the caudal glands in the taxa of the "Enoplida". In the Trefusiidae it was only possible to establish the existence (but not the number and position) of caudal glands. The table has been compiled using data in the literature and personal observations of all families of the "Enoplida" (Except Rhaptothyreidae).

Caudal: the caudal glands are situated completely in the tail (i.e. posterior to the rectum);

Pre-caudal: the part of the caudal gland containing the nucleus lies anterior to the rectum.

3 caudal glands, caudal	3 caudal glands, pre-caudal	2 caudal glands, caudal	No caudal glands
"Tripylinae" "Tobrilinae"		"Cryptonchidae"[1]	"Prismatolaimidae"
"Ironidae": *Ironus* *Parironus bicuspis* Tripyloidae	"Ironidae": *Dolicholaimus* *Parironus* part. *Thalassironus*		"Alaimidae" (except *Adorus*) Raptothyreidae
"Oxystominidae": Halalaiminae	"Oxystominidae": "Oxystomininae"[2] Paroxystomininae Leptosomatidae		
Lauratonematidae Triodontolaimidae Anticomidae	Phanodermatidae		
"Enoplidae": Enoplinae	"Enoplidae"[3]: "Thoracostomopsinae" Chaetonematinae Trileptiinae "Oxyonchinae" "Enoplolaiminae" "Enoploidinae"		
	"Anoplostomatidae": *Anoplostoma*[5] Oncholaimidae Enchelidiidae	Rhabdodemaniidae[4] "Anoplostomatidae": *Pandolaimus*	

1) Personal observations have shown that there is probably only one caudal gland.
2) Personal observations on *Nemanema rotundicaudatum, Oxystomina elongata, Thalassoalaimus pirum, T. tardus.*
3) Personal observations on *Thoracostomopsis barbata, Trileptium sp., Oxyonchus dentatus, Enoplolaimus propinquus, Epacanthion buetschlii, Chaetonema riemanni.*
4) Personal observations on *Rhabdodemania minor.*
5) All three caudal glands, not just two (Jensen 1976 : Fig. 22), are situated pre-caudally.

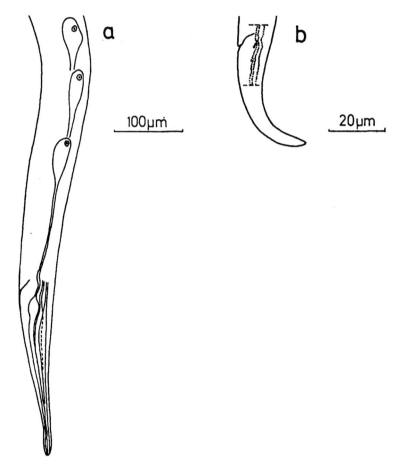

Fig. 18. a: Caudal glands in *Oxyonchus dentatus* ("Enoplida", "Enoplidae"). The part of the cell containing the nucleus of one of each of the three glands is situated well pre-anally. Juvenile from sublittoral muddy sand, Skagerrak, 30th June 1976. b: Tail of *Euteratocephalus palustris* ("Aracolaimida", Teratocephalidae). The short seta at the level of the anus is often thought to be a phasmid, although there is no basis for this assumption. Adult female from the Kaltenhof Moor near Kiel, 17th October 1975.

4.6.2. Phasmids and phasmata

Phasmids are caudal sensory organs which are usually only insignificant; one is situated to the left and one to the right of the tail (they seldom occur anterior to the tail). They have, until now, only been studied electron-microscopically in a few parasitic nematode species (McLaren, 1976 : 149). Phasmids are, in their basic structure, similar to amphids: they are sunk into the body, are connected to the outside through an orifice and have a glandular cell and modified sensory cilia of which only one or two, and not many as in amphids, are present. The phasmids are innervated by pairs of lumbar ganglia (Chitwood & Chitwood, 1950 : 171).

Phasmids are known in species of all former orders of the Secernentea; they may, however, be absent in isolated species. It was thought, until recently, that phasmids were absent in the Adenophorea in the hitherto understood sense (see Chitwood & Chitwood, 1950 : 21, de Coninck, 1965 : 591). Recently Kaiser (1977) found "phasmid-like organs" in juvenile and adult animals from *Hexamermis albicans* and two additional species of the Mermithidae. He was able to show, using a light microscope, that the organs were identical in structure, position and innervation to the phasmids of the Secernentea. Consequently the "phasmid-like organs" of the Mermithidae may be referred to as phasmids. In species from various orders of the Adenophorea structures have been found on the tail which have been referred to as phasmata by Chitwood (1951 : 646). Phasmata were first found by Cobb (1928) in *Syringolaimus striatocaudatus* (syn *S. smarigdus*, "Rhabdolaimida", "Araeolaimida") and were referred to as "phasmids(?)". They were later re-discovered in this species by Chitwood (1951) and Hopper (1969). Chitwood (1951: Fig. 8D, E, H) also found them in *Tricoma* species ("Desmoscolecidae").

Timm (1969) and Decraemer (e.g. 1974a and b, 1978) found phasmata in many species of Desmoscolecinae and Tricominae, and Schrage & Gerlach (1975; referred to as caudal pores) found them in the Greeffiellinae (formerly "Desmoscolecidae"). Schrage & Gerlach found that sometimes a gland opened through the phasmata. It has not, as yet, been possible to demonstrate innervation in phasmata, with the result that one can, at the moment, only suggest a possible homology between phasmids and phasmata.

In *Euteratocephalus crassidens* De Man (1880 : 46) found a left and right lateral "papilla situated posteriorly a short distance from the anus". The papillae, one of which is illustrated in Fig. 19b, were often described as phasmids (e.g. by Goodey 1963 : 1), though there is no evidence in favour of this except for the similarity in position.

4.6.3. Discussion

57. Phylogenetic assessment of the existence of three caudal glands:

The occurrence of three (instead of two) caudal glands is considered holapomorphic in nematodes for the following reasons:

— Outside the nematodes, glands in the tail occur in Nemathelminthes only in a paired arrangement.

— The occurrence of three caudal glands is very common within the Adenophorea.

— Osche (1955–1958) provides a wealth of examples in which species of Secernentea are shown to have three protuberances at the end of the tail; one of these protuberances is always situated dorsally and the remaining two subventrally. They are interpreted as rudiments of caudal gland openings because their arrangement is constant and they are in a position similar to that of the three separate caudal gland openings in *Diplopeltula incisa*. Osche's interpretation is supported by further examples which have since come to light where, in the case of separate caudal gland openings, one lies dorsally and the other two subventrally (Lorenze, 1978c : 528). Thus, the caudal glands are probably absent in the Secernentea at a secondary rather than primary level.

Riemann (1970 : 425) considers it possible that the occurrence of only two caudal glands is plesiomorphic within the nematodes. This interpretation seems improbable in that the presence of three caudal glands must have evolved repeatedly within the nematodes. It is much more likely that one caudal gland has been reduced repeatedly within the nematodes.

58. Phylogenetic assessment of the extension of caudal glands into the pre-caudal body region:

Within the nematodes it is only in all representatives of the "Enoplida" that the caudal glands extend far into the pre-caudal body region (Table 6). On account of the uniqueness of this feature it is considered an apomorphy within the nematodes and particularly within the "Enoplida". On account of the striking frequency of the feature, the genetic potential for its development is, in addition, considered a homology within the "Enoplida", because within two holophyletic subtaxa of the "Enoplida" (Enoploidea and Ironidae, see section 5.3.2.) there are species with caudally situated caudal glands, as well as species in which the caudal glands penetrate far into the pre-anal region. On account of the interpretation of apomorphy and homology the genetic potential for the development of the last-named feature is considered a holapomorphy of a Taxon F which, on the one hand, contains all species with pre-caudally situated caudal glands, and on the other all those species with caudally situated caudal glands which form holophyletic groups with one section of the species first named. Taxon F is described as the sub-order Enoplina in Chapter 5.

59. Phylogenetic assessment of the existence of phasmids:

Phasmids have acquired phylogenetic significance ever since Chitwood (1933) divided the nematodes into two subclasses largely on the basis of this feature: the Secernentea (= Phasmidia) with phasmids and the Adenophorea (= Aphasmida) without phasmids. Nevertheless, Chitwood (1933) and Chitwood & Chitwood (1950 : 196) think it possible that the existence of phasmids may be plesiomorphic within the nematodes and thus cannot be considered in the sense of a holapomorphy of the Secernentea. The discovery of phasmids in three species of Mermithidae, which had hitherto been ascribed to the Adenophorea, seemed to confirm this interpretation. Meanwhile, it has been shown (Lorenzen 1982) that there are no acceptable reasons for classifying the Mermithoidea, Trichuroidea and Dioctophymatoidea in the Adenophorea, so that the three super-families were placed in the Secernentea. From this the phylogenetic conclusion was drawn that the existence of phasmids should be considered a holapomorphy of the Secernentea (in the new sense). This means that phasmids are absent in the Adenophorea probably at a primary and not a secondary level.

60. Phylogenetic assessment of the phasmata:

It seems altogether possible that phasmata are homologous with phasmids on account of their similarity in position laterally on the tail. However, as long as proof of corresponding innervation and of the existence of at least one sensory filament in the phasmata remains unavailable, one cannot be sure that the homology exists.

61. Phylogenetic assessment of the exclusive nature of the occurrence of phasmids and caudal glands:

Chitwood & Chitwood (1950 : 197), Chitwood (1951 : 646), Osche (1952a : 71) and Steiner (1958 : 277) have already discussed the striking fact that caudal glands and phasmids do not occur together in any species of nematodes; phasmids are absent in all nematodes with caudal glands, and caudal glands are absent in all nematodes with phasmids. This observation leads the above authors to interpret the phasmids as transformed caudal glands, the openings of which had been displaced from the tip of the tail to the lateral side of the tail. This interpretaion is not acceptable for the following reasons:
1) It contradicts the interpretation of the three protruberances which were found at the end of the tail of several species of Secernentea, as being rudiments of caudal gland openings (see discussion point 57).
2) Phasmids are sensory organs, caudal glands are not. 3) There is no argument given as to why a phasmid should be a modified caudal gland cell rather than a modified glandular cell in the lateral epidermis.

There are, in fact, no glandular cells in the lateral epidermis of any species which have phasmids. Chitwood & Chitwood (1950 : 197) do, in fact, consider the possibility of a comparison of these glandular cells with phasmids, but do not deal any more closely with the possibility of homology.

Coomans (pers. comm.) does take up this possibility. According to him, a phasmid is exactly like a dereid (see footnote to discussion point 2), a combination of a lateral, modified glandular cell with a sensillum; such a combination can be found along all the lateral epidermal borders of several representatives of the Adenophorea (see Coomans & de Waele, 1979 for *Aphanolaimus* species; personal observations show that similar structures also occur in species of the Plectidae).

The absence of caudal glands is considered a holapomorphy of the Secernentea (including the Mermithoidea, Trichuroidea and Dioctophymatoidea), because the feature occurs in all species of the Secernentea and is linked in most cases with the existence of phasmids.

4.7 The cervical gland

The gland in the Adenophorea which is usually situated ventrally in the cervical region is currently often referred to as the ventral gland, but also occasionally as the excretory gland or "renette cell". The accompanying pore is known as the excretory pore. In accordance with Filipjev (1921 : 36), the gland and its pore are called the cervical gland (neck gland) and cervical pore for the following reasons:
— Experiments have not produced any evidence to indicate that the cervical gland is an excretory organ (see Filipjev, 1921 : 37).
— The cervical gland does not occur in many freeliving Adenophorea, e.g. it is absent in most Xyalidae (see Lorenzen, 1978c) and in many families of the "Enoplida" (see Table 7); in particular, it is absent in many limnic Adenophorea (in Table 7: "Tripylidae", "Prismatolaimidae", *Ironus*, "Cryptonchidae", "Alaimidae" excluding *Adorus*).
— The term "ventral gland" is not always correct, as the gland is situated laterally or subventrally in many cases; at times only the pore is situated ventrally.

In the scope of the present work it was found that within the freeliving Adenophorea the cervical gland is completely restricted to the pharyn-

Table 7: Existence and position of the cervical gland in the taxa of the "Enoplida", based on data in the literature and on personal research on species from all families of the "Enoplida" (except Rhaptothyreidae).

Cervical gland extends into the post-pharyngeal region	Cervical gland is restricted to the pharyngeal region	Cervical gland not observed
Triodontolaimidae Rhabdodemaniidae Oncholaimidae Enchelidiidae	"Oxystominidae" Leptosomatidae Anticomidae Phanodermatidae "Enoplidae"	"Tripylidae"[1] "Prismatolaimidae" "Cryptonchidae" Tripyloididae Trefusiidae[1] Lauratonematidae "Alaimidae" (except *Adorus*) "Anoplostomatidae" Rhaptothyreidae

1) A cervical pore is known in some species; the accompanying cervical gland, however, has not been found.

geal region only in some "Enoplida" and in *Linhystera* ("Monhysterida", Xyalidae), whereas in the remaining freeliving Adenophorea it always extends well into the post-pharyngeal body region.

Discussion:

62. Phylogenetic assessment of the pharyngeal position of the cervical gland in some taxa of the "Enoplida":

The pharyngeal position of the cervical gland in taxa of the "Enoplida" is considered a holapomorphy of a Taxon F, thus establishing the holophyly of this taxon which contains the "Oxystominidae" (excluding *Porocoma* and including *Adorus*), "Ironidae", Leptosomatidae, Anticomidae, Phanodermatidae, "Enoplidae" and "Anoplostomatidae" (excluding *Pandolaimus*). Reasons:

— In cases where a cervical gland is present within Taxon F, it is always restricted to the pharyngeal region.
— The feature hardly ever occurs outside Taxon F.
— No subtaxon of Taxon F forms a holophyletic group with a subtaxon of the remaining Adenophorea on the basis of other features.

Taxon F is named Enoplacea in Chapter 5. From the present phylogenetic interpretation it follows that the extension of the cervical gland into the post-pharyngeal body region, which is common in the Adenophorea in addition to Taxon F, should be considered a plesiomorphic feature at least within the freeliving Adenophorea.

63. Phylogenetic assessment of the position of the cervical gland to the right of the intestine in the Oncholaimoidea:

Personal observations have shown that the cervical gland — when present — is always found to the right of the intestine in species of the Oncholaimoidea. According to personal observations, this feature is unique at least within the Enoplina as they are defined in Chapter 5 of this work. The feature is thus considered a holapomorphy, thereby establishing the holophyly of the Oncholaimoidea. The cervical gland is situated, as is usual, on the ventral line of the body.

4.8 Excretary ducts of epidermal glands in "Enoplidae" and Phanodermatidae

In most freeliving Adenophorea it is typical to find unicellular glands in the lateral epidermal borders. The ducts from these glands show unusual differentiation in the "Enoplidae" and Phanodermatidae. A collective canal leads from each epidermal gland into the vicinity of the pore and widens at the end, forming an ampulla shape. In this ampulliform widening there is a structure shaped like a coffee bean with a thick wall and a narrow lumen. The latter continues into a tubule which leaves the ampulla and opens to the exterior through a pore. The coffee bean-like structure is usually tightly secured to the ampulla wall by a thin ligament. From the epidermal gland the accompanying collective canal in the anterior body section leads towards the anterior and in the posterior body section it leads towards the posterior; and similarly, the outflow tubule from the coffee bean-shaped structure in the anterior body section leads towards the anterior and in the posterior body section it leads towards the posterior. The collective canal and the epidermal gland are not always easy to recognize, and the relative sizes of the coffee bean-like structure and the outflow tubule vary greatly in individual species (Fig. 19a–d, e and f–g). One type of differentiation, as illustrated in Fig. 19a–d, was found by the author in *Enoploides labrostriatus, Enoplolaimus connexus, Epacanthion buetschlii, Fenestrolaimus* sp., *Mesacanthion diplechema, Oxyonchus dentatus, Saveljevia* sp. and *Trileptium* sp. (all "Enoplidae"), as well as in *Phanodermopsis necta* (Phanodermatidae); the differentiation shown in Fig. 19e was found by the author only

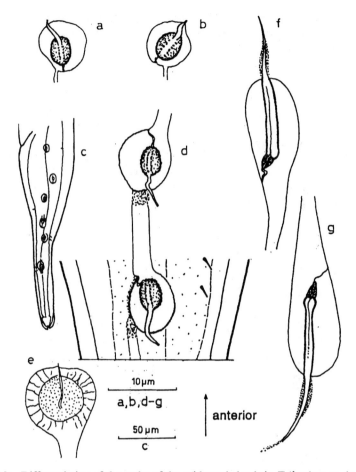

Fig. 19: Differentiation of the outlet of the epidermal glands in *Trileptium* sp. (a – d; "Enoplidae"), *Thoracostomopsis barbata* (e; "Enoplidae") and *Crenopharynx marioni* (f – g; Phanodermatidae) in the pharyngeal (a, b, e, f) and anal body region (c, d, g). a – d: adult females, coarse sand beach on the Island of Chiloé (south Chile), 1st September 1973; e: adult female, from sublittoral sediment in the German Bay, 1st July 1975; f – g: adult male, sublittoral "Schill" sand in the Kattegat, 30th June 1976.

in *Thoracostomopsis barbata* ("Enoplidae") and the differentiation shown in Fig. 19f–g only in *Crenopharynx marioni* (Phanodermatidae); no differentiation could be found in *Enoplus brevis, E. michaelsensi, Enoplolaimus propinquus, Metenoploides* aff. *alatus* (all "Enoplidae"), *Phanoderma campbelli* (Phanodermatidae) and in representatives of the other families of the "Enoplida".

Discussion

64. Phylogenetic assessment of the differentiation found in the excretory ducts of the epidermal glands in species of the "Enoplidae" and Phanodermatidae:

Since the differentiation of the ducts of the epidermal glands is basically the same in structure and orientation in representatives of the "Enoplidae" and Phanodermatidae, it is very probably due to a homology. Since the "Enoplidae" probably represent a holophyletic group on account of the structure of their buccal cavity (discussion point 97), and since the differentiation of the outlet of the epidermal glands in particular also occurs in the Phanodermatidae, the absence of this feature in representatives of the "Enoplidae" must be seen as a secondary condition, i.e. the feature must be reduced. The holophyly of the Phanodermatidae cannot yet be proved. Thus it is not possible to determine whether the absence of the feature in representatives of the Phanodermatidae represents a primary or secondary condition. For this reason one cannot determine the extent of a taxon, the holophyly of which is established by the holapomorphic feature that the outlet of the epidermal glands is differentiated in the way described. On account of the non-holophyletic status of the Phanodermatidae, it may be that the feature is also found outside the present "Enoplidae" and Phanodermatidae.

5. OUTLINE OF A PHYLOGENETIC SYSTEM FOR THE FREELIVING NEMATODES

Previous classifications of freeliving nematodes are of a predominantly diagnostic nature because in general they have not dealt with the holophyly of the taxa. The establishment of holophylies is, however, an essential constituent of a phylogenetic system.

In Chapter 4 new and known features were judged to be holapomorphies; these holapomorphies were then used to establish the holophyly of many taxa. In the present chapter the establishment of these holophylies is used to form a basis on which to outline a phylogenetic system for the freeliving nematodes. It can be no more than an outline, because the freeliving nematodes represent only a non-holophyletic part of the nematodes as a whole, and because not all the features available have been extensively analysed.

In some taxa, from the subfamily to the subclass level, holophyly can be established without changing the extent of the taxon, i.e. without removing known species or adding them from other taxa. However, for the majority of the taxa, from the subfamily to the order level, it is only possible to establish holophyly after changes in the range of the taxa have been made. These changes are often only slight, though in some cases they are considerable. In either case, explicit reference is made to the changes, with the following works serving as the basis for former systems: Gerlach & Riemann (1973/1974, freeliving Adenophorea excluding Dorylaimida), de Coninck (1965, Dorylaimida) and Goodey (1963, freeliving Secernentea).

For a number of taxa a phylogenetic assessment is necessary for those features which have not been analysed in Chapter 4. These phylogenetic assessments have been given numbers which correspond to the numbering used in Chapter 4.

In the following, all taxa, from the subfamily to the subclass level, are cited in the form which is elaborated in the present chapter. The names of these taxa are not quoted in inverted commas as are those taxa in Chapter 4 which do indeed resemble those of the present chapter in a nominal way but differ from them in extent.

5.1 Division of the freeliving nematodes into subclasses

5.1.1. Assessment of currently conflicting concepts

Filipjev (1918/1921, English translation 1968/1970) drew up the first well-thought-out system for freeliving nematodes, based on sound analyses of characteristics. In 1934 he presented it in English in a modified form which included the parasitic nematodes. Filipjev rejected the division of the nematodes into two subclasses (1934 : 6) because the body organization is too homogenous; at the same time, however, he was of the opinion that, if one felt such a division was desirable, then the "Enoplata" (including the Mermithidae), "Chromadorata", "Desmoscolecata" and "Monhysterata" should be placed in one subclass and the "Anguillulata" and the parasitic nematodes in the other. Chitwood (1933) subdivided the nematodes into almost exactly the same two subclasses which he called the Aphasmidia and the Phasmidia. In order to avoid confusion with a group of insects (Phasmida), the Aphasmidia are nowadays generally referred to as the Adenophorea and the Phasmidia as the Secernentea. These names date back to von Linstow (1905).

Chitwood (1933), and later on Chitwood & Chitwood (1950 : 12 and 21), in their diagnoses of the Adenophorea and Secernentea have listed a good twelve features with which both subclasses can be differentiated; however, gonad characteristics are completely absent in these diagnoses. A comparison of the features shows that, for almost every characteristic of one subclass, exceptions exist in the other subclass. Chitwood & Chitwood (1950) do not attempt to establish the holophyly of the two subclasses. In those rather scattered instances where they have made phylogenetic assessments of features they have been more interested in the question of what should be considered plesiomorphic within the entire nematode class. These plesiomorphies have been listed interdependently on page 196 of their 1950 work. The establishment of plesiomorphies in no way automatically proves the holophyly of those taxa in which a non-plesiomorphic expression of the features occurs. What is more, there are no arguments to show why in particular those features which the Plectidae (Adenophorea) and the Rhabditidae (Secernentea) possess in common with one another should occur on the basis of all nematodes and not of only a part of them. Chitwood's division of the nematodes into the Adenophorea and Secernentea has been adopted by de Coninck (1965) and by many other authors.

According to Maggenti (1963, 1970) the earliest division in the nematodes probably did not occur between the Adenophorea and the Secernentea but between the Enoplia and the Chromadoria plus Secernentea. Gadea (1973) has followed through the systematic conclusion of this view and has rejected the two taxa of the Adenophorea and the Secernentea and has instead divided the nematodes into the Enoplimorpha (=Enoplia) and the Chromadorimorpha (=Chromadoria plus Secernentea). The arguments in favour of such a division are to be found in Maggenti (1963). For him it is of major significance that the valvular apparatus in the pharynx of *Plectus parietinus* (a representative of the Adenophorea) and *Rhabditis* (a representative of the Secernentea) differ from one another in structure, method of function and post-embryonic development. Thus it follows, according to Maggenti, that the valvular apparatus in the Plectidae and Rhabditidae must have developed independently from one another from an undifferentiated, posterior section of the pharynx. Therefore, contrary to the view of Chitwood and Chitwood (1950), an undifferentiated, cylindrical pharynx, and not a pharynx with a valvular apparatus, is to be seen as plesiomorphic within the nematodes. Three parts in particular of Magenti's line or argument should be criticised:

a) My own observations indicate that the valvular apparatus of *Plectus parietinus* is simply not representative of the valvular apparatus of the Plectidae: within the Plectidae there are at least two different types of valvular apparatus, the first of which (Fig. 20a) is found in *P. parietinus* and other species and the second (Fig. 20b-c) in *P. parvus* and other species. Maggenti (1961) has dealt with the taxonomy of several of the species named above, but has omitted to include a description of the valvular apparatus. Personal observations show that the valvular apparatus of *P. parvus* (Fig. 20b-c), for example, is more similar in structure and method of function to the valvular apparatus of the Rhabditidae or Panagrolaimidae than to that of *P. parietinus*.

b) No attempt was made to establish the holophyly of the two newly conceived branches of the phylogenetic tree.

c) Since Chitwood & Chitwood (1950) have not established the holophyly of either the Adenophorea or the Secernentea, their view on the presumed plesiomorphic form of the pharynx within the nematodes was unfounded. Consequently, there was no need to reject the view, in order to postulate a cylindrical pharyngeal form as plesiomorphic within the nematodes. On the contrary, if the existence of a valvular apparatus in the Rhabditidae and Plectidae were really

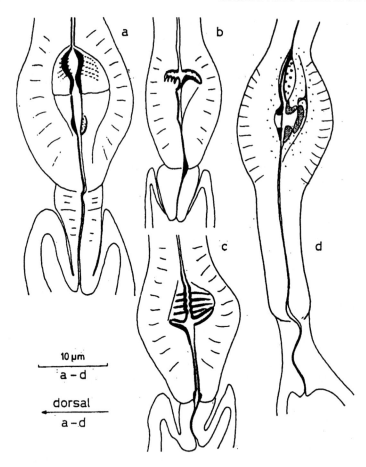

Fig. 20: Different types of valvular apparatus in the end bulb of the pharynx in representatives of the Adenophorea. a) *Plectus parietinus* ("Plectidae"), b – c *P. parvus* (b: inactive, c: active), d) *Chronogaster typica* ("Leptolaimidae"). The drawings are all taken from preserved specimens. a: adult female, freshwater pool near Kiel (Germany), 27th April 1976; b – c: adult female, stream in a wood near Kiel, 26th August 1976; d: adult female, underground water from the banks of a stream in Malaysia, summer 1978.

due to homology and not analogy, then it would confirm rather than discredit Maggenti's (1963) phylogenetic tree.

Andrassy (1976) subdivides the nematodes into three subclasses: the Torquentia (= Chromadoria), the Secernentia (= Secernentea) and the Penetrantia (= Enoplia). His arguments are not sound – he implies that many features are constant within these three groups. These generalizations are often wrong. Using these generalizations Andrassy argues that the Enoplia and Chromadoria each show better characterization as a group than do the Adenophorea which exist out of the union of the Enoplia and Chromadoria. Even if this view were correct, it does not provide either a necessary or a sufficient argument in favour of the rejection of the taxon Adenophorea. Coomans (1977) has written a detailed criticism of Andrassy's work.

Goodey (1963:1) rejects the division of the nematodes into the Adenophorea and Secernentea and instead only accepts a number of orders. Goodey's reasoning relies on two unacceptable arguments: the first, that the phasmids in the Secernentea are "invisible to most people", lacks all strength as a phylogenetic statement. According to the second argument, *Euteratocephalus crassidens*, "which is clearly Aphasmidian in most respects" is meant to possess phasmids; Goodey's interpretation of the caudal papillae (Fig. 18) as phasmids is unfounded and is not accepted.

Recently, Inglis (1983) presented an outline classification of the "phylum Nematoda". He divided it into the three "classes" Rhabditea, Enoplea and Chromadorea (essentially as did Maggenti, Gadea and Andrassy), raised the rank of many taxa by one category (since nematodes were not regarded as a class, but as a phylum), and treated the "order Araeolaimida" in some detail. Essentially, Inglis did not substantiate his views by arguments, but operated with terms like "I believe", "I do not believe", "I prefer", "I disagree", "I consider", etc. Therefore, Inglis' paper is not helpful to a cladistic analysis of nematodes.

Malakhov *et al.* (1982) and Malakhov (1986) divided the class Nematoda into the three subclasses Enoplia, Chromadoria and Rhabditia (essentially as did the above mentioned authors). The Rhabditia are regarded as an offshoot of the order Plectida (Chromadoria), and the Mononchida, Dorylaimida, Mermithida, Trichocephalida and Dioctophymida (all included in the Enoplia) as offshoots of the Enoplida (also included in the Enoplia). These views do not fit cladistic principles. Especially, the dendrogram presented for Enoplia would oblige one to conclude that metanemes had evolved at the origin of the Enoplia and would be secondarily absent in all orders of Enoplia except Enoplida. This view cannot be substantiated.

5.1.2. Criteria for the higher systematic classification of the freeliving nematodes

The holophyly of the nematodes as a class is substantiated by several very reliable holapomorphies:

65. The occurrence of 6+6+4 sensilla at the anterior end (discussion point 2), the existence of a vulva (discussion point 48), the existence of a copulatory apparatus, and post-embryonic development over four juvenile stages.

In striking contrast to these very reliable holapomorphies are the very weak criteria which, as a result of feature analyses, are available for the classification of the freeliving nematodes into subclasses. In the course of the present work the following criteria were compiled in order to establish the holophyly of the subtaxa of the highest possible level in the freeliving nematodes (the numbers correspond to the discussion points):

20: Non-spiral amphids as a holapomorphic feature of the Enoplia.
28: Blister-like form of the amphids as a holapomorphy of the Desmoscolecina.
51: Outstretched ovaries as a holapomorphic feature of the Monhysterida.
52: The holophyly of the Chromadorida cannot be proved; the frequent appearance of the constant position of the anterior gonad to the right and the posterior gonad to the left of the intestine is thought to be the only indication of a possible holophyly of the Chromadorida.
59: The existence of phasmids as a holapomorphy of the Secernentea.
61: Reduction of the caudal glands as a holapomorphy of the Secernentea.
66: Twelve-fold pleated vestibulum as a holapomorphic feature of the Chromadorina.
94: Cephalic setae of the third sensilla circle on peduncles as a holapomorphy of the Desmoscolecina.

There are no criteria available to establish the holophyly of the Leptolaimina. The Chromadorina and Leptolaimina together form the Chromadorida.

According to the results listed above which are illustrated in schematic form in Fig. 21, the classification of the freeliving nematodes into the highest possible holophyletic subtaxa does not quite succeed. A paraphyletic group which consists of the suborder Leptolaimina remains over. The highest possible holophyletic subtaxa of the freeliving nematodes are the Chromadorina, Monhysterida, Enoplia and Secernentea; the ranking of these subtaxa is not a function of the establishment of holophylies and remains open for the time being.

OUTLINE OF A PHYLOGENETIC SYSTEM

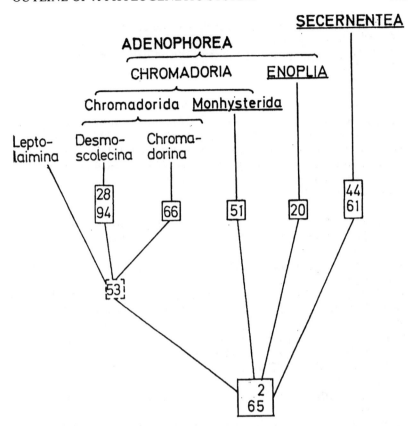

Fig. 21: Attempt to classify the freeliving nematodes into sub-taxa of the highest possible holophyletic rank. The Leptolaimina remain a paraphyletic side-branch. The names of the holophyletic taxa have been underlined and the names of the paraphyletic taxa have not. The system of freeliving nematodes depicted above is used in Chapter 5 of this work.

The numbers in boxes refer to the criteria on which the establishment of the holophylies is based. The dotted box containing the number 53 indicates that only weak circumstantial evidence was found for holophyly. Original.

I know of no criteria in the literature which allow a more complete classification of the freeliving nematodes into fewer than four holophyletic subtaxa. Neither the absence of dereids nor the absence of phasmids in most species of the Adenophorea can be considered holapomorphies with which one may establish the holophyly of the Adenophorea (discussion points 4 and 61).

The Leptolaimina, Desmoscolecina and Chromadorina are placed together to form the order of the Chromadorida on account of the constance of the position of the anterior gonad to the right and the posterior gonad to the left of the intestine. Due to the current state of knowledge the Chromadorida, Monhysterida, Enoplia and Secernentea could be grouped together in any way one chose to form two, three or four subclasses, whereby each subclass would contain at least one and, at the most, three of the four taxa mentioned. The arbitrary nature of the decision is due to the fact that at the moment one cannot establish the holophyly for any of the possible groupings. This arbitrariness shows that the inter-relationships between the four taxa are not yet fully understood. Therefore Chitwood's classification of the nematodes into the Adenophorea and Secernentea is retained in the current work, because the Adenophorea are predominantly aquatic and freeliving, whereas the Secernentea are predominantly terrestrial and, for the large part, parasitic; and because the Adenophorea, in contrast to the Secernentea, are distinguished by several plesiomorphies, the most salient of which is the possession of caudal glands. Likewise, Chitwood's grouping together of the Chromadorida and Monhysterida in the Chromadoria is also retained because in these taxa the ventrally spiral amphids, which is considered a plesiomorphic feature (discussion point 22), occurs particularly frequently, whereas in the remaining taxa its appearance is only sporadic. Only on the basis of these decisions will the currently undecided rank of the holophyletic taxa Chromadorina, Monhysterida, Enoplia and Secernentea be determined as that of a suborder, order, superorder and subclass respectively. Due to the impossibility of establishing holophylies, the Leptolaimina, Chromadorida, Chromadoria and Adenophorea are only paraphyletic. This gross systematic classification of the freeliving nematodes is illustrated in Fig. 21.

5.2 Outline of a phylogenetic system for the Chromadoria

CHROMADORIA Pearse, 1942

The holophyly of the Chromadoria cannot be established at the moment. In most species the cuticle is striated; it is smooth in the Rhadinematidae, Rhabdolaimidae, Monhysteridae, and in one section of the Molgolaiminae, Microlaimidae, Aponchiidae, Camacolaiminae, Aulolaimidae, Tubolaimoididae, Axonolaimidae, Diplopeltidae, Coninckiidae, Sphaerolaimidae, Meyliidae and Linhomoeidae. As regards the length and arrangement of the 6+6+4 cephalic sensilla, two conditions occur with particular frequency: a) the 6+4 cephalic sensilla are arranged in two separate circles, and the posterior 4 are longer than the anterior 6; b) the 6+4 cephalic sensilla form a common circle, and the 6 of one circle are longer than the 4 of the other circle. The amphids are almost always ventrally spiral, occasionally non-spiral (Monoposthiidae, some genera of the Leptolaimidae, *Anaplectus*, Aulolaimidae, Desmoscolecoidea excluding *Meylia*) and only rarely dorsally spiral (Prismatolaimidae, Bastianiidae partim). In instances where teeth are present in the buccal cavity, the dorsal tooth is always larger than, or at most the same size as the two subventral teeth. With the exception of *Glochinema* (Epsilonematidae) and *Ceramonema* partim (Ceramonematidae), the pharynx is never attached to the body wall in the head region. At the posterior end the pharynx often thickens to form a bulb. Where a cervical gland is present, it always extends into the post-pharyngeal region, except in *Linhystera* (Xyalidae). The ovaries are antidromously reflexed or outstretched. One or two testes may be present. There are usually three caudal glands; they are always situated completely in the tail and usually open out through a common passage, seldom through three separate passages (see section 4.6.1.).

Most species of the Chromadoria live in the sea, some live in freshwaters and a few species are terrestrial. Almost all species are freeliving. A few species live as parasites in the gill region of crabs (Riemann, 1970; Sudhaus, 1974). *Theristus (Penzanica) polychaetophilus* (Xyalidae) and the Harpagonchinae live as ectoparasites on the gills and parapodia of polychaetes, while *Domorganus oligochaetophilus* (Ohridiidae) is an endoparasite in oligochaetes.

5.2.1. Classification of the Chromadoria into orders

Within the freeliving nematodes, Chromadoria are the major taxon which has hitherto been subdivided in the greatest variety of ways (Table 8). There are differences in the various names and structures of the individual taxa and, above all, in the assignment of different subtaxa to them. For example, Filipjev (1934), Chitwood & Chitwood (1950) and de Coninck (1965) assign:

- the Leptolaimidae and Plectidae to the "Chromadorata", "Monhysterina" and "Araeolaimida" respectively;
- the Comesomatidae to the "Monhysterata", "Chromadorina" and "Chromadorida" respectively;
- the Microlaimidae to the "Monhysterata"; "Chromadorina" and "Desmodorida" respectively;
- the Desmoscolecoidea to the "Desmoscolecata", "Chromadorina" and "Desmoscolecida" respectively, and
- the Tripyloididae to the "Enoplata", "Chromadorina" and "Chromadorida" respectively.

Table 8: Classification of the Chromadoria into taxa by different authors. The rank of the taxa lies as much as two levels below that of the Chromadoria. The names and positioning of the taxa are identical to those used by the authors concerned.

Filipjev (1934)

Chromadorata
    Camacolaimidae
    Plectidae
    Chromadoridae
 	
Desmoscolecta
    Desmoscolecidae
    Greeffiellidae
 	
Monhysterata
    Linhomoeidae
    Monhysteridae

Chitwood and Chitwood (1950)

Monhysterina
    Plectoidea
    Axonolaimoidea
    Monhysteroidea
 	
Chromadorina
    Chromadoroidea
    Desmodoroidea
    Desmoscolecoidea

de Coninck (1965)

Araeolaimica
    Araeolaimida
    Monhysterida
 	
Chromadorica
    Desmodorida
    Chromadorida
 	
Desmoscolecida
    Desmoscolecoidea
    Greeffielloidea

present study

Chromadorida
    Chromadorina
    Desmoscolecina
    Leptolaimina
 	
Monhysterida
    Axonolaimoidea
    Desmoscolecoidea
    Siphonolaimoidea
    Monhysteroidea

These differences in the gross systematic division of the Chromadoria result from differences in the phylogenetic assessment of ovarian structure, of labial and cephalic sensilla characteristics, shape of the amphids, structure of the vestibulum, form of the cardia lumen and structure of the cuticle. The following is a summary of the various prevailing arguments for a gross systematic classification of the Chromadoria (points a-c). These arguments can, in part, only be inferred from the diagnoses available so far, and inferred from diagnostic keys. Finally (points d-h), the arguments are analyzed critically and the author's decision to divide the Chromadoria into the two orders Chromadorida and Monhysterida is elaborated.

a) Filipjev (1934) did not place together the large number of species which belong to the Chromadoria to form a single taxon. He subdivided them into the "Chromadorata", "Desmoscolecata" and "Monhysterata". A distinction was made between the "Chromadorata" and "Monhysterata" primarily on the basis of their ovarian structure whereby, with uncompromising consistency, only species with reflexed ovaries were placed in the "Chromadorata" and only species with outstretched ovaries were placed in the "Monhysterata". Additional features are given in the diagnoses for both orders, but these features are not generally as valid as the features of ovarian structure. In striking contrast to the uncompromising toleration of only those species with outstretched ovaries within the "Monhysterata" is Filipjev's establishment of the "Desmoscolecata", the females of which in the species known at that time also always possess outstretched ovaries. In his work of 1921 (p. 105), Filipjev suggests that, on account of this feature, they are probably "relatives of a subfamily of the Monhysteridae" (the Monhysteridae in Filipjev 1918/1921 are referred to as the Monhysterata in Filipjev, 1934); in his work of 1934 (p. 24), however, he considers them to be "a very isolated group with somewhat uncertain systematic position". Despite their spiral amphids, Filipjev places the Tripyloididae in the "Enoplata", on account of the conformity in head structure with *Tobrilus* and of the conformity in structure of the spicule apparatus with *Halanonchus*.

b) Chitwood & Chitwood (1950) subdivided the Chromadoria into the "Monhysterina" and "Chromadorina". On page 143 they acknowledge in principle ovarian structure as a systematic criterion, but also list examples which, in their opinion, weaken the phylogenetic strength of this criterion. They suggest that more importance should be attached to other criteria, and on pages 21, 22, 85 and 88 they do in fact introduce the structure of the cardia as a major new criterion in the classification of the Chromadoria into two gross taxa: in the "Monhysterina" the lumen of the cardia is said always to be dorsoventrally flattened or round in cross-section, whereas in the "Chromadorina" it is vertically flattened or triradiate. Further features are specified for both gross taxa, but they are not strongly correlated with cardia structure. Chitwood & Chitwood do not present data for this feature in the Desmoscolecoidea. It is probably only by error that the group was placed in the "Chromadorina", because the ovaries were erroneously described as reflexed. The Tripyloididae were placed in the Chromadoria presumably on account of their spiral amphids and in the "Chromadorina" presumably on account of the cardia's lumen which radiates in three parts.

c) De Coninck (1965) subdivides the Chromadoria into "Araeolaimica", "Chromadorica" and "Desmoscolecida". Stekhoven & de Coninck (1933 : 139) and de Coninck (1965 : 595) consider ovarian structure to be a rather worthless characteristic as a systematic criterion because parallelisms are said to appear. Presumably this observation, which is not substantiated in any way, led de Coninck (1965) to remove the characteristic of ovarian structure from almost all diagnoses and, without considering ovarian structure, to determine the species composition of the individual taxa of the Chromadoria simply using other criteria. De Coninck distinguishes between the three gross taxa of the Chromadoria mainly by the structure of the amphids: in the "Araeolaimica" these are said to be spiral with one loop, round, or shaped like an ear or a question-mark; in the "Chromadorica" they are spiral with one or more loops, or shaped like a kidney, a slit or, occasionally, like an ear; and in the "Desmoscolecida" they are bubble-like. This division is remarkable in that it is based on the assumption (which was never substantiated) by Stekhoven & de Coninck (1933 : Fig. 1) that a closed O-shaped, ear-like amphid, as for example that which occurs in *Ascolaimus* (similar to that in Fig. 7g), is plesiomorphic within the Chromadoria. Filipjev and Chitwood & Chitwood were also well aware of the individual amphid types, but they valued their phylogenetic importance less highly than did de Coninck. De Coninck has not introduced any new systematic criteria. The Tripyloidea are placed among the "Araeolaimica" on account of the shape of their amphids.

Opinion of the criteria listed and elaboration of the author's personal view:

d) The Tripyloidea belong to the Enoplida on account of the fact that they possess metanemes (discussion point 32). Filipjev's placement of this family in the "Enoplata" was thus correct.

e) De Coninck's (1965) outright rejection of ovarian structure and his disregard of cardia structure as criteria need explanation and do not represent, in their current form, any arguments against the principles of classification used by Filipjev and Chitwood & Chitwood. According to postembryological data, a spiral amphid form with only one loop is to be considered plesiomorphic within the Chromadoria, and not a closed O-shaped, ear-like amphid form, as was assumed by Stekhoven & de Coninck, (discussion point 24). The criterion of the amphid form, as used by de Coninck, does not allow a very sharp division of the "Araeolaimica" and "Chromadorica". All in all, de Coninck's gross systematic classification of the Chromadoria is not as well substantiated and more controversial than that of Filipjev (1934) and that of Chitwood & Chitwood (1950).

f) Chitwood & Chitwood's (1950) view that one should not divide all Chromadoria except that Desmoscolecoidea into the "Chromadorata" and "Monhysterata" on the basis of ovarian structure alone – as does Filipjev – is justified. The assessment of other features forces one to assume that without doubt in four instances outstretched ovaries have evolved from antidromously reflexed ovaries within the Chromadoria (discussion point 51). However, only one of the arguments which Chitwood & Chitwood use to substantiate their theory is correct: despite outstretched ovaries the Microlaimidae belong to the Chromadorida. The other two arguments do not hold true: *Halanochus* belongs to the Enoplia and not to the "Monhysteridae", and antidromously reflexed ovaries never occur in the Comesomatidae (discussion point 51). For the following reasons the criterion of cardia structure does not possess the phylogenetic importance attributed to it by Chitwood & Chitwood:

- According to Chitwood & Chitwood (1950: 88 and Fig. 82) the cardia in *Axonolaimus* has a lumen which, in cross-section, is triradiate, although according to the diagnosis of the suborder "Monhysterina", the lumen must be flattened or round.
- The feature of the cardia structure is known in approximately only 20 of the c. 350 genera of the Chromadoria.
- The discriminatory level of the criterion is insufficient since the differences in cross-section (round, flattened or radiating in three parts) of the lumen of the cardia are, in part, only slight.
- There is no analysis to show which cardia structure should be considered plesiomorphic and which apomorphic within the nematodes and particularly within the Chromadoria.

g) The author's own analyses have shown that in a modified form (discussion point 51), Filipjev's use of ovarian structure as a criterion provides the hitherto most reliable way of establishing the holopholy of at least one of the two orders of the Chromadoria: outstretched ovaries are considered to be a holapomorphic feature of the Monhysterida. The criterion is superior to the others for the following reasons:

- Ovarian structure (antidromously reflexed, homodromously reflexed or outstretched) is known in most of the approximately

350 genera of the Chromadoria and in the remaining genera of the freeliving nematodes.
- Outstretched ovaries are considered apomorphic in contrast to antidromously reflexed ovaries (discussion point 51).
- Outstretched ovaries can always be distinguished without doubt from antidromously or homodromously reflexed ovaries with the result that the feature is very selective.
- Ovarian structure shows absolutely constant formation within almost all families of the Chromadoria which have been shown to be holophyletic and within almost all higher rank taxa of the remaining nematodes, so that outstretched ovaries have probably only rarely evolved from antidromously reflexed ovaries within the evolution of the nematodes.
- There is at the moment no other known criterion with which the holophyly of the Monhysterida or of a corresponding taxon could be established.

The extent of the order Monhysterida is determined in discussion point 51 (referred to there as Taxon B).

h) Since antidromously reflexed ovaries are considered plesiomorphic within the nematodes (discussion point 49), this feature cannot be used to establish the holophyly of the Chromadorida. After the removal of the Monhysterida, the Chromadorida represent the rest of the Chromadoria. There is, as yet, no known criterion with which to establish the holophyly of the Chromadorida. The author's own analyses have also failed to produce such a criterion. Only a very insignificant indication of a possible holophyly was found, namely that often within the Chromadorida the anterior gonad is constantly found to the right and the posterior gonad constantly to the left of the intestine (discussion point 53). All in all, only paraphyly, and not holophyly, has as yet been established for the Chromadorida.

5.2.2. Classification of the Chromadorida
CHROMADORIDA Filipjev 1929

No holapomorphy can be cited which establishes the holophyly of the Chromadorida. There is only one weak indication of a possible holophyly: often the anterior gonad is found constantly to the right and the posterior gonad constantly to the left of the intestine within species within this order (discussion point 53); this feature appears in all species of the Chromadoridae and Plectidae, in many species of the Leptolaimidae and Cyatholaimidae and in some species of the Desmoscolecina. Within other taxa of the Chromadorida, the position of the gonads relative to the intestine is only semi-constant (anterior and posterior gonads constantly on different sides or constantly on the same side of the intestine) or totally variable. Further features: the 6+4 cephalic sensilla are often in two separate circles and the posterior 4 cephalic sensilla are in general longer than the anterior 6; less frequently the 6+4 cephalic sensilla are situated at the same level, in which case the 6 cephalic sensilla of the one circle are generally longer than the 4 remaining sensilla (Chromadorina partim). The amphids are almost always ventrally-spiral and only rarely non-spiral. The ventrally-spiral amphids are spiral, inverted U-shaped, O-shaped or round. The buccal cavity contains teeth (most of the Chromadorina) or is toothless (most of the Leptolaimina). The pharynx often bulges out at the end to form a bulb. The ovaries are usually antidromously reflexed and only rarely outstretched (Desmoscolecidae, Tricominae, Microlaimidae, Aponchiidae, Peresianidae, Rhabdolaimidae partim). There are usually two ovaries, seldom only one anterior or only one posterior ovary. There are often two testes, though in part only one anterior testis (Chromadoridae, Desmodoroidea, Desmoscolecidae and individual smaller taxa) and only seldom one posterior testis (some Leptolaimidae, *Pomponema multipapillatum*). The adult males of many species have ventrally situated pre-cloacal papillae or tubules.

Discussion: The Chromadorida are a paraphyletic taxon because the holophyly cannot currently be established and because, after the removal of the holophyletic taxa Monhysterida, Enoplia and Secernentea, the Chromadorida represent the rest of the freeliving nematodes. In their present form the Chromadorida correspond most fully to the "Chromadorata" plus "Desmoscolecata" of Filipjev (1934). Contrary to the checklist of Gerlach & Riemann (1973/1974), the Chromadorida contain species from all 6 of the orders listed there: the Teratocephalidae, "Plectidae", "Rhabdolaimidae" (except *Syringolaimus*), Aulolaimidae, Haliplectidae, "Leptolaimidae", Bastianiidae and part of the "Axonolaimidae" from the "Araeolaimida"; all "Desmoscolecida"; *Cyartonema* (syn. *Southernia*) from the "Monhysterida"; all genera except *Xenella* from the "Desmodorida"; all taxa except the "Comesomatidae" (excluding *Acantholaimus*) from the "Chromadorida"; the Prismatolaiminae, "Monochromadorinae" and *Tobrilia* ("Tobrilinae") from the "Enoplida".

Within the Chromadorida a distinction is made between the suborders Chromadorina, Desmoscolecina and Leptolaimina.

CHROMADORINA Filipjev 1929

Personal analyses of features have not indicated any holapomorphy with which the holophyly of the Chromadorina can be established.

66. Filipjev (1934:22) names the twelve-fold pleated vestibulum which is covered by a soft, flexible cuticle, as a particularly typical feature of this sub-order (which he refers to as the family "Chromadoridae"). This feature is considered to be a holapomorphy of the Chromadorina with which the holophyly of this sub-order can be established. Reason: the feature is unknown in species outside the Chromadorina and occurs in a large number of species within the Chromadorina; it is absent in all Epsilonematidae and in part of the Desmodoridae, Draconematidae, Microlaimidae and Aponchiidae, though these species belong to the holophyletic subtaxa of the Chromadorina on the basis of other features. The absence of the twelve folds in the vestibulum in representatives of the Chromadorina is thus considered a secondary condition.

The presence of teeth in the buccal cavity is also very characteristic of the Chromadorina. There is one tooth situated dorsally and two subventrally. The dorsal tooth is usually the largest; occasionally all three teeth are equal in size, though a single subventral tooth is never the largest of the three, as often occurs within the Enoplia. Occasionally the teeth are absent altogether (Selachinematidae, Stilbonematinae, a few Cyatholaimidae, many Epsilonematidae). The cuticle is often striated. The pharynx often has a posterior bulb.

Discussion: in connection with the twelve-fold vestibulum, Filipjev (1934:22) also refers to "the eversible mouth capsule" as being particularly typical for the Chromadorina. In preserved material this feature can only be detected by way of an exception (see for example Lorenzen, 1972 for *Actinonema pachydermatum*; Boucher, 1976 for *Chromadorella salicaniensis*), and in living animals the feature has not yet been studied. It is also not known whether the possible ability of the teeth in the buccal cavity to be everted *outwards* differs in any way from the same ability which is found in members of the Enoplida. Thus the feature of the "eversible mouth capsule" cannot currently be assessed from a phylogenetic view point.

Chitwood & Chitwood (1950:71) do indeed describe the feature of the twelve-fold vestibulum for taxa of the Chromadorina ("Cheilorhabdions take the form of 12 odontia"), but they ignore it as a systematic criterion, as do de Coninck (1965) and Andrassy (1976). These authors do not give any reason for this. The following may be the reason: Chitwood & Chitwood place the Comesomatidae in the Chromadorina. In both families the vestibulum does not have twelve folds, with the result that the authors were able to view the twelve-fold vestibulum as not being typical of the Chromadorina. In the present work the Tripyloididae are placed in the Enoplida and the Comesomatidae in the Monhysterida.

The Chromadorina correspond most fully to the "Chromadoridae" of Filipjev (1934); with the exception of new species, they are only extended by the Microlaimidae and Aponchiidae, part of the species of which also possess a twelve-fold vestibulum. In the following, the Chromadorina are sub-divided into the holophyletic Chromadoroidea and Desmodoroidea as well as into the non-holophyletic Microlaimoidea. This division shows that within the sub-order, the interrelationships between the larger taxa are not yet understood.

CHROMADOROIDEA Filipjev 1917

67. The cuticle is always striated. In most species the body annules are covered with dots or with adornments which have developed from the dots. Since this feature is not found in the remaining Chromadorina it is considered to be a holapomorphy with which the holophyly of the Chromadoroidea can be established. (In the Desmodoroidea the body striations have at the most vacuoles, longitudinal ridges or little spines.) There are no other criteria available for the holophyly of the super-family.

Personal observations have shown that the body can often be coloured easily with cotton blue; only within the Selachinematidae are there species where cotton blue does not penetrate the cuticle and enter the body; these species have a brownish appearance in glycerine preparations. The labial sensilla are mostly papilliform; they are setiform only in part of the Cyatholaimidae and Selachinematidae. The 6+4 cephalic sensilla are either in two separate circles or they are located at the same level. In the former case the 4 posterior cephalic sensilla are longer than the 6 anterior sensilla, and in the latter case the 6 sensilla of the one circle are usually longer than the 4 sensilla of the other circle (see section 8.1). The ventrally-spiral amphids have one or more loops and are located posterior to or between the 4 posterior cephalic sensilla. Almost all species have two ovaries; they are always antidromously reflexed. There are either two opposed testes, or only one anterior testis; only in one species has a single posterior testis been observed (*Pomponema multipapillatum*). Within species the gonad position relative to the intestine is very constant ($^r_|$), only semi-fixed ($^{lr}_{rl}$, $^{lr}_{lr}$), or occasionally totally unfixed ($^{lrlr}_{rllr}$).

Discussion: Basically the Chromadoroidea correspond in extent to the "Cyatholaiminae", "Choanolaiminae", "Richtersiinae" and "Chromadorinae" of Filipjev (1934). However, from the point of view of nomenclature, Filipjev did not group the 4 taxa together. In the key to his analysis he divided them as a block from the remaining taxa of his "Chromadoridae" on account of the listed features of the cuticular structure. The Chromadoroidea more or less correspond to the "Chromadoroidea" of Chitwood & Chitwood (1950) and to the "Chromadorida" of de Coninck (1965) and of Gerlach & Riemann (1973). In agreement with Filipjev and in contrast to the other authors mentioned, the "Richtersiidae" are put in the Chromadoroidea. In contrast to Chitwood & Chitwood, the Microlaimidae and Tripyloididae are not put into this super-family and, in contrast to de Coninck and to Gerlach & Riemann, the Comesomatidae (excluding *Acantholaimus*) are not included either. The "Richtersiidae" are regarded later on as synonymous with the Selachinematidae.

The following families are put into the Chromadoroidea and are discussed in the following order: Chromadoroidae, Ethmolaimidae, Neotonchidae, Achromadoridae, Cyatholaimidae and Selachinematidae.

Chromadoridae Filipjev 1917

The holophyly of the Chromadoridae is established using the following two holapomorphies.

a) There is always only a single, anterior testis (discussion point 44).
b) The anterior gonad is always situated to the right of the intestine and the posterior gonad always to the left of it (discussion point 53).

The establishment of the holophyly is thus safe-guarded, because no subtaxon of the Chromadoridae forms a holophyletic group with species outside the Chromadoridae on the basis of other features. Additional features: the cuticle is always striated and ornamented. According to personal observations, the intact body can always be coloured using cotton blue after fixation in formalin. The labial sensilla are always minute. There are usually 6 shorter and 4 larger cephalic sensilla situated in two separate circles, except in part of the Euchromadorinae where 6 longer and 4 shorter cephalic sensilla occur in a common circle (see Section 8.1). The ventrally-spiral amphids are usually much wider than they are long and are usually situated between the 4 posterior cephalic sensilla, though in part they are also posterior to them. In part of the Spilipherinae and Chromadorinae all three teeth in the buccal cavity are more or less equal in size, whereas in the other species the dorsal tooth dominates and the subventral teeth are small or absent altogether. In *Megodontolaimus* the largest tooth is situated ventrally (situs inversus). In many species the pharynx swells posteriorly forming a bulb. The cardia is always tiny. There are always two antidromously reflexed ovaries. The males of many species have precloacal, cup-shaped papillae; pre-cloacal tubules are never present. Most species are marine, only a few are limnic.

Discussion: Until now the two holapomorphies named above played no part in the establishment of holophyles and in the diagnoses of the Chromadoridae. By and large however, the family was well characterized because the often typical ornamentation of the cuticle and the transverse oval shape of the amphids hardly ever occur outside the Chromadoridae. In the present work, neither feature is considered to be a holapomorphy of the family for the following reasons:

a) The ornamentation of the cuticle reaches varying levels within the Chromadoridae with the result that a relatively simple ornamentation probably constitutes the starting point for the differentiations. The starting point (the presence of dots or something similar) is probably not significantly different from that of other families of the Chromadoroidea.
b) In the Spilipherinae and a few species of other subfamilies of the Chromadoridae, the amphids are not transversely oval, or only insignificantly so. What is more, in the Spilipherinae the amphids are situated posterior to the 4 cephalic sensilla with the result that, for this sub-family at least, the form and position of the amphids can only be considered as plesiomorphic within the Chromadoroidea (discussion point 18).

At present, the relative length of the 6 anterior and 4 posterior cephalic sensilla can likewise not be used to establish holophyly in the Chromadoridae because the listed characteristics of the cephalic sensilla represent two different apomorphies within the nematodes (discussion point 11).

De Coninck (1965 : 637) and Gerlach & Riemann (1973 : 296) place *Ethmolaimus* in the Chromadoridae. Since *Ethmolaimus* has two testes and the gonad position is only semi-constant, the genus is removed from the Chromadoridae and placed in the Ethmolaimidae. If this were not carried out it would be impossible at present to indicate any holapomorphy for the Chromadoridae.

Gerlach & Riemann (1973) have put the Acantholaiminae in the Comesomatidae. According to personal observations, *Acantholaimus* has only one anterior testis, and the anterior gonad lies to the right and the posterior gonad to the left of the intestine. Furthermore there are similarities in the position of the amphids and in the form and arrangement of the teeth in the buccal cavity between *Spiliphera* and the Acantholaiminae. For these reasons the Acantholaiminae are classed as Spilipherinae in the present study. This view has been confirmed by Gerlach et al. (1979 : 42). Furthermore in contrast to Gerlach & Riemann (1973), *Nygmatonchus* and *Trochamus* have been removed from the Hypodontolaiminae and put into the Euchromadorinae on account of the relaitve size of the cephalic sensilla and the form and position of the amphids.

The holophyly of the currently recognized sub-families cannot be established.

Spilipherinae Filipjev, 1918

Acantholaiminae Gerlach & Riemann, 1973, syn. Lorenzen, 1981: 165[1] nec Ethmolaiminae Filipjev & Steckhoven, 1941

Acantholaimus Allgen, 1933
   Neochromadorina Kreis, 1963, syn. Hope & Murphy 1972 : 18
Spiliphera Bastian, 1865
   Spilophora Bastian, 1865, an impermissible emendation
   Spilophorium Cobb, 1933, an impermissible substitution
Statenia Allgen, 1930, syn. Wieser 1954 : 115
Trichromadorita Timm, 1961
Tridentellia Gerlach & Riemann, 1973
   for *Tridentella* Filipjev, 1946, a homonym

Chromadorinae Filipjev, 1917

Atrochromadora Wieser, 1959
   for *Chromadoropsis* Wieser, 1954, a homonym
Chromadora Bastian, 1865
Chromadorella Filipjev, 1918
Fusonema Kreis, 1928
Prochromadora Filipjev, 1922
Prochromadorella Micholetzky, 1924
   Trichromadora Kreis, 1929, syn. Lorenzen, 1972a : 298
Punctodora Filipjev, 1919
Timmia Hopper, 1961
   for *Parachromadora* Timm, 1952, a homonym

---

[1] All synonymies referred to as 'syn.n." in the original German publication are referred to here as 'syn. Lorenzen, 1981" followed by the appropriate page in the German edition.

OUTLINE OF A PHYLOGENETIC SYSTEM

Euchromadorinae   Gerlach & Riemann, 1973

Actinonema Cobb, 1920
   Pareuchromadora Stekhoven & Adam, 1931, syn. Wieser 1954 : 80
Adeuchromadora Boucher & Bovée, 1972
Austranema Inglis, 1969
Dicriconema Steiner & Hoeppli, 1926
Endeolophos Boucher, 1976
Euchromadora de Man, 1886
Graphonema Cobb, 1898
Nygmatonchus Cobb, 1933
Parapinnanema Inglis, 1969
Protochromadora Inglis, 1969
Rhips Cobb, 1920
Steineridora Inglis, 1969
Trochamus Boucher & Bovée, 1972

Harpagonchinae   Platonova & Potin, 1972

Harpagonchoides Platanova & Potin, 1972
Harpagonchus Platanova & Potin, 1972

Discussion: The Harpagonchinae were established by Platanova & Pontin (1972) as a family of the Chromadorida. The three species known so far live as ectoparasites on polychaetes. A further investigation of the type specimens, which are very squashed, showed that the males probably only possess a single testis. In one female the anterior ovary was found to be on the right and the posterior ovary on the left of the intestine; it was impossible to establish the position of the ovaries in the remaining females. With the exception of the gonad characteristics, the features of the buccal cavity, the pharynx and the male copulatory apparatus suggest the inclusion of the Harpagonchinae in the family of the Chromadoridae.

Hypodontolaiminae   de Coninck 1965

Chromadorissa Filipjev, 1917
Chromadorita Filipjev, 1922
   Allgeniella Strand, 1934 (for *Odontonema* Filipjev, 1929, a homonym), syn. Wieser, 1954 : 69
Deltanema Kreis, 1929
Denticulella Cobb, 1933
Dichromadora Kreis, 1929
Hypodontolaimus de Man, 1886
   Iotadorus Cobb, 1920, syn. Filipjev, 1934 : 20

Innocuonema Inglis, 1969
Megodontolaimus Timm, 1969
Neochromadora Micoletzky, 1924
Panduripharynx Timm, 1961
Parachromadorita Blome, 1974
Ptycholaimellus Cobb, 1920
Spirophorella Fillipjev, 1917

Dubious genera of the Chromadoridae:
Dasylaimus Cobb, 1933, dub. Hope & Murphy, 1972 : 48
Odontocrius Steiner, 1918, dub. Inglis, 1969 : 175

Ethmolaimidae/Neotonchidae

Platt (1982) united Ethmolaimidae, Neotonchidae, *Trichethmolaimus* and *Nannolaimus* into the family Ethmolaimidae which subsequently contains the subfamilies Ethmolaiminae and Neotonchinae. The holophyly of the family Ethmolaimidae *sensu* Platt is based on the single holapomorphy "cup-shaped precloacal supplements with an external articulated flange", that is, "the external part of the cup is surrounded by a longitudinal oval flange with a pore in the middle", and "the anterior and posterior parts of this flange appear to be articulated, allowing the whole supplement to be protruded or retracted." Platt mentioned that such supplements do also occur in *Dichromadora cephalata* (Chromadoridae), and I have seen them in *D. scandula*. Since *Dichromadora* is a typical chromadorid genus, Platt concluded that the special supplements evolved independently in Ethmolaimidae and *Dichromadora*. Since I can not present arguments for the holophyly of Ethomolaimidae and Neotonchidae (both *sensu* Lorenzen), the suggestion by Platt is accepted.

However, the following two families are a translation of Lorenzen (1981) and have not been altered to take Platt (1982) into account.

Ethmolaimidae Filipjev & Stelkhoven, 1941

No holapomorphy is known for this family. The cuticle is striated and is covered with dots. Personal observations show that the prepared body can be coloured well with cotton blue. The 6 papilliform and 4 setiform cephalic sensilla form two separate circles. The ventrally-spiral amphids lie posterior to the cephalic sensilla and are not wider, or hardly wider, than they are long. In the buccal cavity there are three teeth more or less equal in length. The pharynx shows marked muscular thickening at the end and slight muscular thickening in the region of the buccal cavity. Two testes are present, the posterior one of which proceeds anteriorly for a bit and then folds over towards the posterior (personal observations). Past observations on one male and four females show that the anterior gonad lies to the right or to the left of the intestine and the posterior gonad lies to the left of it. The males possess ventrally situated, cup-shaped, pre-cloacal papillae. The family is completely freshwater; males occur regularly in the populations.

The only genus:

Ethmolaimus de Man, 1880
Parachromadora Micoletzky, 1914 (for *Tridontolaimus* Micoletzky, 1913, a homonym), syn. Micoletzky, 1922a : 393

Discussion: *Ethmolaimus* resembles the Chromadoridae in the structure of the cuticle, of the labial region, of the posterior pharyngeal bulb, of the spicule apparatus, of the pre-cloacal papillae and of the tail, so it seems that the two families are probably closely related. However, if on account of this one were to place *Ethmolaimus* in the Chromadoridae, as de Coninck (1965) and Gerlach & Riemann (1973) have done, one would no longer be able to demonstrate any holapomorphy for the family "Chromadoridae" as they understand it, because *Ethmolaimus* has two testes and only a semi-fixed gonad position. For this reason the Ethmolaimidae are recognised as a family in their own right. *Ethmolaimus* is put into the "Cyatholaiminae" by Filipjev (1934 : 20) and into the "Microlaimidae" by Chitwood & Chitwood (1950 : 71).

Neotonchidae   Wieser & Hopper, 1966

No holapomorphy is known at present for the Neotonchidae. The family is well characterized by the combination of the following features:

a) The buccal cavity has one distinct dorsal tooth which is always larger than the subventral teeth when they are present.
b) The pharynx possesses a muscular end bulb.
c) The amphids are spiral, they have several loops (multi-spiral) and they are situated posterior to the cephalic sensilla.
d) Cup-shaped, ventrally situated cloacal papillae occur which are similar to those of the Chromadoridae.

Additional features: The cuticle is always striated; the annules are always punctated. Personal observations show that the intact body, prepared in formalin, can be coloured well with cotton blue. The posterior 4 cephalic sensilla are always longer than the anterior 6 sensilla; the two circles of cephalic sensilla are usually separated from one another. Usually there are two opposed testes, though there may also be only a single, anterior testis. The females have two antidromously reflexed ovaries. Within species the anterior gonad lies to the left or to the right of the intestine and the posterior testis lies on the opposite side of the intestine to that of the anterior testis. The species are marine.

Discussion: Wieser & Hopper (1966 : 520) established the Neotonchinae as a sub-family and placed them in the "Cyatholaimidae" on account of their "cyatholaimid affinity". The "cyatholaimid affinity" was not explained any further, but it is presumably connected with the presence of a distinct dorsal tooth in combination with multispiral amphids which are situated posterior to the cephalic sensilla. On the other hand, the development of the buccal cavity, the pharyngeal bulb, the cephalic sensilla characteristics and the pre-cloacal papillae is similar to that of the Chromadoridae and Ethmolaimidae, and not to that of the Cyatholaimidae. The Neotonchidae cannot currently be placed either in the Cyatholaimidae or in the Chromadoridae, without thereby destroying the holophyly, as it is now understood, of the families concerned. It is thus preferable, as in the case of the Ethmolaimidae, to give the Neotonchinae the rank of a family despite their paraphyletic status.

The genera of the Neotonchidae:
Gomphionema Wieser & Hopper, 1966
Neotonchus Cobb, 1933
   Comesa Gerlach, 1956, syn. Wieser & Hopper, 1966 : 521

Achromadoridae   Gerlach & Riemann, 1973

The holophyly to the Achromadoridae is established by the following two holapomorphies, both of which are unique to the Chromadorina.

a) The anterior and posterior ovaries are situated on the same side of the intestine (both on the left or both on the right of the intestine, discussion point 53).

68. b) Males are found only extremely rarely, so that parthenogenetic reproduction must be the rule; this characteristic is connected with the purely limnic-terrestrial occurrence of the family.

Additional features: Most of the species are small (0.5–0.6 mm long, only *Paradoxolaimus* is said to be over 2 mm long). The cuticle is striated; each body annule has dots on it. The 6+4 cephalic sensilla are usually situated in a single circle and seldom form two separate circles; the 6 cephalic sensilla of the second circle are usually longer than those of the fourth and third circles. The amphids are spiral and are situated well posterior to the cephalic sensilla. the buccal cavity contains one distinct dorsal tooth; subventral teeth are either small or absent altogether. The pharynx has an obvious posterior bulb. The females have two anti-dromously reflexed ovaries. Pre-cloacal organs are absent in the males.

Discussion: *Achromadora* has hitherto always been put into the Cyatholaimidae, presumably on account of its spiral amphids which are situated posterior to the cephalic sensilla. However, as far as the amphids are concerned, *Achromadora* also resembles the Spilipherinae (Chromadoridae) and Ethmolaimidae, with the result that this feature does not provide a valid argument for placing the *Achromadora* in the Cyatholaimidae. Ever since their introduction, the Kreisonematinae have always been placed in the "Leptolaimidae", but in the present work they are put into the Achromadoridae on account of the structure of the buccal cavity (longitudinal ribs in the vestibulum, dorsal tooth present), the presence of a pharyngeal end bulb in *Kreisonema* and the position of the two gonads on the same side of the intestine (see Fig 1a in Khera, 1969). It has not been decided to which family of the Chromadoroidea the Achromadoridae are most closely related. Its apparently isolated status and the discovery of the two holapomorphies justify the rank of family. The holophyly of the two sub-families cannot be established.

Achromadorinae   Gerlach & Riemann, 1973

Achromadora Cobb, 1913

Kreisonematinae   Khera, 1969

Kreisonema Khera, 1969
Paradoxolaimus Kreis, 1924

Cyatholaimidae   Filipjev, 1918

The holophyly of the Cyatholaimidae is established by the holapomorphy that the two cephalic sensilla circles are situated at the same level and are not separated (discussion point 15). Reasoning:

a) The characteristic is unique within the Chromadorina in combination with the following characteristics:

— The 6 cephalic sensilla of the second sensilla circle are always longer than the 4 cephalic sensilla of the third sensilla circle.

— The amphids are always multi-spiral.
— The buccal cavity usually contains one distinct dorsal tooth, whereas the subventral teeth are very small or absent altogether.
— The anterior and posterior gonads always lie on different sides of the intestine (Section 8.2.1).

b) No subtaxon of the Cyatholaimidae forms a holophyletic group with species outside the family on the basis of other characteristics.

Additional features: The cuticle is always striated and is covered with dots. The labial sensilla are often setiform. The pharynx never has a posterior bulb. There are almost always two ovaries (exceptions: *Pomponema multipapillatum, P. syltense, Dentatonema* all have only one ovary); the ovaries are always antidromously reflexed. There are usually two opposed testes, rarely only one anterior or only one posterior testis (Section 8.2.2. of the appendix). Pre-cloacal tubules or papillae are often present. The Cyatholaimidae are marine except in a few cases (e.g. *Paracyatholaimus truncatus*).

Discussion: In contrast to previous sytems, the Achromadorinae and Neotonchinae are not listed as members of the Cyatholaimidae, but as families in their own right. Otherwise no holapomorphy could be demonstrated for the Cyatholaimidae. The arrangement of the 6+4 cephalic sensilla in two separate circles in *Nannolaimus volutus* is interpreted as atavism because on the basis of the other characteristics the species clearly belongs to the Cyatholaimidae. The four established sub-families have hitherto been distinguished on the basis of purely diagnostic new points; the establishment of a holophyly is still outstanding in their case. Very probably nothing more than paraphyly can be established at least for the Cyatholaiminae in its present extent because, according to Gerlach & Riemann (1973:383), this sub-family contains all those genera which could not be placed in the other sub-families.

Pomponematinae   Gerlach & Riemann, 1973

Longicyatholaiminae Andrassy, 1976, syn. Lorenzen 1981 : 173
Kraspedonema Gerlach, 1954
  Craspodema Gerlach, 1956, an unjustifiable substitution since *Kraspedonema* does not represent a homonym of *Craspedonema* Richters, 1908
Minolaimus Vitiello, 1970
Nannolaimoides Ott, 1972
Nannolaimus Cobb, 1920
Pomponema Cobb, 1917
  Anaxonchium Cobb, 1920, syn. Lorenzen, 1972 : 285

Cobbiacanthonchus Allgen, 1953, syn. Hope & Murphy, 1972 : 23
Endolaimus Filipjev, 1922, syn. Wieser, 1954 : 6
Haustrifera Weiser, 1954, syn. Wieser & Hopper, 1967 : 264
Nummocephalus Filipjev, 1946, syn. Lorenzen, 1972 : 286
Parapomponema Ott, 1972, syn. Lorenzen, 1981 : 173
Propomponema Ott, 1972, syn. Lorenzen, 1981 : 173

Discussion: Andrassy (1976 : 131) also puts, among others, *Pomponema*, the type genus of the Pomponematinae, in the "Longicyatholaiminae". Both sub-families are thus synonymous with one another. Andrassy has not listed the Pomponematinae as a sub-family of the Cyatholaimidae. *Parapomponema* and *Propomponema* do not differ from the revised diagnosis of the genus *Pomponema* (Lorenzen 1972c : 286). They are now called the *Parapomponema* — species *Pomponema hastatum* (Ott, 1972) and *P. macrospirale* (Ott, 1972) and the single *Propomponema* — species *Pomponema foeticola* (Ott, 1972) (erroneously called *foeticolum* in the original description).

Raracanthonchinae de Coninck, 1965

Acanthonchus Cobb, 1920
  Subgen. *Acanthonchus* Cobb, 1920
  Subgen. *Seuratiella* Ditlevsen, 1921 — for *Seuratia* Ditlevsen, 1918, a homonym
Biarmifer Wieser, 1954
Paracanthonchus Micoletzky, 1924
  Harveyjohnstonia Mowson, 1953, syn. Wieser, 1959 : 39
  Paraseuratiella Stekhoven, 1950, syn. Hope & Murphy, 1972 : 24
Paracyatholaimoides Gerlach, 1953
Paracyatholaimus Micoletzky, 1922

Xenocyatholaiminae Gerlach & Riemann, 1973

Xenocyatholaimus Gerlach, 1953

Cyatholaiminae Filipjev, 1918

Cyatholaimus Bastian, 1865
  Necticonema Marion, 1870, syn. Inglis, 1961 : 73
Longicyatholaimus Micoletzky, 1924
Marylynnia Hopper, 1977
  for *Marylynia* Hopper, 1972, a homonym
Metacyatholaimus Stekhoven, 1942
  Metachoniolaimus Stekhoven, 1950, syn. Wieser, 1954 : 14
Paralongicyatholaimus Stekhoven, 1950

Phyllolaimus Murphy, 1964
Praeacanthonchus Micoletzky, 1924
Xyzzors Inglis, 1963

Dubious genera of the Cyatholaimidae:
Boreomicrolaimoides Allgen, 1954, dub. Hope & Murphy, 1974 : 48
Dentatonema Kreis, 1928, dub. Hope & Murphy, 1972 : 48
Dispira Cobb, 1933, dub. Hope & Murphy, 1972 : 48
Dispirella Cobb, 1933, dub. Hope & Murphy, 1972 : 48
Heterocyatholaimus Allegne, 1935, dub. Hope & Murphy, 1972 : 49
Micranthonchus Allgen, 1959, dub. Hope & Murphy, 1972 : 50

Selachinematidae   Cobb, 1915

Choniolaimidae Stekhoven & Adam, 1931, syn. Lorenzen, 1981 : 715
Choanolaimidae de Coninck & Stekhoven, 1933, syn. Gerlach, 1964 : 26
Richtersiidae Kreis, 1929, syn. Lorenzen, 1981 : 175

The holophyly of the Selachinematidae is established by the following holapomorphies:

69. The buccal cavity is spacious and either contains no teeth at all, or only teeth which are not homologous with those of the other Chromadorina and which are referred to as jaws. Reasoning: a) a spacious buccal cavity with no teeth otherwise occurs within the Chromadorina only in a few species of the Cyatholaimidae; b) no subtaxon of the Selachinematidae forms a holophyletic group with other species of the Chromadorina on the basis of other characteristics.

Additional features: In some species the pharynx has muscular swelling posteriorly and, in part, also anteriorly. The cuticle is always striated, and is always dotted. Personal observations have shown that, in species of several genera (e.g. *Choniolaimus, Halichoanolaimus, Latronema, Richtersia*), intact animals preserved in formalin do not, or hardly, become coloured with cotton blue, and vary from a yellowish to a brownish colour in glycerine preparations; *Synonchiella*, however, colours well with cotton blue. The sensilla of one of the three circles may be longer than the sensilla of the other two circles (Scetion 9.1 of the appendix). In many cases the sensilla are jointed. The amphids are spiral with several loops. Most males have two testes and the females always have two antidromously reflexed ovaries. The position of the gonads relative to the intestine is only semi-fixed ($^{lr}_{rl}$) or totally variable ($^{lrlr}_{rllr}$), see section 8.2.1. of the appendix). Pre-cloacally the males of many species have cup-shaped, ventrally situated pre-cloacal papillae; pre-cloacal tubules are always absent. The family is purely marine.

Discussion: Filipjev (1934 : 22) mentions the large intestinal cells as an additional, very characteristic feature of this family. According to personal observations this feature does seem to hold good, but it is in need of closer investigation and a comparison should be made with the intestinal cells of the remaining Chromadoroidea. As regards the structure of the cuticle (dots, pores), many species of the Selachinematidae are similar to the Cyatholaimidae, but where the resistance to colouring with cotton blue is concerned, however, they resemble the Desmodoroidea. This finding cannot currently be considered phylogenetic. Toothless species of Cyatholaimidae clearly belong to the Cyatholaimidae on the basis of other characteristics which indicates that the teeth have been reduced independently at least twice within the Chromadoroidea.

Gerlach (1964) has revised the "Selachinematidae" and "Choniolaimidae" and on analysis of the buccal cavity has come to the conclusion that the presence of jaws should be considered apomorphic within the species set of the two families. However, the jaw structures could not be considered homologous to one another, with the result that the "Selachinematidae" could only be regarded as a "collective group for genera with jaw structures" (Gerlach 1964 : 26); in other words: the "Selachinematidae" were seen as a polyphyletic taxon. The polyphyly is upheld by the synonymisation of the Selachinematidae and Choniolaimidae (see above).

Latronema (Selachinematidae) and *Richtersia* ("Richtersiidae") resemble each other to such an extent in the structure of the labial region of the buccal cavity, of the setiform labial and cephalic sensilla and of the longitudinal rows of cuticular adornment, that the similarities are probably due to homology. The uniqueness in the combination of the homologies is taken to be an indication of a very close relationship between *Latronema* and *Richtersia*. Since *Latronema* is also related on the other hand to *Halichoanolaimus* (Selachinematidae), the Richtersiidae are seen as synonymous with the Selachinematidae. While Filipjev (1934 : 23) regarded the Richtersiidae to be closely related to the Cyatholaimidae, Chitwood & Chitwood (1950 : 22), de Coninck (1965) and Gerlach & Riemann (1973) all placed the family in the "Desmodorida" where it seemed like a foreign body on account of its spacious, toothless buccal cavity and the setiform structure of its labial sensilla. Boucher (1976 : 59) states that *Latronema* and *Richtersia* resemble one another in numerous characteristics, but he did not draw any systematic conclusion from this observation.

The genera of the Selachinematidae:

Cheironchus Cobb, 1917
  Dignathonema Filipjev, 1918, syn. Filipjev in Kreis, 1926 : 157
Choanolaimus de Man, 1880
Choniolaimus Ditleusen, 1918
  Bulbopharyngiella Allgen, 1929, syn. Gerlach, 1964a : 28
Cobbionema Filipjev, 1922
Demonema Cobb, 1894
  Selachinema Cobb, 1915, syn. Gerlach, 1964a : 47
Gammanema Cobb, 1920
Halichoanolaimus de Man, 1886
  Smalsundia Allgen, 1929, syn. Stekhoven & Adam, 1931 : 32
Kosswigonema Gerlach, 1964
Latronema Wieser, 1954
Richtersia Steiner, 1916
  Richtersiella Kreis, 1929, syn. Stekhoven, 1935 : 99
Synonchiella Cobb, 1933
Synonchium Cobb, 1920
Trogolaimus Cobb, 1920

Dubious genera of the Selachinematidae:

Nunema Cobb, 1933, dub. Hope & Murphy, 1972 : 50
Pteronium Cobb, 1933, dub. Hope & Murphy, 1972 : 51

DESMODOROIDEA Filipjev, 1922

The holophyly of the Desmodoroidea is established by the holapomorphy that only the anterior testis is present (discussion point 44). Reasoning:

a) Within the Chromadorina the feature is unique in combination with the following features:

— The intact body preserved in formalin cannot be coloured using cotton blue (personal observation) and it appears yellowish to brownish in glycerine preparations.
— In the buccal cavity there is often one distinct dorsal tooth, while the subventral teeth are always smaller or absent altogether; in cases where the buccal cavity contains no teeth it is very small indeed.
— The position of the gonads relative to the intestine is variable.

b) As far as is known no subtaxon of the Desmodoroidae forms a holophyletic group with species outside the Desmodoroidea on the basis of other features.

Additional features: The cuticle is almost always striated (only in

Molgolaimus partim is it unstriated); the body annules are not dotted and, at the most, have vacules, longitudinal ridges or little spines on them. The head region is usually striated. In many species the labial region is covered in soft skin and can be everted. The labial sensilla are always very short and papilliform. The 6+4 cephalic sensilla are usually situated in two separate circles; the posterior 4 cephalic sensilla are always longer than the anterior 6 sensilla. Within families the amphids, which are always ventrally-spiral, are variable in shape. In many species the pharynx swells at the end forming a muscular bulb; in addition, a musuclar swelling occurs in the anterior region in many species. The females always have two antidromously reflexed ovaries. The males of many species have ventrally situated pre-cloacal papillae or tubules. With the exception of *Prodesmodora*, the Desmodoroidea are marine.

Discussion: The Desmodoroidea correspond extensively to the "Desmodoroidea" of Chitwood & Chitwood (1950 : 22) and to the "Desmodorida" of de Coninck (1965 : 623) and Gerlach & Riemann (1973). Chitwood & Chitwood and de Coninck, in accordance with Filipjev (1934: diagnostic key on p.21), have used the absence of cuticular ornamentation as the fundamental criterion in the separation of the "Desmodorida" from the "Chromadorida". As the ornamentation of the cuticle is considered a holapomorphic feature of the Chromadoroidea (discussion point 67), its absence in the "Desmodorida" can be considered only as plesiomorphic and is consequently not suitable as a criterion for the establishment of the holophyly of this "order". Within the existing "Desmodorida", the Desmodoroidea form the largest number of species for which the holophyly can be established. This species pool is thus given an equal rank to that of the Chromadoroidea. From the remaining taxa of the existing "Desmodorida", the Microlaimidae, Aponchiidae and Monoposthiidae are placed in the non-holophyletic Microlaimoidea, the Ceramonematidae and Ohridiidae are placed in the Leptolaimina, and the Xenellidae in the Trefusiida.

The families Desmodoridae, Epsilonematidae and Draconematidae are included in the Desmodoroidea. So far, only the last two families named above can be shown to be holophyletic with the result that, as yet, only paraphyly can be established for the extensive family of the Desmodoridae. The paraphyly of the Desmodoridae shows that the relationship of the three families to one another is, as yet, only incompletely understood.

<u>Desmodoridae</u> Filipjev, 1922

There is no known holapomorphy for the Desmodoridae. The family stands out largely because it does not possess the features typical of the Epsilonematidae and the Draconematidae. It should be pointed out that, within the Desmodoridae, and in particular in many *Desmodora* species, there is a difference in thickness between the various regions of the body, the ovaries and the vulva lie well posterior to the middle of the body, and the cuticle has very coarse annules, at least in the anterior body region. These features appear in a more strongly derived form in the Epsilonematidae and Draconematidae (see also Lorenzen, 1976b : 71).

Desmodorinae   Filipjev, 1922

Acanthopharynginae Filipjev, 1918, syn. Gerlach & Riemann, 1973 : 235, nomen oblitum
Acanthopharyngoides Chitwood, 1936
Acanthopharynx Marion, 1870
   Xanthodora Cobb, 1920, syn. Gerlach, 1963 : 94
Amphispira Cobb, 1920
Desmodora de Man, 1889
   Mastodes Steiner, 1921, syn. Gerlach, 1963 : 77
   Subgen. *Bolbonema* Cobb, 1920
   Subgen. *Croconema* Cobb, 1920
     Aculeonchus Kreis, 1928, syn. Kreis in Wieser, 1954 : 170
   Subgen, *Desmodora* de Man, 1889
     Desmodorella Cobb, 1933, syn. Lorenzen, 1976b : 71
   Subgen. *Pseudochromadora* Daday, 1899
     Bradylaimoides Timm, 1961, syn. Gerlach, 1963 : 83–84
     Micromicron Cobb, 1920, syn. Andrassy, 1959 : 51
   Subgen. *Xenodesmodora* Wieser, 1951
     Bla Inglis, 1963, syn. Wieser & Hopper, 1967 : 277
   Subgen. *Zalonema* Cobb, 1920
     Heterodesmodora Micoletzky, 1924, syn. Gerlach, 1963 : 92
Metadesmodora Stekhoven, 1942
Paradesmodora Stekhoven, 1950
Pseudodesmodora Boucher, 1975

Spiriniinae   Gerlach & Murphy, 1965

Metachromadorinae Chitwood, 1936, syn. Hope & Murphy, 1972 : 2–3, 26
Alaimonema Cobb, 1920
Chromaspirina Filipjev, 1918
   Mesodorus Cob, 1920l syn. Filipjev in Kreis, 1926 : 157
Metachromadora Filipjev, 1918
   Ichthyosdesmodora Chitwood, 1951, syn. Timm, 1961 : 62

Subgen. *Gradylaimus* Stekhoven, 1931
Subgen. *Chromadoropsis* Filipjev, 1918
Subgen. *Metachromadora* Filipjev, 1918
Subgen. *Metachromadoroides* Timm, 1961
Subgen. *Metonyx* Chitwood, 1936
Subgen. *Neonyx* Cobb, 1933
Onyx Cobb, 1891
   Oistolaimus Ditlevsen, 1921, syn. Gerlach, 1951 : 61
Parallelocoilas Boucher, 1975
Polysigma Cobb, 1920
Pseudometachromadora Timm, 1952
Sigmophoranema Hope & Murphy, 1972
   for *Sigmophora* Cobb, 1933, a homonym
   Parachromadora Schulz, syn. Gerlach, 1951 : 61
Spirinia Gerlach, 1963
   for *Spira* Bastian, 1865, a homonym
   for *Spirina* Filipjev, 1918, a homonym
Subgen. *Spirina* Gerlach, 1963
Subgen. *Perspiria* Wieser & Hopper, 1967

Discussion: Until now, *Alaimonema* was always put in the "Diplopeltinae" (see Gerlach & Riemann, 1973: 58). In the structure of the head the only species of the genus greatly resembles the species of *Spirinia (Spirinia)*; in addition, the structure of the pharynx, the existence of only one anterior testis and the structure of the precloacal organs all indicate the inclusion of *Alaimonema* in the Spiriniinae. *Bolbolaimus* was seen as synonymous with *Chromaspirina* by Luc & de Coninck (1959 : 130–131). The idea that the two genera are synonymous must be rejected because, among other things, *Bolbolaimus* species have outstretched ovaries (Jensen, 1978a : 162).

Pseudonchinae Gerlach & Riemann, 1973

Pseudonchus Cobb, 1920
   Cheilopseudonchus Murphy, 1964, syn. Warwick 1969 : 381

Stilbonematinae Cobb, 1936

Eubostrichus Greeff, 1869
   Catanema Cobb, 1920, syn. Gerlach, 1963 : 96
Laxus Cobb, 1894
Leptonemella Cobb, 1920
Robbea Gerlach, 1956
Squanema Gerlach, 1963
Stilbonema Cobb, 1920

Molgolaiminae Jensen, 1978

The only genus: *Molgolaimus* Ditlevsen, 1921

Discussion: Until very recently the species of the Molgolaiminae were classed as Microlaimidae. Jensen (1978a : 165) erected the family Molgolaimidae and characterized it above all by the presence of a single anterior testis in the males and by reflexed ovaries in the females. He sub-divided the family into Molgolaiminae and Aponematinae. *Molgolaimus* and *Prodesmodora* were put in the Molgolaiminae and the montypical genus *Aponema* was put in the Aponematinae. In the current study, the Molgolaiminae are only seen as a sub-family of the Desmodoridae and are listed in extent to species of the genus *Molgolaimus* for the following reasons:

a) In *Aponema* the ovaries are outstretched – a feature which Jensen considered to be an exception within the Molgolaimidae. Almost all holophyletic families of freeliving nematodes have only one type of ovary, whereas on the other hand in many holophyletic families there are species with two and species with only one testis. It is therefore more likely that the posterior testis has become reduced independently in *Aponema* and *Molgolaimus*, than that outstretched ovaries have developed independently in *Aponema* and in the remaining Microlaimidae. For this reason *Aponema* is again placed in the Microlaimidae.

b) There are no reasons to accept the assumption that *Prodesmodora* should be more closely related to the Molgolaiminae than to the other taxa of the Desmodoridae (see Prodesmodorinae).

c) *Molgolaimus* corresponds to the Desmodoridae in that it only has an anterior testis, antidromously reflexed ovaries, its body does not become coloured with cotton blue, and it differs from the remaining taxa of this family mainly because it has round amphids. As long as the Desmodoridae are only a paraphyletic taxon, it is preferable not to give their subfamilies the rank of family.

Jensen extended the genus *Molgolaimus* by several species which had hitherto been considered as belonging to *Microlaimus*. These changes are acceptable with one exception: personal observations showed that *Microlaimus tenuispiculum* de Man, 1922 has two outstretched ovaries and two testes. This contradicts its diagnosis as belonging to the Molgolaiminae and it is thus put back into *Microlaimus* again. Consequently, *Molgolaimus demani* Jensen, 1978 becomes a synonym for *Microlaimus tenuispiculum* de Man, 1922; *Molgolaimus demani* was created because *Molgolaimus tenuis* (de Man 1922) was a subjective homonym for *Molgolaimus tenuispiculum* Ditlevsen, 1921.

Prodesmodorinae Lorenzen, 1981

Type and only genus: *Prodesmodora* Micoletzky, 1923.

Discussion: The ovaries are antidromously reflexed, their position relative to the intestine is variable, only one testis is present, the body appears a brownish colour in glycerine preparations, and the dorsal tooth is prominent. It is on the basis of these features that the Prodesmodorinae are put in the Desmodoridae. The following features are considered holapomorphies of the Prodesmodorinae:
70. a) Reproduction is almost exclusively parthenogenetic (males are extremely rare, see Gagarin, 1978).
71. b) The species are purely limno-terrestrial.
Both features are closely connected to one another and are unique within the Desmodoroidea. Jensen has already classified *Microlaimus arcticus* and *Melepturus* as *Prodesmodora* on the basis of these features.

Doubtful genera of the Desmodoridae:

Laxonema Cobb, 1920, dub. Hope & Murphy, 1972 : 49
Parathalassoalaimus Allgen, 1929, dub. Hope & Murphy, 1972 : 51
Xenonema Cobb, 1920, dub. Hope & Murphy, 1972 : 49

Epsilonematidae   Steiner, 1927

The following features are considered holapomorphies of the Epsilonematidae on account of their uniqueness within the nematodes (from Lorenzen, 1974b):
72. a) In the epsilon-shaped bodies which are the commonest, and in the S-shaped bodies which are rare, the ovaries are situated posterior to the dorsal curvature of the body (i.e. in the lower branch of the epsilon or the base of the S respectively).
73. b) The rigid subventral setae (which are never adhesive) are situated anterior to, or at the same level of the body region as the ovaries are situated in the females. The rigid setae may be absent at a secondary level (*Perepsilonema*).
The animals move like looper caterpillars, i.e. on the ventral side of the body. The cuticle is coarsely striated. The buccal cavity is tiny. The pharynx always has a substantial end bulb and in *Metepsilonema* it is also swollen in the middle. Postembryological studies on *Epsilonema, Bathyepsilonema* and *Metepsilonema* have shown (Lorenzen, 1973b : Clasing, 1980), that rigid setae are completely absent in juvenile stage I and first appear in the course of postembryonic development. The family is marine.
Discussion: Forward progression on the ventral side of the body in a way similar to that of the looper caterpillar occurs only in the

Epsilonematidae and Draconematidae within the nematodes, and is therefore taken to be an indication of the close relationship between the two families. The Epsilonematidae have recently been revised by Lorenzen (1973 and 1974b).
Within the Epsilonematidae, Gourbault & Decraemer (1986) have erected the new subfamily Keratonematinae with the only genus *Keratonema*. The new subfamily resembles strongly the Glochinematinae because of the presence of horn-like spines at the anterior body end and of the body shape. Unique within Epsilonematidae is that glands open through the ambulatory subventral setae. This character is common within the related family Draconematidae.

Epsilonematinae Steiner, 1927

Archepsilonema Steiner, 1931
Bathyepsilonema Steiner, 1931
 Epsilonella Steiner, 1931, syn. Lorenzen, 1973b : 41
Epsilonema Steiner, 1927
 for *Rhabdogaster* Metschinkoff, 1867, a homonym; *Prochaetosoma* Baylis & Daubney, 1926 is also a new name for *Rhabdogaster*, but is also itself a homonym
 Epsilonematina Johnston, 1938, syn. Lorenzen, 1973b : 39
 Epsilonoides Steiner, 1931, syn. Lorenzen, 1973b : 39
Metepsilonema Steiner, 1927
Perepsilonema Lorenzen, 1973

Glochinematinae Lorenzen, 1974

Glochinema Lorenzen, 1974

Draconematidae Filipjev, 1918

Chaetosomatidae Pagenstecher, 1881; this name is not valid because the type genus *Chaetosoma* Claparede, 1963 is a homonym
Drepanonematidae Johnston, 1938, a later and thus invalid substitution for Chaetosomatidae
Claparediellidae Allgen, 1954, a later and thus invalid substitution for Chaetosomatidae.
The following features are considered holapomorphides of the Draconematidae on account of their uniqueness within the nematodes (from Lorenzen, 1974b):
74. a) In the S through Z-shaped (only occasionally epsilon-shaped) curved bodies, the ovaries are situated anterior to the dorsal curvatures of the body, i.e. in the middle section of the S or Z, or in the upper curve of the epsilon respectively.

75. b) The subventrally located stilt setae (through which adhesive glands open in all species except those of *Dracognomus*) are situated posterior to that region in which the ovaries are found in females.
76. c) Situated on the dorsal side of the anterior end there are adhesive setae through which adhesive glands open in all species except those of *Dracognomus*.

The animals move in the same way as the Epsilonematidae, like the caterpillar of the looper moth, on the ventral side of the body. The cuticle has coarse or fine striatims. Postembryological research on *Draconema* and *Dracograllus* (Allen & Noffsinger, 1978; Clasing, 1980), has shown that adhesive setae are completely absent in juvenile stage I and only appear during the course of postembryological development. The family is marine.

Discussion: Allen & Noffsinger (1978) have produced an extensive monograph on the family Draconematidae. The authors have given them the rank of super-family because de Coninck (1965) had combined the Draconematidae and Epsilonematidae to form the suborder of the Draconematina. The high ranks are not accepted because the Draconematidae and Epsilonematidae together with the Desmodoridae form a holophyletic group which in the present work is referred to as the super-family Desmodoroidea. The idea of combining the Epsilonematidae and the Draconematidae to form a single taxon is rejected on account of the paraphyly of the Desmodoridae.

Within the Draconematidae a distinction is made between the Draconematinae and the Prochaetosomatinae. This division orginates from Allen & Noffsinger (1978). However, they gave both sub-families the rank of family and in addition they further divided the "Prochaetosomatidae" into four sub-families. The latter are synonymous with the Prochaetosomatinae because the "Prochaetosomatidae" already have the rank of sub-family and because the phylogenetic status of all taxa of the Draconematidae has not been debated. The Prochaetosomatinae show much greater physical variation than do the Draconematinae.

Draconematinae Filipjev, 1918

The pharynx always has a swelling in the middle and at the end. The buccal cavity is always very small and contains no teeth.
 Dracograllus Allen & Noffsinger, 1978
 Draconema Cobb, 1913
  Drepanonema Cobb, 1933 (for *Chaetosoma* Claparede, 1863, a homonym), syn. Allen & Noffsinger, 1978 : 27
  Claparediella Filipjev, 1934, a later and thus invalid substitution for *Chaetosoma* Claparede
  Tristicochaeta Panceri, 1876, syn. Schepotieff, 1907 : 157

Dracotoranema Allen & Noffsinger, 1978
Paradraconema Allen & Noffsinger, 1978

Prochaetosomatinae   Allen & Noffsinger, 1978

Cygnonematinae Allen & Noffsinger, 1978, syn. Lorenzen, 1981 : 188
Dracognominae Allen & Noffsinger, 1978, syn. Lorenzen, 1981 : 188
Notochaetosomatinae Allen & Noffsinger, 1978, syn. Lorenzen, 1981 : 188

The pharynx always has a posterior swelling and, in *Dracognomus* alone, a central swelling too. The buccal cavity is very narrow and usually contains one small dorsal tooth; this tooth is also particularly common in *Dracognomus*.

Apenodraconema Allen & Noffsinger, 1978
Cygnonema Allen & Noffsinger, 1978
Dracogalerus Allen & Noffsinger, 1978
Dracognomus Allen & Noffsinger, 1978
Draconactus Allen & Noffsinger, 1978
Notochaetosoma Irwin-Smith, 1918
Prochaetosoma Micoletzky, 1922

MICROLAIMOIDEA   Micoletzky, 1922

The holophyly of the Microlaimoidea cannot be established. In the three member families (Microlaimidae, Aponchiidae, Monoposthiidae) there are species with a twelve-fold pleated vestibulum and a prominent dorsal tooth which is larger than any subventral teeth which may occur. Thus the Microlaimoidea are classified as Chromadorina. However, in the species of the Microlaimoidea, the cuticle is not punctated as in the Chromadoroidea, nor is the presence of a single testis a constant feature as in the Desmodoroidea. Therefore the Microlaimoidea form the paraphyletic remains of the Chromadorina after the latter has been divided into the two holophyletic superfamilies Chromadoroidea and Desmodoroidea. The paraphyly of the Microlaimoidea indicates that, within the Chromadorina, the interrelationships between the larger taxa are not fully understood.

Microlaimidae   Micoletzky, 1922

Aponematinae Jensen, 1978, syn. Lorenzen, 1981 : 189
Bolbolaiminae Jensen, 1978, syn. Lorenzen, 1981 : 189

The cuticle is usually striated; only in *Ixonema* is it smooth. Personal observations have shown that animals of most of the species, when preserved in formalin, cannot be coloured using cotton blue and are a yellowish to brownish colour in glycerine preparations. The labial sensilla are very short. The 6+4 cephalic sensilla are always in two separate circles; the posterior 4 sensilla are always longer than the anterior 6 sensilla. The amphids are round. The buccal cavity contains one larger dorsal tooth and two, smaller sub-ventral teeth which are situated further back. The pharynx swells posteriorly, forming a muscular end bulb. The females always have two outstretched ovaries and the males usually have two opposed testes; rarely is only one anterior testis present (*Aponema*). Within species the anterior gonad usually lies to the left or the right of the intestine and the posterior gonad on the opposite side to the anterior one; it is less common to find the anterior gonad constantly to the left and the posterior gonad constantly to the right of the intestine. The family is marine; only a few species penetrate brackish water which contains a lot of freshwater.

Discussion: Despite the presence of outstretched ovaries, the Microlaimidae are put in the Chromadorida not in the Monhysterida on account of the twelve-fold vestibulum found in many species of the Microlaimidae (mostly in those species with one large dorsal tooth), and on account of the teeth in the buccal cavity which do not occur with this structure and in this arrangement anywhere else besides in the Chromadorida (discussion points 51 and 66). It is currently undecided whether the outstretched condition of the ovaries can be considered as a holapomorphy of the Microlaimidae alone or of the Microlaimidae and the Aponchiidae together (discussion point 51).

Molgolaimus, Prodesmodora, Paramicrolaimus and *Ohridius* have hitherto been classed as Microlaimidae. The species of the four genera possess antidromously reflexed ovaries. In the present work the four genera are put in the Molgolaiminae, Prodesmodorinae (both Desmodoridae), Paramicrolaimidae and Ohridiidae (both Leptolaimina) respectively. Jensen (1978a) has already removed the four genera from the Microlaimidae; they were, in part, given a different systematic classification from that of the current work.
The distinction between the "Microlaimidae" and the "Molgolaimidae", and the division of the two families into sub-families by Jensen (1978a) depends partly on an incorrect series of arguments and is therefore not accepted. This all resulted in the Aponematinae and Bolbolaiminae being regarded as synonymous with the Microlaimidae. According to Jensen, the pharynx in the "Microlaimidae" has a small end bulb with weak cuticularization of the inner wall, and in the "Molgolaimidae" it has a marked end bulb with a strongly cuticularized inner wall. However, the type species of *Microlaimus, M. globiceps,* has a marked end bulb with a very thick, cuticularized inner

wall (de Man, 1922 : 240; personal observation), with the result that the two families cannot be separated using the criterion. According to Jensen's diagnosis the cervical pore (referred to as the excretion pore) may be situated anterior or posterior to the nerve ring both in the Microlaimidae and in the Mologolaimidae, and it therefore cannot be used to distinguish between the two families, as Jensen has done in his Fig. 1. The marked cuticularization of the buccal cavity, the inner wall of the pharynx and the copulatory apparatus, as well as the large end bulb of the pharynx in *Bolbolaimus*, show no more than gradual differences from the remaining Microlaimidae and therefore do not justify the introduction of the Bolbolaiminae. It cannot be concluded that the sub-family represents a connecting link between the Microlaimidae and Desmodoridae because species without the peculiarities described also exist within the Desmodoridae.

Jensen's classification of species in the individual genera of the Microlaimidae is accepted, with the exception of *Microlaimus tenuispiculum* de Man, 1922, which is kept in the genus *Microlaimus* and not put into *Molgolaimus* on account of the occurrence of outstretched ovaries.

The genera of the Microlaimidae:
Aponema Jensen, 1978
Bolbolaimus Cobb, 1920
   Pseudomicrolaimus Sergeeva, 1976, syn. Jensen, 1978a : 162
Calomicrolaimus Lorenzen, 1976
Cinctonema Cobb, 1920
Crassolaimus Kreis, 1929
Ixonema Lorenzen, 1971
Microlaimus de Man, 1980
   Microlaimoides Hoeppli, 1926, syn. Andrassy, 1960 : 211
   Paracothonolaimus Schulz, 1932, syn. Stekhoven & de Coninck, 1933 : 7

Doubtful genera of the Microlaimidae:
Ungulilaimella Allgen, 1958, dub. Hope & Murphy, 1972 : 51

<u>Aponchiidae</u>   Gerlach, 1963

The holophyly of the Aponchiidae is established by the holapomorphy that only one anterior ovary is present (discussion point 46). This feature in combination with outstretched ovaries is unique within the Chromadorida.

Additional features: The cuticle is only slightly striated or smooth. Of the 6+4 cephalic sensilla, which are arranged in two separate circles, the posterior 4 are more prominent. The amphids form an 0-shaped loop or a spiral with one loop. The buccal cavity contains three teeth, which appear to be homologous with those of the remaining Chromadorida. In *Synonema ochrum*, at least, the vestibulum is folded (Gerlach, 1963). The cervical gland consists of 4–7 cells. As far as is known, the males have only one testis (Cobb, 1920 for *Aponchium cylindricolle*). The adult males have ventrally situated precloacal papillae. The family is purely marine.

Discussion: In agreement with Gerlach (1963 : 164), the Aponchiidae are put in the Chromadorina on account of the presence of teeth in the buccal cavity, and somewhere near the Microlaimidae on account of the outstretched ovary. The multicellularity of the cervical gland cannot yet be assessed phylogenetically.

The genera of the Aponchiidae:
Aponchium Cobb, 1920
Synonema Cobb, 1920
Microlaimella Wieser, 1954 (for *Microlaimoidae* Allgen, 1942, a homonym), syn. Gerlach, 1963 : 160
Synonemoides Chitwood, 1951, syn. Wieser, 1954 : 207

Monoposthiidae Filipjev, 1934

The holophyly of the Monoposthiidae is established by the following holapomorphies:
77. The amphids are non-spiral. This characteristic is very rare within the Chromadorina and is unique in combination with the presence of two testes.

Additional features: The cuticle is strongly striated and has longitidinal ornamentation. Personal observations have shown that, when fixed in formalin, the body does not become coloured with cotton blue, or only slightly so, and looks grey in glycerine preparations. The labial sensilla are very short. The 6+4 cephalic sensilla are arranged in two separate circles; the posterior 4 cephalic sensilla are longer than the anterior 6. According to current data, the males have two opposed testes. Two ovaries (*Rhinema*) or only one anterior ovary (the remaining genera) are present; the ovaries are always antidromously reflexed. The anterior gonad is situated to the left and the posterior gonad to the right of the intestine. The family is marine.

Genera of the Monoposthiidae:
Monoposthia de Man, 1889
Monoposthioides, Hopper, 1963
Nudora Cobb, 1920
Rhinema Cobb, 1920

DESMOSCOLECINA Filipjev, 1934

The diagnoses and extent of the Desmoscolecina and Desmoscolecoidea are identical.

Discussion: Until recently the females of the Desmoscolecina were only known to have outstretched ovaries. For this reason the Desmoscolecoidea were put in the Monhysterida and not in the Chromadorida in the German edition of the current work. Decraemer & Jensen (1982) have found antidromously reflexed ovaries in all the species of the Meyliinae which they investigated. From this it follows that outstretched ovaries must have developed from antidromously reflexed ovaries independently from one another within the Desmoscolecoidea and in the Monhysterida. On account of this, the Desmoscolecoidea are herewith placed in the Chromadorida where they form the suborder Desmoscolecina. Jensen & Decraemer (1982) even confer the rank of order to suborder. This suggestion is not taken up here because the remaining Chromadorida and the Monhysterida are much more heterogenous amongst themselves than are the Desmoscolecoidea.

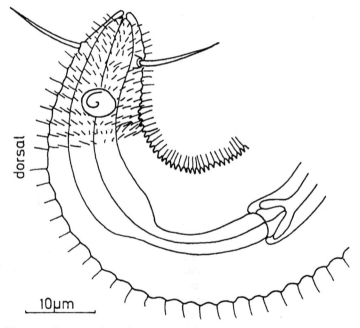

Fig. 22: Anterior end of *Meylia spinosa* (Meyliidae). The peduncles, on which the cephalic setae are located, the spines on the body ridges and the structure of the buccal cavity and pharynx all suggest the inclusion of this species in the Desmoscolecoidea. Adult female, sublittoral coarse sand, North Sea, 30th March 1977. Original.

DESMOSCOLECOIDEA Shipley, 1896

94. The holophyly of the Desmoscolecoidea is established by the two following holapomorphies:
a) The 4 cephalic setae of the third sensilla circle are set on peduncles; this feature is unique within the nematodes. In very few species of the Desmoscolecoidea where the feature has not developed, an analysis of the other characteristics shows, without doubt, that this must be a secondary condition.
b) The non-spiral amphids have a blistered corpus gelatum with a very strong outer wall (discussion point 28); this feature is, at the most, only rudimentary in other nematodes (e.g. in *Paramonohystera elliptica*, a species of the Xyalidae). Only in *Meylia* is the corpus gelatum spiral and only slightly blistered (Fig. 22). The amphids are always situated between the 4 cephalic setae or slightly behind them.

The following feature is unique within the Chromadoria:

c) The adult animals of many species have rings or warts made up of their own body secretions or out of foreign bodies on their cuticles which are almost always striated (the cuticle is smooth only in *Gerlachius*); these adornments are absent in all juvenile animals as far as is known (Lorenzen, 1971 : Decraemer, 1978), in the Greeffiellinae and in some species of the Meyliidae. An analysis is still necessary to establish whether the absence of the coatings represents a primary or secondary condition within the Desmoscolecoidea. Ring-shaped coatings of secretion and foreign bodies are otherwise known to the author only in *Criconemoides* (Tylenchida, Criconematidae; personal observations).

Additional features: In many species the body setae are on peduncles. The 6 labial and 6 anterior cephalic sensilla are hardly, or not at all, perceptible. The buccal cavity is extremely small. The pharynx is only slightly muscular and in some species it bulges out like a bubble at the end through the pharyngeal glands (Decraemer, 1978). In some species at least, the pharyngeal glands open dorsally subventrally in the proximity of the buccal cavity (Decraemer, 1976, 1978); this occurs very rarely within the Chromadoria (otherwise only in some Siphonolaimoidea, according to Riemann, 1977a, and in the Tubolaimoididae, personal observations, Fig. 24). Subventral to the anterior intestine a large number of species have two yellowy-brown pigment particles which look like droplets of fluid. The Meyliinae have antidromously reflexed ovaries, the Tricominae and Desmoscolecidae have outstretched ovaries. The males have two testes (Meyliidae) or only one anterior testis (Desmoscolecidae). Many species have phasmata on the tail. The majority of species are marine; only within the Desmoscolecinae are there a few freshwater and terrestrial species.

Discussion: The two pigment particles situated subventrally to the anterior intestine are still often referred to as ocelli in the literature, although the following arguments contradict this interpretation:
a) In ocelli the pigment is always arranged in cup-like shapes, so that the rays of light can reach the photoreceptors through the open side of the cup. A massive, bubble-shaped array of pigment, as occurs in the pigment particles of the Desmoscolecoidea, cannot provide this effect of a shield and thus contradicts the underlying structure of ocelli.
b) The shape and size of the pigment particles shows intraspecies variation; sometimes pigment particles may be present in a species or they may be absent (e.g. in *Tricoma similis*, according to Decraemer 1978). However, where ocelli do occur in nematodes, they always have the same structure within species.
c) The pigment particles occur independently from the biotope where the animal lives, whereas they occur, above all, in phytal-living nematode species and they are always absent in nematodes from greater depths (from approximately 30m to deep seas). However, many Desmoscolecoidea with pigment particles live in greater depths. The Meyliidae were classed as "Desmoscolecida" by de Coninck (1965) and Gerlach & Riemann (1973), while Timm (1970 : 1), Freudenhammer (1975 : 20) and Decraemer (1977 : 233) regarded them as members of the "Monhysterina". The reason behind this is given by Timm: the body is extended, the head delineated, the "prominent annulation" and the "tubular setae" are absent, and the amphids are "cryptospiral", they are not blistered and they are not situated in the cephalic region. Timm arrived at his assessment of the Meyliidae, among other things, by studying *Meylia spinosa*. The same species, from the sublittoral region of the North Sea, was also available to the author. After studying this species (Fig. 22) and assessing the original description of the type species of *Meylia, M. alata*, the author places the Meyliidae in the Desmoscolecoidea, for the following reasons:

a) The four cephalic setae are located on peduncles.

b) The body rings of *M. spinosa* carry little rods which are, as yet, known only in species of the Desmoscolecoidea and not in other nematodes.

c) The spiral structure which is detectable within the amphids is dorsally-spiral and not ventrally-spiral (beginning from the periphery of the amphids). It thus only represents a spiraling of the corpus gelatum, as is also known to the author in representatives of the Xyalidae. Therefore the spiral structure does not represent an obstacle for the classification of the Meyliidae in the Desmoscolecoidea.

d) The body proportions correspond to those of many Tricominae.

Desmoscolecidae Shipley, 1896

The holophyly of the Desmoscolecidae is established by the following two holapomorphies:
95. a) The body setae in adult animals are arranged according to the following pattern:

1st	3rd	5th	7th	9th	11th	13th	16th	17th
2nd	4th	6th	8th	10th	12th	14th	15th	

The above is interpreted as follows: the 1st pair of setae is subdorsal, the 2nd subventral, the 3rd subdorsal etc. Individual pairs may be reduced, in particular on the ventral side and only very rarely on the dorsal side of the body (see Lorenzen 1969 : 239, Freudenhammer 1975). As far as is known, juvenile animals have no subventral setae, whereas the number of subdorsal setae increases during the course of postembryonic development (Lorenzen, 1971). In many species of the Desmoscolecinae, at least, adhesive glands open through the subdorsal setae so that the animals move along on the subdorsal setae as if on stilts (Staufer, 1924).
b) Only the anterior testis has developed (discussion point 44); this feature is unique in combination with the setae arrangement just described.
Discussion: In the current work the Tricominae, species of which have hitherto always been classified as Desmoscolecidae, are not put in the Meyliidae (the reasoning behind this is in the new systematic position).

Desmoscolecinae Shipley, 1896

Adult animals almost always have annular or wart-like layers of secretion and foreign bodies. As far as is known these are always absent in juvenile animals (Lorenzen, 1971). The number of body annuli is always greater than the number of rings made out of secretion and foreign bodies. The latter are referred to as desmen by Freudenhammer (1975 : 6).

Desmolorenzenia Freudenhammer, 1975
Desmoscolex Claparède, 1863
  Eudesmoscolex Steiner, 1916, syn. Lorenzen, 1971 : 344
  Eutricoma Allgen, 1939, syn. Lorenzen, 1971 : 344
  Heterodesmoscolex Stammer, 1935, syn. Lorenzen, 1969 : 242
  Prodesmoscolex Stauffer, 1924, syn. Lorenzen, 1971 : 344
  Protodesmoscolex Tim, 1970, syn. Freudenhammer, 1975 : 29
Pareudesmoscolex Weischer 1962
Protricomoides Timm, 1970

Greeffiellinae Filipjev 1929

Calligyridae Andrassy, 1974, syn. Lorenzen, 1981 : 247
Hapalominae Andrassy, 1976, syn. Lorenzen, 1981 : 247

Coverings of secretion and foreign bodies are never present; the body rings are adorned with spines (*Greeffiella, Greeffiellopsis, Progreeffiella*), smooth warts (*Calligyrus*) or with fine fringes (*Hapalomus*). Several pairs of setae are always reduced subventrally, and almost always subdorsally, from the setal pattern typical of each family (Schrage & Gerlach, 1975).
Discussion: The "Calligyridae" and "Hapalominae" were characterized by the absence of body setae and by the type of cuticular ornamentation. Schrage & Gerlach (1975) were able to show that considerable remains of setae trimmings typical of the family are definite and were simply overlooked in the original descriptions. The differences in ornamentation of the cuticle certainly justify the introduction of genera, but not of sub-families or families. For this reason the above synonyms were adopted.
In contrast to Timm (1978 : 233), the author does not put *Antarcticonema* in the Greeffiellinae, but in the Tricominae on account of its tricomoid complement of body bristles.

The genera of the Greeffiellinae:

Calligyrus Lorenzen, 1969
Greeffiella Cobb, 1922
    for *Trichoderma* Greeff, 1869, a homonym
Greeffiellopsis Schrage & Gerlach 1975
Hapalomus Lorenzen, 1969
Progreeffiella Timm, 1970

Meyliidae de Coninck, 1965

96. The holophyly of the Meyliidae is established by the following feature which has been judged to be a holapomorphy: there are more body setae on the ventral half of the body than on the dorsal half. This feature is unique within the Desmoscolecoidea because in the Desmoscolecidae there are always more body setae on the dorsal half of the body than on the ventral half. The feature of setae distribution is unknown for *Gerlachius* and the type species of *Meylia, M. alata*.

It is assumed that only very rarely do ahesive glands open through the body setae in the Meyliidae, this seems to be the case in *Usarpnema*. It has only recently become known that the Tricominae, at least, have two opposed testes (Decraemer, 1977, 1978, personal observations on two *Tricoma* species).

The Tricominae have outstretched ovaries, the Meyliinae, according to Decraemer & Jensen (1982), have antidromously reflexed ovaries.

Discussion: The Tricominae are transferred from the Desmoscolecidae to the Meyliidae in the present work because it is possible to establish the holophyly of both the Desmoscolecidae as they now stand and the Meyliidae as they now stand. For this reason, in agreement with Freudenhammer (1975 : 20) and Decraemer (1977 : 234), the synonymization of the Tricominae with the Desmoscolecinae, taken up by Timm (1970 : 14) and by Gerlach & Riemann (1973 : 82), is upheld here.

Meyliinae   de Coninck, 1965

Gerlachiinae Andrassy, 1976, syn. Decraemer & Jensen, 1982

The amphids have a dorsally-spiral corpus gelatum (Fig. 28). The body does not have any coverings of secretion and foreign bodies. In *Meylia spinosa* at least, the pharynx consists of two dissimilar sections (Fig. 28). The distribution of body setae (if any are present) and the number of testes are unknown. The ovaries are antidromously reflexed.

Boucherius Decraemer & Jensen, 1982
Gerlachius Andrassy, 1976
Meylia Gerlach, 1956
Noffsingeria Decraemer & Jensen, 1982

Tricominae   Lorenzen, 1969

In almost all species the ridged body has ring-shaped or wart-like coverings of secretion and foreign bodies. Except in *Usarpnema*, and unlike the Desmoscolecinae, the number of ring-shaped coverings (desmen) is presumably always identical to the number of body rings, for as far as is known, within species the number of body rings in juvenile animals is approximately the same as the number of desmen in adults. The corpus gelatum never has a spiral structure.

Antarcticonema Timm, 1978
Desmogerlachia Freudenhammer, 1975
Desmotimmia Freudenhammer, 1975
Demanema Timm, 1970
Haptotricoma Lorenzen, 1977
Paratricoma Gerlach, 1956
   Protricoma Timm, 1970, syn. Freudenhammer, 1975 : 22

Prototricoma Timm, 1970
Quadricoma Filipjev, 1922
  Neoquadricoma Kreis, 1963, syn. Timm 1970 : 38
Quadricomoides Decraemer, 1976
Tricoma Cobb, 1893
Usarpnema Timm, 1978

<u>LEPTOLAIMINA</u>   Lorenzen 1981

All those taxa of the Chromadorida which do not belong in the Chromadorina and Desmoscolecina are collected together in the Leptolaimina. The Leptolaimina consist predominantly of taxa from the former "Araeolaimida" and, to a small extent, from taxa of the former "Desmodorida", "Monhysterida" and "Enoplida". No holapomorphy can be found for the Leptolaimina. The suborder is paraphyletic because it forms the non-holophyletic remains when the free-living nematodes are divided into the holophyletic sub-taxa of the highest possible ranks (Section 5.1.2. and Fig. 21).

Features: The cuticle is almost always striated and in many species it is also strongly refractive. The body rings only have punctations in the Metateratocephalinae. The labial sensilla are tiny or not detectable at all. The 6+4 cephalic sensilla usually stand in two separate circles and the 4 posterior cephalic sensilla are usually longer than the 6 anterior sensilla (exception: Ceramonematidae partim). Dereids are occasionally present (Plectidae). The amphids are usually ventrally-spiral and rarely non-spiral (*Anaplectus, Aulolaimus*, some Leptolaimidae, Chronogasteridae). The buccal cavity is often tubular, in part, however extremely small. Teeth are only rarely found in the buccal cavity (Camacolaiminae, *Diodontolaimus, Dagda, Stephanolaimus* partim); their form and arrangement suggests that they are probably not homologous with those of the Chromadorina. The pharynx often has an end bulb. A valvular apparatus is present in the pharynx in the Chronogasteridae, Plectidae and Teratocephalidae.

In most species the males have two testes and the females have two ovaries. The ovaries are almost always antidromously reflexed (outstretched only in the Peresianidae and part of the Rhabdolaimidae). The adult males of many species have pre-cloacal, ventrally situated tubules (Leptolaimidae, Chronogasteridae, Plectidae, Peresianidae) or papillae.

No criteria were found to establish holophyletic super-families within the Leptolaimina. The following 18 families are distinguished and discussed in the order listed below: Leptolaimidae, Chronogasteridae,

OUTLINE OF A PHYLOGENETIC SYSTEM 167

Plectidae, Teratocephalidae, Peresianidae, Haliplectidae, Aulolaimidae, Rhadinematidae, Tarvaiidae, Aegialoalaimidae, Ceramonematidae, Tubolaimoididae, Paramicrolaimidae, Ohridiidae, Bastianiidae, Prismatolaimidae, Odontolaimidae and Rhabdolaimidae.

Leptolaimidae   Örley, 1880

No holapomorphies were found with which the holophyly of the Leptolaimidae can be established. The family is therefore at most paraphyletic.

Features: Except in *Ionema, Nemella* and *Onchium* partim, the cuticle is always striated and often has a non-striated lateral field. The 6 anterior cephalic sensilla are usually invisible or hardly detectable; they are setiform only in some *Stephanolaimus* species. The 4 posterior cephalic sensilla are usually setiform. The amphids are usually ventrally-spiral and have only one turn; more than one turn is rare. The amphids are non-spiral in *Anomonema, Leptolaimoides, Leptolaimus* partim and *Stephanolaimus*. The buccal cavity is usually narrow and tubular and contains tooth-like structures in the Camacolaiminae and some Leptolaiminae (*Diodontolaimus, Dagda, Stephanolaimus* partim). The pharynx often shows muscular or glandular swelling at the end and does not have a valvular apparatus. The outlet of the cervical gland is cuticularized for a long stretch in the genera *Assia* (personal observation), *Anonchus* (see Riemann, 1970 : 380 for *A. maculatus*), *Paraphanolaimus* (see Riemann, 1970 : 377 for *P. anisitsi*), *Pakira* and *Paraplectonema* and in *Anonchus maculatus* and *Paraplectonema* it is even coiled in the cardia region, similar to the condition found in the Plectidae. The adult males of almost all species have pre-cloacally situated tubules and/or papillae; in addition, the males of many species have subventral sensory setae anterior and posterior to the cloaca. Usually there are two opposed testes; the occurrence of two forward facing testes or only one anterior or only one posterior testis is rare (see Section 9.2.2.). Usually two ovaries are present; it is rare to find only one anterior ovary (*Anonchus* partim) or only one posterior testis (*Alaimella, Stephanolaimus* partim, *Camacolaimus monohystera*). The ovaries are always antidromously reflexed. Within species the anterior gonad almost always lies constantly to the right of the intestine and the posterior gonad constantly to the left of it. Occasionally the tail end has a capsule-like thickening (*Cynura, Diodontolaimus*) like that which is otherwise only found in some Oxystominidae (Enoplida). The majority of the species are marine, only a few are limnic or terrestrial (*Anonchus* partim, *Aphanolaimus, Bathyonchus, Pakira, Paraphanolaimus*).

Discussion: Vis-a-vis Gerlach & Riemann's checklist (1973/1974), the extent of the Leptolaimidae has been changed in the following ways: the Peresianinae receive the rank of family; *Rhadinema* and *Chronogaster* are each put in their own family; the Kreisonematinae are put in the Achromadodoridae and *Anthonema* in the Plectidae; new additions to the Leptolaimidae are *Polylaimium* (formerly "Axonolaimidae") and *Pakira* (formerly Plectidae). The last named genera correspond to the diagnosis of the Leptolaimidae and seemed like foreign bodies within the former families. *Polylaimium* corresponds so well with *Leptolaimus* that the two genera are seen as synonymous with one another. Thus the only *Polylaimium* species in now called *Leptolaimus exilis* (Cobb, 1920).

There are currently no available criteria by which to divide the Leptolaimidae into holophyletic sub-families.

Leptolaiminae Örley,1880

Aphanolaiminae Chitwood, 1936, syn. Hope & Murphy, 1972 : 43
Halaphanolaiminae de Coninck & Stekhoven, 1933, syn. de Coninck, 1965 : 610
Alaimellinae Andrassy, 1976, syn. Lorenzen, 1981 : 197

Discussion: Presumably on account of the absence of cuticularized pre-cloacal organs, *Alaimella* was put into the "Araeolaimoidea" and from here into the "Cylindrolaimidae" by Andrassy (1976 : 109, 115); within this family Andrassy erected the Alaimellinae with the single genus *Alaimella*. The absence of cuticularized pre-cloacal organs is not accepted as a reason for removing *Alaimella* from the Leptolaimidae because the genus fits into the Leptolaimidae well on the basis of its other features and because within this family *Leptolaimus minutus, Leptolaimoides thermastris, Assia* and part of the Camacolaiminae also do not have cuticularized pre-cloacal organs.

Andrassy's suggestion (1973a : 248) that *Plectolaimus* and *Cynura* are synonymous is not accepted because the tail end in *Plectolaimus*, in contrast to *Cynura*, is not surrounded by a cuticularized capsule.

The genera of the Leptolaiminae:
Alaimella Cobb, 1920
Anomonema Hopper, 1963
Antomicron Cobb, 1920
  Eutelolaimus de Man, 1920, syn. de Coninck, 1965 : 611
Aphanolaimus de Man, 1880
Bathyonchus Kreis, 1936

Caribplectus Adrassy, 1973
Cricolaimus Southern, 1914
Cynura Cobb, 1920
Dagda Southern, 1914
Diodontolaimus Southern, 1914
Halaphanolaimus Southern, 1914
Leptolaimoides Vitiello, 1971
Leptolaimus de Man, 1876
  Aplectus Cobb, 1914, syn. Hope & Murphy, 1972 : 44
  Dermatolaimus Steiner, 1916, syn. de Coninck, 1965 : 611
  Polylaimium Cobb, 1920 syn. Lorenzen, 1981 : 198
  Subgen. *Alveolaimus* Alekseev & Rassadnikova, 1977
  Subgen. *Boveelaimus* Alekseev & Rassadnikova, 1977 (from Alekseev, 1979 : 1300 promoted to the rank of genus)
  Subgen. *Tubulaimus* Alekseev & Rassadnikova, 1977
Pakira Yeates, 1967
Paraphanolaimus Micoletzky, 1923
Paraplectonema Strand, 1934
  for *Paraplectus* Filipjev, 1929, a homonyn
Plectolaimus Inglis, 1966
Stephanolaimus Ditlevsen, 1918

Anonchinae   Andrassy, 1973

Anonchus Cobb 1913
  Pseudobathylaimus Filipjev 1918, syn. Filipjev & Stekhoven, 1941 : 125; the name was introduced for *Bathylaimus* Daday, 1905, a homonym *Dadaya* Micoletzky, 1922, a later and therefore invalid substitution for *Bathylaimus*, Daday
Assia Gerlach, 1957
Haconnus Andrassy, 1973

Camacolaiminae   Micoletzky, 1924

Procamacolaiminae de Coninck, 1965, syn. Lorenzen, 1981 : 199

The Camacolaiminae always have amphids which are ventrally-spiral and look like those of *Plectus* (spiral with one loop); they always lie between the posterior 4 cephalic sensilla or may even be anterior to them. There is usually a baccal cavity spear which is inserted in the dorsal wall of the buccal cavity. The pharynx usually has a glandular swelling at the end. The cuticle is usually striated and is seldom smooth. Some of the males have pre-cloacal tubules.

Discussion: The only difference between the Procamacolaiminae and

Camacolaiminae is the presence of pre-cloacal tubules. Within the Chromadorida there are several families and sub-families in which pre-cloacal tubules or papillae occur in one part of the species and not in the other (e.g. Chromadoridae, Cyatholaimidae, Leptolaiminae, Anonchinae, Plectidae, Peresianidae), with the result that the presence or absence of these organs alone is not sufficient for the establishment of a sub-family. Since the Camacolaiminae and Procamacolaiminae are similar to one another in the features listed above, they are judged to be synonymous. The genus *Deontolaimus* is removed from the Leptolaiminae and put in the Camacolaiminae, because it is exactly the same as *Camacolaimus* in the structure of the anterior end (including the buccal cavity), of the pharynx and of the spicule apparatus. At present it is equally difficult to establish the holophyly of the Camcolaiminae as it is in the case of the other sub-families of the Leptolaiminae; in particular it should still be considered whether *Dagda* and *Diodontolaimus* (both Leptolaiminae) can be classified as Camacolaiminae on account of their tooth formation and similarities in other features.

The genera of the Camacolaiminae:

Anguinoides Chitwood, 1936
Camacolaimoides de Conick & Stekhoven, 1933
Camacolaimus de Man, 1889
   Acontiolaimus Filipjev, 1918, syn. Hope & Murphy, 1972 : 45
     Digitonchus Cobb, 1920, syn. Wieser, 1956 : 26
   Ypsilon Cobb, 1920, syn. Wieser, 1956 : 26
Deontolaimus de Man, 1880
Ionema Cobb, 1920
Nemella Cobb, 1920
Neurella Cobb, 1920
Onchium Cobb, 1920
   Onchulella Cobb, 1920, syn. Wieser, 1956 : 29
Procamacolaimus Gerlach, 1954

Doubtful general of the Leptolaimidae:

Donsinemella Allgen, 1949, dub. Hope & Murphy, 1972 : 49
Necolaimus Allgen, 1959, dub. Hoep & Murphy, 1972 : 50

Chronogasteridae   Gagarin, 1975

78. The holophyly of the Chronogasteridae is established by the holapomorphy that the females always have a single anterior ovary. This feature otherwise occurs within the Leptolaimina only in *Teratocephalus* (Teratocephalidae) and in a few other isolated species. The

ovary is antidromously reflexed, and personal observations have shown that it is situated constantly to the right of the intestine in *Chronogaster typica* and *C. magnifica*.

Additional features: The cuticle is always striated and in some species it has longitudinal ornamentation. Four cephalic setae are clearly recognizable. The amphids are generally non-spiral and pocket-shaped, but in *Chronogaster boettgeri* they are said to be spiral with one loop. The cylindrical buccal cavity has no teeth. Posterior to the buccal cavity the pharynx has one ventral and two subdorsal pharyngeal tubes. The pharynx swells at the end forming a bulb, the inner wall of which is differentiated to form a valvular apparatus (Fig. 20d); in the anterior part of the valvular apparatus only the two subventral walls are rasp-like in that they are covered in small teeth. The outlet of the cervical gland is cuticularized only distally on a small area. Males are rare. They have two opposed testes and several pre-cloacal tubules. The family is limno-terrestrial; *Chronogaster alata* has been found in coastal ground water.

Type and only genus: *Chronogaster* Cobb, 1913
                         Walcherenia de Man, 1921, syn. de Coninck, 1935 : 226–4227

Discussion: *Chronogaster* was put in the Plectidae by Filipjev (1934 : 19), Chitwood & Chitwood (1950 : 110), Andrassy (1957 : 4), Goodey (1963 : 300) and Loof & Jairajpuri (1965), and in the Leptolaimidae by de Coninck (1965 : 611), Gerlach & Riemann (1973 : 37) and Andrassy (1976 : 112). In both families the genus seemed misplaced because unlike the Plectidae the outlet of the cervical gland in *Chronogaster* is not coiled up, and unlike the Leptolaimidae, *Chronogaster* has a valvular apparatus in the pharyngeal bulb. A valvular apparatus otherwise only occurs within the Adenophorea in the Plectidae and Teratocephalidae. There has been no research to show whether the existence of a valvular apparatus in the three named families is the result of homology or analogy, and whether the valvular apparatus is absent at a primary or secondary level in the remaining taxa of the Leptolaimina. Furthermore, it is still unclear whether the valvular apparatus of the three named families is homologous with that of the Rhabditida or not.

Plectidae   Örley, 1880

There is no known holapomorphy to establish the holophyly of the Plectidae.

The cuticle is always striated and has a non-striated lateral field. The

4 cephalic sensilla and 3 sensilla circles are clearly recognizable. Dereids are present; they are situated at approximately the same level as the nerve ring. The amphids are non-spiral in *Anaplectus* and in the other genera they are ventrally spiral with one turn. They are located posterior to the cephalic sensilla. The buccal cavity is tubular. Connected to these there are three tubes in the pharynx, one of which lies ventrally and the other two subdorsally. At the end of the pharynx there is a muscular bulb with a valvular apparatus which, however, does not have a homogenous structure in the family (Fig. 20a–c). The cuticularized outlet of the cervical gland is strikingly twisted in the region anterior to the pharyngeal bulb; this characteristic gave the genus *Plectus* its name. The females have two antidromously reflexed ovaries and the males have two opposed testes. Within species the position of the gonads is constant: the anterior gonad lies to the right and the posterior gonad to the left of the intestine. The adult males usually have ventrally situated pre-cloacal tubules and subventrally situated pre-cloacal and post-cloacal papillae. In many species no males occur, or only very few, with the result parthenogenetic reproduction is common. The family is limno-terrestrial.

Discussion: The Plectidae and Rhabditoidea are very similar to one another in the structure of the pharynx. This and other similarities have led Chitwood & Chitwood (1933, 1950: 196) to consider the common characteristics of the two families as being particularly plesiomorphic within the nematodes. Maggenti on the other hand, considered the similarities in the structure of the valvular apparatus in both families as anologous. Both views are discussed in section 5.1.1. of this work.

Dereids which resemble those of the Rhabditoidea in form and position occur within the Adenophorea only in the Plectidae. Since at present the holophyly of the Adenophorea cannot yet be established, the author prefers not to consider the feature as a holapomorphy of the Plectidae, for it is possible that the Plectidae and other taxa of the Adenophorea are more closely related to the Secernentea than to the remaining Adenophorea.

Pakira was put in the Plectidae by Yeates (1967a) and Gerlach & Riemann (1973). In accordance with Andrassy (1976: 112), the current work places the genus in the Leptolaimidae because the cuticularized outlet of the cervical gland is not twisted, the pharynx only has a glandular swelling at the end and because dereids are absent.

In accordance with de Coninck (1965: 614), *Anthonema* is put in the Wilsonematinae on account of the existence of a valvular apparatus and on account of the structure of the anterior end. The structure and position of the amphids, as well as the dots on the body annules are, however, suggestive of the Metateratocephalinae.

Plectinae   Örley, 1880

Anaplectus de Coninck & Stekhoven, 1933
   Marinoplectus Kreis, 1963, syn. Allen & Noffsinger, 1968 : 80
Oligoplectus Taylor, 1935 (fossil)
Perioplectus Sanwal in Gerlach & Riemann, 1973
   for *Periplectus* Sanwal, 1968, a homonym
Plectus Bastian, 1865
   Plectoides de Man, 1904, syn. Micoletzky, 1914 : 450
   Proteroplectus Paramonov, 1964, syn. Gerlach & Riemann, 1973 : 11

Wilsonematinae   Chitwood, 1951

Anthonema Cobb, 1906
Ereptonema Anderson, 1966
Pyenolaimus Cobb, 1920
Tylocephalus Crossman, 1933
Wilsereptus Chawla, Khan & Saha, 1975
Wilsonema Cobb, 1913
   Bitholinema de Coninck, 1931, syn. Filipjev & Stekhoven, 1941 : 126
Wilsotylus Chawla, Khan & Prasad, 1970
   Neotylocephalus Ali, Farooqui & Tejpal, 1969, syn. Andrassy, 1976 : 119

Teratocephalidae   Andrassy, 1958

79. The holophyly of the Teratocephalidae is established by the holapomorphy that the unusual labial region has 6 deep cuticularized incisions and 6 extensions ("cephalic plicae" and "cephalic ribs" respectively in Anderson, 1969 : 829 and 830); this feature is unique within the Denophorea.

Additional features: The cuticle is striated and in *Euteratocephalus* it is punctated. At most it is possible to detect only the posterior 4 labial and cephalic sensilla, and in a few species they are short and setiform. The amphids are round (*Euteratocephalus*) or pore-shaped (*Teratocephalus*). The buccal cavity has no teeth. The pharynx widens at the end to form a muscular bulb with a valvular apparatus. There are two ovaries (*Euteratocephalus*) or only one anterior ovary (*Teratoce-*

phalus). The ovaries are antidromously reflexed. The males only have one testis which folds over at the tip. In *Euteratocephalus* both gonads lie to the right or both lie to the left of the intestine; as far as is known the anterior gonad always lies to the right of the intestine in *Teratocephalus*. The adult males in *Euteratocephalus* have no precloacal tubules or one ventrally situated pre-cloacal tubule and additional subventrally situated pre- and post-cloacal papillae, whereas in *Teratocephalus* they only have one pre-cloacal and one post-cloacal, sometimes ventrally situated, papilla. In some *Euteratocephalus* there is a pair of lateral papillae on the tail (Fig. 19b). Males are found only very rarely, so that parthenogenetic reproduction is common. The family is purely limnic/terrestrial.

Discussion: The systematic position of the Teratocephalidae has been assessed in very different ways. Filipjev (1934 : 30), Ritter & Theodorides (1965 : 776) and Andrassy (1976 : 140) put the family in the Secernentea; Chitwood & Chitwood (1950 : 71), de Coninck (1965 : 614) and Gerlach & Riemann (1973 : 5) on the other hand, put it in the Adenophorea. The caudal papillae in *Euteratocephalus*, which Goodey (1963 : 1) believed to be phasmids, led him to reject the division of the nematodes into the Adenophorea and Secernentea, and to give the Teratocephalidae the rank of order. The interpretation of the pair of papillae as phasmids cannot be sustantiated and is therefore not accepted. The existence of a single anteriorly reflexed testis conforms with the diagnosis of the Rahbditida, and the labial region and oral cavity are similar in structure to those of the Cephalobidae (Rhabditoidea). The last-named similarities are regarded as convergences for the following two reasons:

a) In the Rhabditoidea, to which the Cephalobidae also belong, the ovaries are homodromously reflexed. This feature is considered circumstancial evidence for the holophyly of the Rhabditoidea (discussion point 50).

b) In many taxa within the Rhabditoidea the labial region and oral cavity are not similar in structure to those of the Cephalobidae.

The Teratocephalidae cannot be classed as Diplogasteroidea because of the varying structure of the pharynx. All in all, the Teratocephalidae would, at the most, occupy an isolated position within the Rhabditida. Since phasmids have not yet been found in the Teratocephalidae, and since a pre-anal tubule occurs in some *Euteratocephalus*, the family is put in the Adenophorea in the current work. It is further classified as Leptolaimina on account of the round amphids in *Euteratocephalus*, the presence of a valvular apparatus and the absence of 12 ribs (rugae) in the vestibulum.

In the following, *Metateratocephalus* is regarded as synonymous with *Euteratocephalus*, because the absence of the posterior four cephalic setae and the absence of the pre-anal tubule in males is not sufficient to distinguish them from *Euteratocephalus*; both genera correspond completely in all the remaining features.

Teratocephalinae   Andrassy, 1958

Teratocephalus de Man, 1876
   Mitrephoros Linstow, 1877, syn. de Coninck, 1965 : 614

Metateratocephalinae   Eroshenko,1973

Euteratocephalus Andrassy, 1958
   Meteratocephalus Eroshenko, 1973, syn. Lorenzen, 1981 : 207

Peresianidae   Vitiello & de Coninck 1968

Manunematinae Andrassy 1973, syn. Lorenzen 1981 : 207

The holophyly of the Peresianidae is established by two holapomorphies:

a) The ovaries are outstretched (discussion point 51); this feature is unique within the Leptolaimina.

b) Only one posterior testis is present (discussion point 45); within the Leptolaimina this feature otherwise only occurs sporadically in the Leptolaimidae and has therefore probably developed independently in the two families.

Additional features: The cuticle is striated. The body setae stand on socles. Only the posterior 4 cephalic setae are recognisable. The amphids are round and are situated posterior to the cephalic setae. The buccal cavity is narrow and tubular. In the anterior section the pharynx is not very muscular but in the posterior section it is very muscular and inside it has a thick lining of cuticle. The adult males have two pre-anal tubules or none at all. This family which has few species, is purely marine.

The only genus:

Manunema Gerlach 1957
   Peresiana Vitiello & de Coninck 1968, syn. Rieman,, Thun & Lorenzen 1971 : 148

Discussion: Despite their outstretched ovaries, the Peresianidae are put in the Leptolaimina and not in the Monhysterida because of the existence of pre-anal tubules, the existence of only one posterior testis and the structure of the pharynx. These features are never found in the Monhysterida but do certainly occur in the Leptolaimina. On account of the socles, on which the cephalic setae are inserted, and on account of the position of the anus on a papilla, Riemann et al. (1971 : 151) referred to *Peresiana* as playing the role of "a connecting link between the Araeolaimida and Desmoscolecida". This evaluation is dubious because, at a primary level, the body setae probably do not insert on socles and the anus probably does not lie on a papilla in many species of the Desmoscolecoidea (above all in the Meyliidae). Riemann et al. (1971) and Gerlach & Riemann (1973) have taken the Peresianidae to be a sub-family of the Leptolaimidae.

The sub-family Manunematinae, introduced by Andrassy (1973a : 243), is a synonym for the Peresianidae because *Peresiana* is a synonym for *Manunema*. Andrassy also includes *Anomonema* in the sub-family. In accordance with Gerlach & Riemann (1973) and other authors, the genus is further classified as belonging to the Leptolaiminae on account of the antidromously reflexed ovaries.

Haliplectidae Chitwood 1951

80. The holophyly of the Haliplectidae is established by the holapomorphy that the pharynx is only slightly muscular in the anterior part, has a small bulb in the middle section and a large bulb with a well cuticularized inner wall but no valvular apparatus in the posterior section; a pharynx of this description is unique within the nematodes.

Two further features are very characteristic of the Haliplectidae: the amphids are circular and personal observations have shown that the body appears a brownish colour in glycerine preparations and cannot be coloured with cotton blue. Both these features are otherwise rare within the Leptolaimina.

Additional features: the cuticle is striated; there is no differentiation in the lateral field. At most it is possible to detect only the posterior 4 of the labial and cephalic sensilla. The buccal cavity is narrow and tubular. In the females there are two antidromously reflexed ovaries and in the males there are two testes which face in opposite directions to one another. The anterior gonad lies to the right or left of the intestine and the posterior gonad lies on the opposite side to that of the anterior gonad. The adult males have ventrally situated pre-anal papillae. The family is marine; most species live in the coastal regions.

The genera of the Haliplectidae:
Haliplectus Cobb 1913
Aegialospirina de Coninck 1943, syn. Gerlach 1957 : 150
Setoplectus Vitiello 1971

Aulolaimidae   Jairajpuri & Hooper 1968

81. The holophyly of the Aulolaimidae is established by the holapomorphy that in its anterior, longer section, the pharynx consists of a cuticularized tube without surrounding muscle tissue and in its posterior, shorter section, it is strongly muscular; such a pharynx is unique within the Chromadoria and only occurs in anything like a similar form in the Peresianidae and Haliplectidae (both Leptolaimina) and in the Cryptonchidae and Isolaimiidae (both Dorylaimida).

Additional features: The cuticle is smooth or weakly striated and, except in *Mehdilaimus*, has thin longitudinal stripes on it. The labial and cephalic sensilla can hardly be detected or are totally invisible. The amphids are either non-spiral and have a diagonal, slit-like opening (*Aulolaimus*), or they are ventrally-wound with one loop (*Mehdilaimus, Pseudoaulolaimus*). *Aulolaimus* and *Pseudoaulolaimus* have two ovaries, *Mehdilaimus* has only one anterior ovary. The ovaries are antidromously reflexed and in *Aulolaimus* they lie anteriorly to the left or right of the intestine and posteriorly on the opposite side of it. The males have two testes and in *Aulolaimus* they also have ventrally situated pre-anal papillae and in some cases a ventrally situated cervical papilla as well. Caudal glands were not observed. The family is purely limnic/terrestrial.

Discussion: Only the presence of ventrally-wound amphids in *Mehdilaimus* and *Pseudoaulolaimus* supports the classification of the Aulolaimidae in the Chromadoria. On account of the antidromously reflexed ovaries and the narrow opening of the mouth, the family is put in the Leptolaimina. Prabha (1974 : 479) put *Mehdilaimus* in the "Cylindrolaiminae" ("Axonolaimidae") on account of the spiral amphids and the cylindrical oral cavity. Andrassy (1976 : 114) has adopted this classification. However, in this work, the genus is put in the Aulolaimidae, because the ovary is anti-dromously reflexed and because the oral cavity and pharynx have exactly the same structure as in the remaining Aulolaimidae.

The only *Pseudoaulolaimus* species, *P. anchilocaudatus*, is said to occur abundantly in damp soil in Japan, and is distinguished by a unique tail which ends in four curved points like an anchor. In accordance with Andrassy (1976 : 199), but in contrast to the checklist of Gerlach & Riemann (1973/1974), *Gymnolaimus* is removed from the Aulolaimidae and put in the Cryptonachidae.

On account of the similarity in body shape and in the structure of the pharynx, the limnic species *Thalassoalaimus aquaedulcis* is removed from the Oxystominidae and placed in *Aulolaimus*. The species is therefore now called *Aulolaimus aquaedulcis* (W. Schneider 1940).

The genera of the Aulolaimidae:
  Aulolaimus de Man 1880
    Pandurinema Timm 1957, syn. Jairajpuri & Hooper 1968 : 43
  Mehdilaimus Prabha 1974
  Pseudoaulolaimus Imamura 1931

Rhadinematidae   Lorenzen 1981

82. The holophyly of the Rhadinematidae is established through the holapomorphy that the vestibulum is reinforced by 6 cuticularized ribs and the adjacent section of the buccal cavity by a cuticularized ring; this feature is unique within the freeliving nematodes, and at most resembles the buccal cavity of the Kreisonematinae (Achromadoridae) in which however, the number of longitudinal, cuticular reinforcements in the vestibulum is probably twelve. The smooth cuticle and the almost round amphids which still have the original ventral loop are also very characteristic of the family.

Additional features: The pharynx does not become wider at the end. The four posterior cephalic sensilla are setiform; the remaining 6+6 sensilla are not detectable. The males have two testes facing towards the anterior, and the females have antidromously reflexed ovaries. The anterior and posterior gonads lie on the same side or on different sides of the intestine; the gonad position is thus completely variable. The adult males have ventrally situated pre-anal tubules. The family is marine.

Type and only genus: *Rhadinema* Cobb 1920

Discussion: Riemann (1966 : 53) has removed *Rhadinema* from the Linhomoeidae and put it in the Leptolaimidae on account of the pre-anal tubules. However, because of the completely variable position of the gonads relative to the intestine, this genus differs from all other species of the Leptolaimidae, in which this characteristic has been studied so far (see Section 9.2.1. of the appendix).

Tarvaiidae Lorenzen 1981

The holophyly of the Tarvaiidae is established by the following holapomorphy: the loops of the ventrally spiral amphids, beginning at the canalis, run from the inside to the outside (discussion point 26, Fig. 8b), instead of from the outside to the inside, as is otherwise always the case.

Additional features: in the cephalic region the cuticle is smooth (Fig. 8b) and on the rest of the body it is striated; there is no ornamentation of the body rings, and differentiation in the lateral field is absent. Only the 4 cephalic setae of the third sensilla circle can be detected. The buccal cavity is extremely small. The pharynx is only slightly muscular. The females have two anti dromously reflexed ovaries, and the males have two testes which face in opposite directions to one another. Personal observations have shown that the testes produce fine, thread-like sperm. Pre-anal supplements are absent. The family is marine.

Type and only genus: *Tarvaia* Allgen 1934

Discussion: In his re-description of the type species of *Tarvaia, T. donsi*, Gerlach (1950 : 150) has already pointed out the peculiarity of the amphid loops. *Tarvaia* was put into the "Diplopeltidae" by de Coninck (1965 : 607) and into the "Campylaiminae" ("Axonolaimidae") by Gerlach & Riemann (1973 : 81) both "Araeolaimida", whereas Vitiello (1974 : 145), on account of the non-striated cephalic region, was in favour of putting the genus into the "Desmodorida", where it would have occupied an intermediate position between the "Desmodoroidea" and "Ceramonematoidea". In the current work, *Tarvaia* is not put in the Ceramonematidae, because of the absence of ornamentation in the cuticle, or in the Axonolaimoidea, because of the antidromously reflexed ovaries, or in the Desmodoroidea, because of its tiny oral cavity and because of the existence of two testes. Since there is no valid reason for putting *Tarvaia* into any other family, and since holophyly can be proved, the family Tarvaiidae has been created. In agreement with Vitiello (1974 : 146), the similarity in the structure of the anterior end in *Tarvaia* and *Pselionema* (Ceramonematidae) is seen as circumstantial evidence of a close relationship between the two families.

Aeglialoalaimidae Lorenzen 1981

83. The holophyly of the Aegialoalaimidae is established by the following holapomorphy: the pharynx has extremely thin walls and

no muscle at least in the middle section and, in part also (*Aegialoalaimus*), in the anterior section, and at the end it swells out to form a bulb (Fig. 23); the feature does not occur in this form anywhere else within the freeliving nematodes, except in an approximate form in *Robbea* (Desmodoridae).

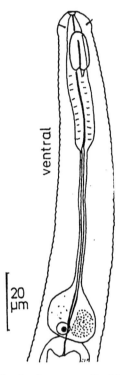

Fig. 23: Anterior end of *Diplopeltoides ornatus* (Aegialoalaimidae). It is characteristic of this family that the pharynx has hardly any tissue in the middle section and that it swells to form a bulb at the end. Adult female, sublittoral coarse sand, Kiel Bay, 27th January 1971.

Additional features: The cuticle is striated, in part only indistinctly so. Only the 4 cephalic sensilla of the third sensilla circle are visible. The large amphids are round or (*Diplopeltoides*) bent into an inverted U-shape. The buccal cavity is very narrow and toothless. The females have two anti dromously reflexed ovaries and the males have two testes, or, less frequently (*Cyartonema* partim), only one anterior testis. In *Aegialoalaimus* the two testes face towards the anterior, and in *Cyartonema* they face in opposite directions. The position of the

OUTLINE OF A PHYLOGENETIC SYSTEM   181

gonads relative to the intestine is completely variable (*Aegialoalaimus*) or semi-fixed, with the anterior gonad lying to the left or right of the intestine and the posterior gonad on the opposite side to that of the anterior one (*Cyartonema*). The family is marine.

Type genus: *Aegialoalaimus* de Man 1907
  Tubuligula Boucher & Helleouet 1977, syn. Jensen 1978b : 229
  nec *Kreisia* Allgen 1929 (doubtful genus of unknown affiliation, see Section 5.4)
Additional genera:
Cyartonema Cobb 1920
  Southernia Allgen 1929, syn. Juario 1973 : 85
Diplopeltoides Gerlach 1962
Paraterschellingia Kreis in Stekhoven 1935

Discussion: In Gerlach & Riemann's checklist (1973/1974) *Aegialoalaimus* and *Diplopeltoides* are put in the "Axonolaimidae" (specifically "Cylindrolaiminae" and "Campylaiminae" respectively), and *Cyartonema* and *Paraterschellingia* are put in the "Siphonolaimidae". Such groupings seemed provisional because in the families concerned species with outstretched ovaries predominate and the pharynx has a different structure. The presence of outstretched ovaries was recorded for *Paraterschellingia*; this information is queried because the fold in the ovaries is difficult to detect in some instances.

Ceramonematidae   Cobb 1933

Dasynemellidae de Coninck 1965, syn. Lorenzen 1981 : 211
Leptodasynemellinae Haspeslagh 1973, syn. Lorenzen 1981 : 211
Metadasynemellinae de Coninck 1965, syn. Lorenzen 1981 : 211
Pselionematinae de Coninck 1965, syn. Lorenzen 1981 : 211

84. The holophyly of the Ceramonematidae is established by the holapomorphy that the body rings are ornamented in a unique way: the ornamentation of each body ring overlaps onto the neighbouring body ring. This is unique within the Leptolaimina, and is not comparable with the longitudinal striations found in the Aulolaimidae.

Additional features: The anterior end is not striated. In *Dasynemella* there are 3 lips. Labial sensilla are not discernable. In most cases there are 6+4 cephalic sensilla, less frequently (*Pselionema, Pterygonema*) only four; the 6+4 cephalic sensilla are usually situated in two separate circles and less frequently (*Metadasynemoides, Metadasynemella*) in one common circle; the 6 anterior cephalic sensilla may be longer, shorter or in equal length to the posterior 4 sensilla. The ventrally-spiral amphids are usually an inverted U-shape. The buccal cavity is very small. In *Dasynemella sexalineata* and *Pselionema dissimile*, the pharynx is divided into an anterior, muscular and a posterior glandular part; the pharynx in the remaining species has not been described in detail or has not been described at all. In some species at least, two testes, facing in opposite directions, have been observed (*Pristionema, Ceramonema chitwoodi*). The anterior and posterior gonads usually lie on different sides of the intestine (Haspeslagh 197 : 71). The family is marine.

Discussion: Until now the Ceramonematidae, together with the Xenellidae, formed the super-family "Ceramonematoidea". In the present study the Xenellidae are put in the Trefusiida because of the non-spiral, pocket-shaped amphids and the single posterior ovary (females were hitherto unknown). The Ceramonematidae therefore remain alone in the super-family. Since the Leptolaimina have not yet been satisfactorily divided up from a gross systematic point of view, the Ceramonematidae are only given the rank of family in the current work. The family is not put in the Desmodoroidea, because the vestibulum does not have twelve folds in it and because, in part, two testes occur. It is possible that the Ceramonematidae are closely related to the Tubolaimoididae (see discussion on this family).

Both de Coninck (1942b) and Haspeslagh (1973) made invalid use of the structure of the body cuticle as the main criterion for the systematic division of the Ceramonematidae, while other features played an insignificant role or were not used at all. The synonyms, listed at the beginning of the section, are adopted for the following reasons:

a) The differences between the taxa listed are only slight.

b) The phylogenetic assessment of the structures of the body cuticle does not exclude alternative systematic divisions.

c) The former classification contradicts the phylogenetic interpretation of the cephalic sensilla characteristics.

On account of the arguments put forward, of the 14 genera which Haspeslagh (1973) lists, only the 8 specified below are accepted, and further synonymization is not considered impossible. In the following division of the Ceramonematidae, the relative importance of the two features, the body cuticle and the cephalic sensilla, is taken into consideration. The remaining features are either largely homogenous, or (structure of the pharynx, structure of the buccal cavity and labial region, number of testes, position of the gonads relative to the intestine) only known in isolated species.

OUTLINE OF A PHYLOGENETIC SYSTEM

The genera and species of the Ceramonematidae:

Ceramonema Cobb 1920
  Ceramonemoides Haspeslagh 1973, syn. Lorenzen 1981 : 218
  Cyttaronema Haspeslagh 1973, syn. Lorenzen 1981 : 218
  Species: *C. attenuatum* Cobb 1920, *C. carinatum* Wieser 1959, *C. chitwoodi* de Coninck 1942, *C. filipjev* de Coninck 1942, *C. pisanum* Gerlach 1953, *C. racovitzai* Andrassy 1973, *C. rectum* Gerlach 1957, *C. reticulatum* Chitwood 1936, *C. salsicum* Gerlach 1956, *C. sculpturatum* Chitwood 1936, *C. undulatum* de Coninck 1942.
Dasynemella Cobb 1933
  for *Dasynema* Cobb 1920, a homonym
  Species: *D. phalangida* Chitwood 1936, *D. sexalineata* (Cobb 1920)
Dasynemoides Chitwood 1936
  Dasynemelloides Haspeslagh 1973, syn. Lorenzen 1981 : 218
  Leptodasynemella Haspeslagh 1973, syn. Lorenzen 1981 : 218
  Species: *D. albaensis* (Warwick & Platt 1973) (=*Dasynemella a.*), *D. cinctus* (Gerlach 1957) (=*Dasynemella c.*), *D. conicus* (Gerlach 1956) (=*Dasynemella c.*), *D. falciphallus* (Vitiello & Haspeslagh 1972) (=*Metadasynemella f.*, lapsus *falciphalla*), *D. filum* (Gerlach 1957) (=*Ceramonema f.*), *D. pselionemoides* (Gerlach 1953) (= *Ceramonema p.*), *D. rhombus* (Andrassy 1973) (=*Ceramonema r.*), *D. riemanni* (Haspeslagh 1973) (=*Leptodasynemella r.*), *D. setosus* Chitwood 1936, *D. spinosus* Gerlach 1963.
Metadasynemella de Coninck 1942
  Dictyonemella Hspeslagh 1973, syn. Lorenzen 1981 : 218
  Species: *M. cassidiniensis* Vitiello & Haspeslagh 1972, *M. elegans* Vitiello 1973, *M. macrophallus* de Conick 1942 (lapsus *macrophalla*), *M. picrocephala* (Haspeslagh 1973) (=*Dictyonemella p.*).
Metadasynemoides Haspeslagh 1973
  Species: *M. cristatus* (Gerlach 1957) (=*Dasynemoides c.*), *M. latus* (Gerlach 1957) (=*Dasynemoides l.*), *M. longicollis* (Gerlach 1952) (=*Dasynemoides l.*).
Pselionema Cobb 1933
  for *Steineria* Filipjev 1922, a homonym
  Pselionemoides Haspeslagh 1973, syn. Lorenzen 1981 : 219
  Species: *P. annulatum* (Filipjev 1922), *P. beauforti* Chitwood 1936, *P. deconincki* Vitiello & Haspeslagh 1972, *P. detriticola* Vitiello 1973, *P. dissimile* Vitiello 1973 (lapsus *dissimilis*), *P. longiseta* Ward 1973, *P. longissimum* Gerlach 1953, *P. minutum* Vitiello & Haspeslagh 1972, *P. parasimplex* Vitiello 1971, *P. richardi* de Coninck 1942, *P. rigidum* Chitwood 1936 (*P. hexalatum* Chitwood 1936, syn. Haspeslagh 1973), *P. simile* de Coninck 1942, *P. simplex* de Coninck 1942.

Pterygonema Gerlach 1954
Specis: *P. alatum* Gerlach 1954, *P. cambriense* Ward 1973 (lapsus *cambriensis*), *P. ornatum* Timm 1961.
Doubtful genus of the Ceramonematidae: *Pristionema* Cobb 1933, dub. Lorenzen 1981 : 219

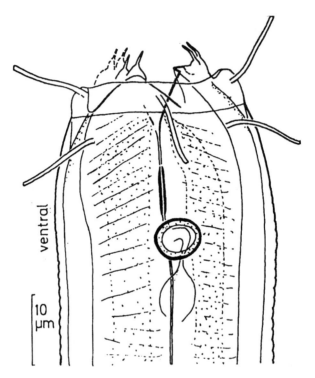

Fig. 24: *Tubolaimoides* sp. (Tubolaimoididae). It was hitherto unknown that only three lips are present and that pharyngeal glands open in all three sectors of the buccal cavity. Adult female, sublittoral sand, North Sea near Helgoland, 9th May 1970.

Tubolaimoididae   Lorenzen 1981

There is no known holapomorphy with which to establish the holophyly of the Tubolaimoididae. The combination of the two following features is unique within the Chromadoria:

a) There are three lips, on each of which there are two, closely packed labial papillae (personal observations; Fig. 24); this feature was hitherto unknown for species of this family; presumably it was

OUTLINE OF A PHYLOGENETIC SYSTEM 185

overlooked in most cases. Within the Chromadoria the existence of three lips was, until now, known only in *Dasynemella* (Ceramonematidae).

b) Pharyngeal glands open dorsally and subventrally in the oral cavity (Fig. 24); this feature was hitherto also unknown in the family and was presumably overlooked. Within the Chromadoria it was hitherto known only in species of the Siphonolaimoidea and Desmoscolecoidea, that the pharyngeal glands may also open subventrally in or near the buccal cavity.

Additional features: The cuticle is striated or smooth. The 6+4 cephalic setae are arranged in two separate circles, they are more or less the same length and hardly taper off towards the tip. The 6 anterior cephalic setae stand on an annular pad which has probably been overlooked until now in most cases. The amphids are ventrally spiral. The buccal cavity does not contain teeth; only the tips of the lips may appear tooth-like (Fig. 24). A pharyngeal gland opens out into each of the three buccal cavity sectors. In most animals the pharynx hardly becomes wider at the end; only *Tubolaimoides bullatus* has a bulb. The females have two anti dromously reflexed ovaries and the males, in *Tubolaimoides* at least, have two testes. Personal observations on *Tubolaimoides* sp. showed that the position of the gonads relative to the intestine is variable. The outstretched tapering tail has three caudal glands. The family is marine.

Type genus: *Tubolaimoides* Gerlach 1963
  for *Tubolaimella* Allgen 1934, a homonym
Additional genus: *Chitwoodia* Gerlach 1956

Discussion: The systematic position of the Tubolaimoididae is difficult to assess. On the one hand the existence of three lips and the fact that the pharyngeal glands open into all three buccal cavity sectors supports the classification of the family in the Enoplida, because both features are common within this order, whereas they never occur together — as far as is known — within the Chromadoria. On the other hand, the ventrally spiral amphids, the absence of metanemes (personal observation on *Tubolaimoides* sp.) and the annulation of the cuticle in some of the species are all arguments for putting the family in the Chromadoria. The author is in favour of the second possibility and puts the Tubolaimoididae specifically in the Leptolaimina, because both the Chromadoria and the Leptolaimina are paraphyletic and are thus catch-alls for taxa of uncertain systematic status. The great similarity between *Tubolaimoides* and *Dasynemella* (Ceramonematidae) in the structure of the labial region, the buccal cavity and the amphids as well as in the structure and position of the cephalic setae, also supports this decision. This similarity could be circumstantial evidence for the Tubolaimoididae and Ceramonematidae being closely related.

Formerly *Tubolaimoides* was placed in the "Linhomoeidae" ("Monhysterida") and *Chitwoodia* in the "Axonolaimidae" ("Araeolaimida").

<u>Paramicrolaimidae</u>   Lorenzen 1981

85. The holophyly of the Paramicrolaimidae is established by the holapomorphy of the unique form of the teeth in the oral cavity such as are unknown in any other nematodes (Fig. 25).

Additional features: The cuticle is striated. An intact body preserved in formalin does not become coloured using cotton blue and appears a brownish colour in glycerine preparations. The labial sensilla are

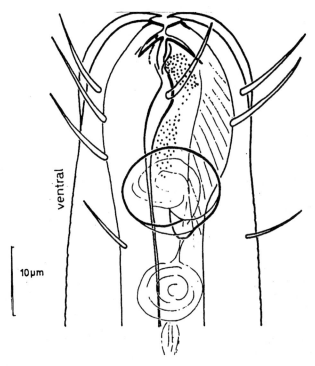

Fig. 25:   *Paramicrolaimus spirulifer* (Paramicrolaimidae). The structure of the buccal cavity is characteristic of this family. Adult male, sublittoral sand, Patagonia (south Chile), 21st March 1972.

hardly visible. The 6+4 cephalic setae are equal in length and in two separate circles. The ventrally spiral amphids form a spiral with one loop. The pharynx becomes only slightly thicker at the end. The females have two antidromously reflexed ovaries and the males have two testes, facing in opposite directions, and ventrally situated pre-anal papillae. The family is marine.

Type and sole genus: *Paramicrolaimus* Wieser 1954

Discussion: Formally, *Paramicrolaimus* was usually put in the Microlaimidae (e.g. by Gerlach & Riemann 1973). Jensen (1978a : 167) puts the genus in the Stilbonematinae (Desmodoridae) on account of "the reduced buccal cavity", the equal length of the 6+4 cephalic setae and the cuticular structures which are interpreted as attached bacteria. This classification is not possible for the following reasons:

a) In the Stilbonematinae only one testis is present; Jensen does indeed record only one testis for *Paramicrolaimus* as well, but in all probability this is an error; personal observations on animals from southern Chile and from the Öresund (Jensen's animal came from the Öresund) reveal the presence of two testes.

b) In general, the posterior 4 cephalic setae in the Stilbonematinae are considerably longer than the anterior 6 cephalic sensilla; the subcephalic setae which are about the same length as the posterior 4 cephalic setae and which are situated nearby can easily be mistaken for cephalic setae.

c) The buccal cavity in the Stilbonematinae is tiny and has no teeth.

The family of the Paramicrolaimidae is created because *Paramicrolaimus* cannot legitimately be put in any of the existing families and because the holophyly of the family can be established. The extent of the genus *Paramicrolaimus* is described by Jensen (1978a : 167).

Ohridiidae Andrassy 1976

There is no known holapomorphy with which to establish the holophyly of the Ohridiidae. The cuticle is striated and has an unstriated lateral field. Only the 4 cephalic setae of the third sensilla circle are recognisable. The amphids are round in outline. The buccal cavity is tiny and does not contain teeth. The pharynx is slightly swollen in the middle and at the end (only in *Domorganus oligochaetophilus* is the swelling more marked). The cervical pore lies at the end of the pharynx or posterior to it. The females have two antidromously reflexed ovaries; the males have two testes facing in opposite directions to one another, and one pre-anal papilla. In two animals the anterior gonad was observed to be on the right of the intestine and the posterior gonad on the left of it.

Domorganus oligochaetophilus lives endoparasitically in the intestine of the oligochaete *Lumbricillus lineatus* (see von Thun 1967), and *D. macronephriticus* was found during the preparation of earthworms, from which the possibility of endoparasitism was concluded (Morgan in T. Goodey 1947). It is possible that all species of the Ohridiidae live endoparasitically in oligochaetes. Endoparasitism is not known in any other species of the Chromadoria. All known species come from limnic, terrestrial and brackish habitats.

Only genus:

Domorganus T. Goodey 1947
  Ohridius Gerlach & Riemann 1973 (for *Ohridia* W. Schneider 1943, a homonym), syn. Lorenzen 1981 : 225 [1] (*Ohridius* is the type genus of the family.)
  Leoberginema Tsalolikhin 1977, syn. Lorenzen 1981 : 225
  Mikinema Tschesunov 1978, syn. Lorenzen 1981 : 225

Discussion: The synonymy of the four genera arises from the similarity in the structure of the amphids, the oral cavity, the pharynx, in the position of the cervical pore (not discribed in *Leoberginema*), the spicule apparatus and the existence of a pre-anal papilla in adult males. The only *Ohridius* species is now called *Domorganus bathybius* (W. Schneider 1943), the only *Leoberginema* species is now called *Domorganus acutus* (Tsalolikhin 1977) and the only *Mikinema* species *Domorganus supplementatus* (Tschesunov 1978). The spiral nature of the amphids, reported in *Leoberginema*, very probably only applies to the spiral nature of the corpus gelatum. Formerly *Domorganus* and *Leoberginema* were put in the "Cylindrolaiminae" ("Araeolaimida", "Axonolaimidae"), *Ohridius* in the "Microlaimidae" ("Desmodorida") and *Mikinema* in the "Linhomoeidae" ("Monhysterida"). This arrangement is not possible because of the antidromously reflexed ovaries. Since the 12 ribs in the vestibulum are also absent the Ohridiidae are put in the Leptolaimina.

---

[1] My thanks to Dr F. Riemann in Bremerhaven for drawing my attention to the synonymy of *Domorganus, Ohridius* and *Mikinema*, and also for the opportunity to work on the syntypes of the type species of *Ohridius, O. bathybius*.

OUTLINE OF A PHYLOGENETIC SYSTEM

Domorganus bathybius (W. Schneider 1943)
(Fig 26a–e)

After studying the four syntypes (δ_1, juvenile $\female_1$, an additional juvenile and an additional adult), the original description is extended in the following points: the striated cuticle has an unstriated lateral field. It is not possible to interpret the exact structure of the vestibulum. The pharynx has a central and a posterior swelling. The pharynx is surrounded by little cells behind the nerve ring (Fig 26e), as is also known to occur in *D. acutus*. The cardia is very large. The intestinal wall is packed with a large number of little, eliptical to round particles (Fig 26b). On the ventral side of the body, 3–4 times the length of the pharynx posterior to the front end, there are a few large cells (coelomocyctes); in the δ_1, it was possible to establish with certainty the presence of two subventral nuclei in the anterior cells only, but not in the posterior ones. The δ_1 has two testes, facing in opposite directions to one another. The anterior testis lies on the right of the intestine, the posterior testis on the left. The sperm are thread-like. The position of the ovaries could not be established with certainty. The δ_1 is 2110 μm long, the juvenile $\female_1$ is 1990 μm long.

Bastianiidae de Coninck 1935

There is no known holapomorphy with which to establish the holophyly of the Bastianiidae. One feature, discovered during research for the current study, is unique within the Chromadoria, and is that the amphids are either ventrally spiral or dorsally spiral within a species (Fig. 9a–c). In ventrally spiral amphids the canalis lies dorsal to the lateral line of the body and in dorsally spiral amphids it lies ventral to it; from this it has been concluded that the dorsally spiral condition has probably evolved directly from the ventrally spiral condition (discussion point 23). As a result, the Bastianiidae are put into the Chromadoria, and from here into the Leptolaimina because of their antidromously reflexed ovaries and the absence of a twelve-fold folded vestibulum. The fusus of the amphids is markedly dilated (Fig. 9a–c).

Additional features: The cuticle is striated; neither ornamentation nor a lateral field are present. The labial sensilla are papilliform. There are 6 longer, jointed and 4 shorter, unjointed cephalic setae in one common circle or in two circles situated extremely close to one another. The buccal cavity is small and conical. The pharynx has a slight, conical swelling at the end. The cardia is short and strikingly thick. The males have two testes, facing in opposite directions to one another, and the females have two antidromously reflexed ovaries.

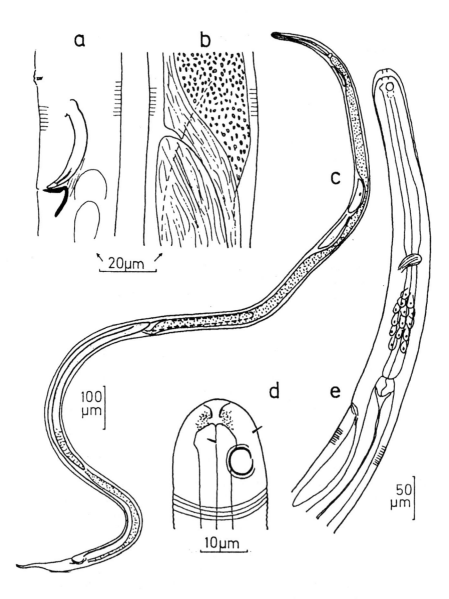

Fig. 26: *Domorganus bathybius* (Ohridiidae). a) Spicular apparatus ♂₁; b) region of the body of ♂₁ where the posterior testis folds over (the thread-like structure of the sperm is indicated by dashes); c) ♂₁ entire specimen; d) head of ♂₁; e) anterior end of juvenile ♀₁ (in moult). The drawings are taken from syntypes.

The position of the gonads relative to the intestine is variable. The adult males have ventrally situated pre-anal papillae which may even be found as far as the cervical region. Females are found considerably more often than males which suggests that parthenogenetic reproduction is probably widespread. The family is terrestrial.

Discussion: The relationship between the Bastianiidae and Prismatolaimidae is dealt with in the discussion of the Prismatolaimidae.

The genera of the Bastianiidae:

Bastiania de Man 1876
Dintheria de Man 1921; family uncertain; the only species has never been found since the original description and, in the author's opinion, may belong to the Odontolaimidae.

Prismatolaimidae Micoletzky 1912

The holophyly of the Prismatolaimidae is established by the holapomorphy that the amphids are always dorsally spiral, with the canalis always ventral to the lateral line of the body (Fig. 9d, discussion point 23); this combination of features is unique within the freeliving nematodes. The fusus of the amphids is broadened.

Additional features: The cuticle is striated; neither ornamentation nor a lateral field are differentiated. The labial sensilla are papilliform. The 6 longer, jointed and 4 shorter unjointed cephalic setae are arranged in a single circle. The buccal cavity is spacious and has a small dorsal tooth at the end; at the same level pockets which are round subventrally are set into the wall of the buccal cavity (Fig. 9d). The pharynx hardly becomes larger at the end. According to Chitwood & Chitwood (1937 : 519) there are 5 pharyngeal glands (one dorsal, two in every subventral pharyngeal sector), one of which, at the end of the buccal cavity, opens into each of the three pharyngeal sectors. The cardia is short and strikingly thick. There is no cervical gland. There are two testes facing in opposite directions and two antidromously reflexed ovaries which constantly lie either to the right or to the left of the intestine within a given species. The adult males have ventrally situated pre-anal papillae which may extend as far as the cervical region (Gagrin & Kuzmin 1972 for *Prismatolaimus intermedius*). Males are found only very rarely, so parthenogenetic reproduction must be the norm. The family is limno-terrestrial.

Only genus: *Prismatolaimus* de Man 1880
Takakia Yeates 1967, syn. Andrassy 1969 : 22

Discussion: The amphids in the Prismatolaimidae were hitherto thought to be non-spiral and pocket-shaped, and for this reason the family was put in the "Enoplida". This position can now no longer be accepted because the dorsally spiral amphids of the Prismatolaimidae have probably evolved directly from ventrally spiral and not from non-spiral amphids (discussion point 23). As a result the family is put in the Chromadoria. On account of the current knowledge of the structure of the amphids the Prismatolaimidae can no longer be grouped together with the Onchulidae and Odontolaimidae to form a single family as was commonly the case before. The Prismatolaimidae are probably more closely related to the Bastianiidae, for both families show similarities in the structure of the cuticle, the structure and position of the cephalic setae, the structure and position of the pre-anal papillae, the structure of the cardia and the presence of dorsally spiral amphids. In 1935 de Coninck (1935 : 223) had already stressed the relationship between the two families.

Odontolaimidae   Gerlach & Riemann 1974

The holophyly of the Odontolaimidae is established by the following feature which is considered to be holapomorphic:

86. The dorsal tooth can be extended forwards like a spear (Fig. 27); within the freeliving nematodes this feature is known otherwise only in *Kinonchulus* (Onchulidae, Enoplia) (see Riemann 1972). The structure of the rest of the buccal cavity is so different in the two taxa that the characteristic very probably developed independently.

Additonal features: The cuticle is striated. The labial sensilla are hardly visible. The 6+4 cephalic setae are situated at the same level; the 6 setae of one circle are longer or shorter than the 4 setae of the other circle. The non-spiral amphids have a round aperture. There is a muscular swelling at the end of the pharynx. It is unknown where the pharyngeal glands open. There is, as yet, no proof of the existence of a cervical gland; personal research indicates that it is very probably absent. The females have two ovaries or only one posterior ovary; the ovaries are antidromously reflexed. Males have not yet been found. In all probability caudal glands are absent for no trace of them has been detected. The family is limnic/terrestrial.

Only genus: *Odontolaimus* de Man 1880
   Neonchus Cobb 1893, syn. Micoletzky 19221a : 419

Discussion: The Odontolaimidae were introduced as a sub-family of the "Prismatolaimidae". On account of differences in the structure

OUTLINE OF A PHYLOGENETIC SYSTEM 193

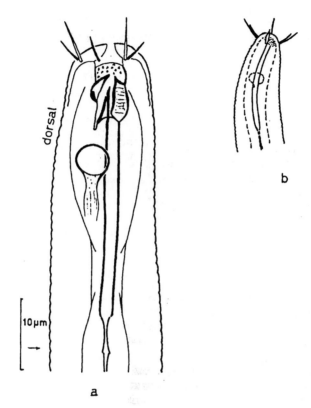

Fig. 27: *Odontolaiums aquaticus* (Odontolaimidae). The dorsal tooth can be everted like a spear. a) dorsal tooth retracted; b) dorsal tooth everted. a: juvenile, underground water from the banks of the River Oconaluftee (U.S.A.), 3rd August 1973. b: after Andrassy (1963: 257).

of the buccal cavity, the amphids, the pharynx and the arrangement of the cephalic setae, the Odontolaimidae cannot be joined together to form a holophyletic group either with the Prismatolaimidae, as they are now understood, or with the Onchulidae. In the discussion on the Bastianiidae it was already indicated that *Dintheria* de Man 1921 possibly belongs to the Odontolaimidae. Andrassy (1976 : 196) listed *Odontolaimus* under the Bastianiidae, which may have meant that the Odontolaimidae were synonymous with the Bastianiidae. Arguments against this synonymization are the considerable differences in the structure of the amphids and the buccal cavity.

Despite their non-spiral amphids, the Odontolaimidae are placed in the Chromadoria and not in the Enoplia because of the striated nature of the cuticle and the small size of the body (0.6–0.9mm). Within the Chromadoria the family is placed in the Leptolaimina on account of the antidromously reflexed ovaries and the absence of the 12 ribs (rugae) in the vestibulum.

Rhabdolaimidae Chitwood 1951

Monochromadorinae Andrassy 1958, syn. Lorenzen 1981 : 231
Rogerinae Andrassy 1976, syn. Lorenzen 1981 : 231
Tobriliinae Andrassy 1976, syn. Lorenzen 1981 : 231

There is no known holapomorphy with which to establish the holophyly of the Rhabdolaimidae. The combination of the following features is characteristic of the family:
a) All species are diminutive — they are only about 500μm in length.
b) The amphids are always non-spiral and pocket-shaped.
c) The small buccal cavity contains a small, distinct dorsal tooth in the central or anterior section; subventrally there are no teeth or, at the most, just rudiments of teeth.
d) Except in *Tobrilia* the pharynx has a marked end bulb which usually has a thick, cuticularized inner wall.

Additional features: The cuticle is smooth. The 6+4 cephalic sensilla are papilliform and rarely setiform; they are arranged in two separate circles or in one common circle. It is not known whether pharyngeal glands exist. Within the genera *Monochromadora, Rogerus* and *Udonchus*, spherical anterior curvatures of the cardia were observed; these are thought to be cardial glands and probably also occur in *Rhabdolaimus*, according to personal observations on *R. terrestris*. There are two ovaries (*Rhabdolaimus, Rogerus, Tobrilia*) or only one anterior ovary (*Monochromadora, Sinanema, Udonchus*); the ovaries are either outstretched (*Rhabdolaimus* partim, *Rogerus, Sinanema*) or antidromously reflexed (the remaining taxa). Personal research has shown that, in *Rhabdolaimus* at least, both ovaries are always on the same side of the body (both on the right or both on the left of the intestine, see Section 9.2.1.). Males occur only extremely rarely, so that parthenogenetic reproduction must be widespread. The family is limnic/terrestrial.

Discussion: In its present form the Rhabdolaimidae consists of *Rhabdolaimus, Rogerus* (both formerly Chromadoria, "Araeolaimida", "Rhabdolaimidae"), *Monochromadora, Sinanema, Udonchus* (all formerly "Enoplidae", "Tripylidae", "Tobrilinae" respectively). The genera therefore originate from two different sub-orders. If the former "Rhabdolaimidae" were put in the Chromadoria on account of their pharyngeal bulb and in spite of their non-spiral amphids, then it was illogical to put the former "Monochromadorinae" in the Enoplia on account of their non-spiral amphids and in spite of their pharyngeal bulb.

In the species descriptions of the former "Monochromadorinae", reference is often made to the curvatures of the cardia which are also found in the same form in representatives of the Tripylidae and Tobrilidae; however, they also occur in *Rogerus rajasthanensis* (formerly "Rhabdolaimidae") and, according to personal observations, probably also in *Rhabdolaimus*. Therefore, this feature likewise fails to provide an argument for the variable grouping of the genera now classed together as the Rhabdolaimidae. Previously hardly any attention has been paid to the feature of the curvatures in the cardia in representatives outside the former "Tripylidae", with the result that, for this reason if for no other, it can hardly be currently considered phylogenetic.

Tobrilia was initially put in the "Tobrilinae" within the former "Tripylidae", and then in the "Tobriliinae" by Andrassy (1976 : 198). The genus differs completely from the Tobrilidae and Tripylidae in the structure of the buccal cavity and resembles the Rhabdolaimidae in this feature. For this reason the genus is put in the Rhabdolaimidae.

Syringolaimus was put in the "Rhabdolaimidae" by Chitwood (1951 : 646), Riemann (1970a : 382) and Gerlach and Riemann (1973 : 23). However, the structure of the buccal cavity in *Syringolaimus* is completely different and therefore the author, in agreement with other authors, places the genus in the Ironidae (Enoplida).

The above synonymization with the Rhabdolaimidae is adopted because, at present, there are no criteria to support a valid division of the family into sub-families and because on account of the small number of species within the family a provisional, artificial division into sub-families is superfluous.

Despite the non-spiral amphids and the smooth cuticle, the Rhabdolaimidae are put into the Chromadoria and not into the Enoplia on account of the smallness of the body (approximately 0.5mm), the dominance of the dorsal tooth in the buccal cavity and the presence of the pharyngeal bulb which is usually lined with a thick, cuticularized inner wall. The family is put in the Leptolaimina because some of the species have antidromously reflexed ovaries and because the vestibulum does not have twelve ribs (rugae).

The genera of the Rhabdolaimidae:
Monochromadora Goodey 1951
Rhabdolaimus de Man 1880
Rogerus Hoeppli & Chur 1934
for *Greenia* Hoeppli & Chu 1932, a homonym
Greenenema Andrassy 1959 is a later and thus invalid substitution
Sinanema Andrassy 1960
Tobrilia Andrassy 1967
Udonchus Cobb 1937

5.2.3. *Classification of the Monhysterida*

MONHYSTERIDA Filipjev 1929

The holophyly of the Monhysterida is established by the holapomorphy that the ovaries are always outstretched (discussion point 51 and section 5.2.1.). There are, at present, no other holapomorphies. An apomorphic feature in all Monhysterida, in contrast to the Chromadorida, is the constant absence of pre-anal tubules in all adult males (discussion point 54). Furthermore, the buccal cavity never contains teeth which could be seen as homologous with those in the Chromadorina. Neither feature can be used to establish the holophyly of the Monhysterida because they also occur in many taxa of the Chromadorida.

Additional features: The cuticle can be smooth or striated. The 6 labial sensilla are often recognizable and the 6+4 cephalic sensilla are recognizable in most cases; seldom are only the posterior 4 cephalic sensilla recognizable. In some cases the 6+4 cephalic sensilla are in two separate circles and in other cases in one common circle. The 6 anterior cephalic sensilla are longer, equal in length or shorter than the 4 posterior sensilla; the sensilla in both circles may be jointed. The ventrally spiral amphids are spiral, loop-shaped or round. Ocelli with lenses are present in most species with a smooth or only weakly striated cuticle. There are often two ovaries or only one anterior ovary; the presence of only one posterior ovary is rare (Axonolaimidae partim). The males possess two testes or only one anterior testis; the presence of only one posterior testis has not yet been observed. In several families the adult males of at least some species possess ventral situated pre-anal papillae; in many species these are absent. The position of the gonads relative to the intestine is seldom completely variable, is often semi-fixed (the anterior gonad on the left or right and the posterior gonad on the opposite side of the intestine) and in many cases is completely constant. The caudal glands are always situated completely in the tail. Within the Diplopeltidae partim,

Xyalidae and Sphaerolaimidae they open separately, in the remaining taxa they open through a common pore.

Discussion: The Monhysterida, as they are understood here, contain the species with outstretched ovaries from the "Monhysterida", "Araeolaimida" and "Chromadorida". Within the Chromadoria only the following also possess outstretched ovaries: the Peresianidae (formerly "Araeolaimida"), Microlaimidae and Aponchiidae (formerly "Desmodorida") and Rhabdolaimidae (formerly partly "Araeolaimida" and partly "Enoplida"); in the current work they are placed in the Chromadorida (discussion point 51).

No criteria have been found with which to divide the Monhysterida into two holophyletic sub-orders or, in some appropriate form, into a holophyletic and paraphyletic sub-order. A better solution might be to accept the former "Monhysterida" as a sub-order because, according to current diagnoses, they are said to possess round amphids as a particular characteristic. However, not all 'Monhysterida" have this feature: in *Anticyclus junctus, Didelta, Disconema* and *Halinema* (all Linhomoeidae) the amphids are loop-shaped. According to Gerlach (1963 : 604) and according to personal analyses (discussion point 25), this amphid type is considered plesiomorphic in comparison with the round amphid form. Since round amphids also occur within several families of the Chromadorina and Leptolaimina as well as within the Desmoscolecoidea, the feature cannot be considered a holapomorphy of the former "Monhysterida".

On the basis of the arguments presented, only the three super-families Monhysteroidea, Siphonolaimoidea and Axonolaimoidea are recognized within the Monhysterida; they are discussed in the order in which they are listed. According to the criteria available so far, it is only possible to establish the holophyly of the first super-family named but not that of the other two. The Axonolaimidea and Siphonolaimoidea are not joined together to form a single paraphyletic group because there is a chance, for the Siphonolaimoidea at least, that holophyly can be established.

MONHYSTEROIDEA de Man 1876

The system of the Monhysteroidea has only recently been revised according to the principles of phylogenetic systematics (Lorenzen 1978c). The most important points are dealt with in the following report.

The holophyly of the Monhysteroidea is established solely by the holapomorphy that all species have only a single, anterior ovary (discussion point 46); within the Monhysteridae this feature occurs regularly only in the Siphonolaimidae which, on the basis of other characteristics however, presumably form a holophyletic group with a part of the Linhomoeidae. The ovary is always outstretched.

Additional features: The cuticle is smooth (Monhysteridae, Sphaerolaimidae) or striated. The labial sensilla are short and setiform (Xyalidae partim) or papilliform. With a few exceptions (section 8.1), the 6+4 cephalic sensilla are always arranged in a single circle. The 6 cephalic sensilla of the second sensilla circle are longer (Xyalidae, Monhysteridae partim) or shorter than 4 of the third sensilla circle. The amphids are always round. The lips are almost always pointed and are turned towards the central axis of the body. The buccal cavity is funnel-shaped and occasionally contains tooth-like formations which are grouped into three different types (Lorenzen 1978c : 523). The pharynx never thickens to form a muscular bulb. There are two testes (Xyalidae partim, Sphaerolaimidae partim) or only one anterior testis. The position of the gonads is typical of each family respectively. The caudal glands have separate openings (Xyalidae, Sphaerolaimidae) or one single opening.

Within the Monhysteroidea the three holophyletic families Monhysteridae, Xyalidae and Sphaerolaimidae are discerned. The two last-named families form a holophyletic group on the basis of the following holapomorphies (from Lorenzen 1978c):

87. a) The cuticle is striated, at least at a primary level.

88. b) The tail has 2 or 3 terminal setae, at least at the primary level.

89. c) There exists the genetic potential for the formation of 8 groups of subcephalic setae, whereby the longest subcephalic setae are always longer than the longest cephalic setae. This feature is unique within the nematodes.

Monhysteridae de Man 1876

Diplolaimelloidinae de Coninck 1965, syn. Timm 1967 : 120

The holophyly of the Monhysteridae is established by the two following holapomorphies:

a) The presence of only a single anterior testis is constant (discussion point 44); within the Monhysterida this feature is also constant within families only in the Desmoscolecidae and Siphonolaimidae, neither of which however, on the basis of other features, belong to the Monhysteroidea, with the result that the posterior testis must have become reduced independently in each case in the three families.

b) The anterior gonad always lies to the right of the intestine (the posterior gonad is absent) (discussion point 53); in combination with a single, anterior gonad this feature otherwise only occurs within the Monhysterida in very few species of the Xyalidae.

Additional features: The cuticle is always smooth. The length of the cephalic sensilla of the second sensilla circle relative to the length of those of the third circle is variable within families. Many species have ocelli with lenses. The buccal cavity is always completely surrounded by pharyngeal tissue. The cervical gland is usually present; personal observations have shown that it only occurs to the right of the anterior part of the intestine. The caudal glands open through a common pore. Caudal terminal setae are always absent. The family is predominantly limnic and terrestrial, though there are a few marine species. *Gammarinema, Monhystrium* and *Tripylium* live as commensals or ectoparasites in the gill region of crabs of the sub-class Malacostraca; *Odontobius* on the other hand, lives in the plates of baleen whales (the exact form of relationship is not known for all cases). The remaining species are freeliving.

Discussion: In agreement with Sudhaus (1974 : 372), *Branchinema* is seen as synonymous with *Gammarinema* because in both genera the structure of the oral cavity and the tail, as well as the life style, are identical. The type species and only *Branchinema* species are therefore now called *Gammarinema paratelphusae* (Farooqui 1967) (lapsus *paratelphusi* in the original description). When Farooqui (1967) created *Branchinema* he did in fact discuss the differences in *Tripylium* and *Monhysterium*, but not those in *Gammarinema*. Riemann (1968 : 40) accepted *Branchinema* because of the perianal papillae in the adult males of the type species. The author considers this feature to be no more than a species characteristic.

The species of the Monhysteridae:

Diplolaimella Allgen 1929
  Diplolaimita Chitwood & Murphy 1964, syn. Timm 1967 : 115
Diplolaimelloides Meyl 1954
Gammarinema Kinne & Gerlach 1953
  Branchinema Farooqui 1967, syn. Sudhaus 1974 : 372
Monhystera Bastian 1865
  Helalaimus de Cillis 1917, syn. Meyl 1960 : 96
  Tachyhodites Bastian 1865, syn. Bütschli 1874 : 26–27

Monhystrella Cobb 1918
Monhystrium Cobb 1920
Odontobius Roussel de Vauzème 1834
Sitadevinema Khera 1971
Tripylium Cobb 1920

Xyalidae   Chitwood 1951

Cobbiinae de Coninck 1965, syn. Lorenzen 1978 : 531
Rhynchonematinae de Coninck 1965, syn. Lorenzen 1978c : 531
Scaptrellidae de Coninck 1965, syn. Lorenzen 1978c : 531

The holophyly of the Xyalidae is established by the following holapomorphy: within species the anterior gonad is constantly to the left of the intestine and the posterior gonad constantly to the right of it (discussion point 53); this feature is unique within the Monhysteroidea. Only 2 out of the 98 species studied deviate from this pattern: within species in *Hofmaenneria niddensis* and *Steineria pilosa* the anterior gonad lies constantly on the right and the posterior gonad (only present in *Steineria*) constantly to the left of the intestine; in these cases other characteristics support their membership of the Xyalidae.

Additional features: The cuticle is always striated. The 6+4 cephalic setae are almost always situated at the same level (exception — *Rhynchonema* partim). The 6 cephalic setae of the second sensilla circle are always longer than or, at the most, equal in length to the 4 of the third sensilla circle and are thus more frequently jointed than the latter. There are often additional cephalic setae, the lateral ones almost always situated ventrally to the lateral cephalic setae of the second sensilla circle (dorsally only in *Gonionchus inaequalis*, see Lorenzen 1971a : 215). Occasionally 8 groups of subcephalic setae develop. The oral cavity is often surrounded by pharyngeal tissue, though in part this is not the case (see Lorenzen 1978c :523). The cervical gland is absent in most cases (it is present in *Hofmaenneria, Lynhystera* and *Steineira*). Two testes are often present, but in many cases there is only one anterior testis (section 8.2.1.). The majority of the species are marine, only a few are limnic or terrestrial. *Therisus (Penzancia) polychaetophilus* lives ectoparasitically on the polychaete *Scolelepsis squamata*; no further instances of parasitism are known in the family.

The genera of the Xyalidae:

Ammotheristus Lorenzen 1977
Amphimonhystera Allgen 1929
Amphimonhystrella Timm 1961

OUTLINE OF A PHYLOGENETIC SYSTEM

Cenolaimus Cobb 1933
Cobbia de Man 1907
Dactylaimus Cobb 1920
Daptonema Cobb 1920
  Cylindrotheristus Wieser 1956, syn. Lorenzen 1977a : 222
  Mesotheristus Wieser 1956, syn. Lorenzen 1977a : 222
    Allomonhystera Micoletzky 1923, Gerlach & Riemann 1973 : 140, nomen oblitum
  Pseudotheristus Wieser 1956, syn. Lorenzen 1977a : 222
  Spirotheristus Timm 1961, syn. Lorenzen 1977a : 222
  Tubolaimus Allgen 1929, syn. Lorenzen 1977a : 222
Echinotheristus Thun & Riemann 1967
Elzalia Gerlach 1957
  Megalolaimus Timm 1961, syn. Hope & Murphy 1972 : 39
Filipjeva Ditlevsen 1928
Gnomoxyala Lorenzen 1977
Gonionchus Cobb 1920
Hofmaenneria Gerlach & Meyl 1957
Linhystera Juario 1974
Megalamphis de Coninck 1965
  for *Macramphis* Timm 1961, a homonym
Metadesmolaimus Stekhoven 1935
Omicronema Cobb 1920
Paramonohystera Steiner 1916
  Subgen *Leptogastrella* Cobb 1920
  Subgen *Paramonohystera* Steiner 1916
Promonhystera Wieser 1956
Pseudosteineria Wieser 1956
Retrotheristus Lorenzen 1977
Rhynchonema Cobb 1920
Scaptrella Cobb 1917
Sphaerotheristus Timm 1968
Spiramphinema Wieser 1956
Steineria Micoletzky 1922
Stylotheristus Lorenzen 1977
Theristus Bastian 1865
  Subgen *Penzancia* de Man 1889
  Subgen *Theristus* Bastian 1865
Trichotheristus Wieser 1956
Valvaelaimus Lorenzen 1977
Wieserius Chitwood & Murphy 1964
Xenolaimus Cobb 1920
Xyala Cobb 1920
  Neotheristus Schulz 1938, syn. Gerlach 1951 : 402
Zygonemella Cobb 1920

Dubious genera of Xyalidae:

Austronema Cobb 1914, dub. Hope & Murphy 1972 : 48
Buccolaimus Allgen 1959, dub. Hope & Murphy 1972 : 48
Pulchranemella Cobb 1933, dub. Hope & Murphy 1972 : 51

Sphaerolaimidae   Filipjev 1918

90. The holophyly of the Sphaerolaimidae is established by the following holapomorphy: the 4 cephalic setae of the third sensilla circle are always longer than the 6 of the second sensilla circle, whereby both circles are situated at the same level (discussion point 14); this feature does not appear in the Xyalidae, the sister group of the Sphaerolaimidae.

Additional features: The cuticle is ridged or smooth. There are always 8 groups of subcephalic setae. The buccal cavity is not completely surrounded by pharyngeal tissue and is usually barrel-shaped. The inner wall of the pharynx is lined with an unusually thick layer of cuticle. The cervical gland is always present. Two testes are usually present. The anterior gonad lies on the right or left of the intestine and the posterior gonad lies on the side opposite the anterior testis. The caudal glands have separate openings. The tail always has a cylindrical end part to it and 2 or 3 terminal setae. The family is marine.

Sphaerolaiminae   Filipjev 1918

The holophyly of the Sphaerolaiminae is established by the holapomorphy that the number of groups of subcephalic setae in the 4 juvenile stages are 0, 0, 4 or 6, and 8 respectively; this sequence is unique within the Monhysteroidea (Lorenzen 1978e : 76).

Doliolaimus Lorenzen 1966
Sphaerolaimus Bastian 1865
   Euthoracostomopsis Sergeeva 1974, syn. Lorenzen 1978e : 67
Subsphaerolaimus Lorenzen 1978

Parasphaerolaiminae   Lorenzen 1978

The holophyly of the Parasphaerolaiminae is established by the following three holapomorphies (after Lorenzen 1978c and e):

91. a) The number of groups of subcephalic setae is 8 in all four juvenile stages.

OUTLINE OF A PHYLOGENETIC SYSTEM 203

92. b) In all four juvenile stages the longitudinal ribs of the labial region are significantly different in structure from those of the adult animals.
93. c) The buccal cavity is surrounded by 6 anterior and 3 posterior plates arranged in rings around it, and the connecting, anterior-most part of the inner wall of the pharynx is divided into 6 plates.

All three features are unique within the Sphaerolaimidae.

Only genus: *Parasphaerolaimus* Ditlevsen 1918

SIPHONOLAIMOIDEA Filipjev 1918

In the present study, in agreement with Filipjev (1918, 1934) and Riemann (1977a), but in contrast to de Coninck (1965), the Siphonolaimidae and Linhomoeidae are classed together as one sub-family because many species of the two families show agreement in the following features (after Riemann 1977a : 112–114):

a) The pharyngeal glands open right by the buccal cavity, one dorsally and two subventrally.
b) In their own characteristic way the intestinal cells are turgescent and possibly act like a backbone.
c) The amphids have a very long canal which extends backwards and is gently curved.
d) The tail is, in part, shortened by histolysis.
e) The body surface is smooth.
f) The lateral field is very wide.
g) Anteriorly the head is round.

These features only apply to some of the species. Extensive analyses are still needed in order to decide which of the features listed can be used to establish the holophyly of the Siphonolaimoidea.

Analyses by the author did not produce any holapomorphies.

Siphonolaimidae Filipjev 1918

The holophyly of the Siphonolaimidae is established by the following two holapomorphies:

97. a) The wall of the buccal cavity comes to a point like a spear at the anterior in its own characteristic way. According to all the current observations available, this spear-like formation cannot be extended forwards.

b) Only the anterior ovary is present (discussion point 46); this feature otherwise occurs at most sporadically within the Siphonolaimoidea. The ovary is outstretched.

It is possible that the following two features represent holapomorphies in the Siphonolaimidae:

c) Only the anterior testis is present.
d) Within species the gonad lies constantly to the left of the intestine.

Insufficient observations of the two features are available for a phylogenetic assessment to be made.

Additional features: The cuticle is striated. The 6+4 cephalic setae are situated at more or less the same level. The 6 of the second sensilla circle are always shorter than the 4 of the third sensilla circle. The amphids are round. The pharynx shows thickening of the muscles behind the buccal cavity and at the end. The family is marine.

Discussion: Formerly, *Cyartonema* (syn. *Southernia*), *Paraterschellingia* and *Tubuligula* were put in the Siphonolaimidae. They are now put in the Aegialoalaimidae (Leptolaimina) (see new systematic position for reasons).

Genera of the Siphonolaimidae:

Siphonolaimus de Man 1893
  Anthraconema zur Strassen 1904, syn. Steiner 1916 : 631
  Chromagaster Cobb 1894, syn. Filipjev 1918 : 307
Solenolaimus Cobb 1894

Linhomoeidae   Filipjev 1922

There is no known holapomorphy with which to establish the holophyly of the Linhomoeidae. The following two features are characteristic of many species of the Linhomoeidae, but more extensive analyses are still necessary before they can be assessed phylogenetically:

a) In most cases the cardia is noticably lengthened. Gerlach (1963 : 606) considers this feature to be primitive because it is also said to occur in a similar form in *Plectus* and related forms, but in the Linhomoeidae the cardia is always clearly set-off from the intestine, in contrast to the cardia of *Plectus* and its relatives, in which this division is not detectable.
b) Personal research has shown that in species from very different genera (*Desmolaimus, Metalinhomoeus, Paralinhomoeus, Eleuthero-*

laimus), the inner side of the labial region forms an annular, soft-skinned pad or bulge which produces a marked narrowing to the entrance of the buccal cavity. Such a pad is not known in other nematodes and is often not described for species of the Linhomoeidae, even in cases where, in all probability, the pad is present.

A further feature has hitherto often been overlooked in species of the Linhomoeidae — personal observations have shown that this feature in fact appears very frequently within the family:

c) The base of the buccal cavity forms one dorsal and two subventral tooth-like arches which point towards the body axis and which are more markedly cuticularized frontally than they are abfrontally. There are only a very few illustrations of this type of buccal cavity structure (e.g. by Inglis 1963 for *Linhomoeus timi*, by Lorenzen 1973a for *L. filaris* and by Riemann 1977a for *Paralinhomoeus lepturus* and *Anticyclus* sp.). The buccal cavity of some representatives of the Xyalidae is formed in a very similar way (Lorenzen 1978c : 524), so the phylogenetic interpretation of this feature as a holapomorphy of the Linhomoeidae alone is probably not justifiable.

Additional features of the Linhomoeidae: the cuticle is often striated and seldom smooth. The labial sensilla are small or not recognizable at all. The 6+4 cephalic sensilla are arranged in two separate circles, or both circles are combined to form a common circle. The 6 cephalic sensilla of the one circle are longer or shorter than the 4 of the other circle (section 9.1 of the appendix). The amphids are round in most cases; they are curved into a round, bow-like shape in *Anticyclus junctus, Didetta, Disconema, Halinema* and *Sphaerocephalum*. There are usually two outstretched ovaries and two testes facing in opposite directions to one another; only seldom are a single ovary and a single testis present (Section 9.2 of the appendix). Usually the anterior gonad lies on the left or right of the intestine and the posterior gonad lies on the opposite side; within species the anterior gonad seldom lies constantly to the left and the posterior gonad constantly to the right (Section 9.2.1. of the appendix). The family is marine.

Discussion: The enumeration of the features reflects the heterogenous nature of the Linhomoeidae. The following classification of the family was established largely by Gerlach (1963). From amongst the former "Linhomoeidae" *Coninckia* is put in the Coninckiidae (Axonolaimoidea), *Tubolaimoides* is put in the Tubolaimoididae (Leptolaimina), *Pandolaimus* is put in the Pandolaimidae (Enoplida; Jensen 1976 has already put the genus in this order, specifically into the Anoplostomatidae) and *Mikinema* is put in the Ohridiidae (see new systematic position for reasons).

Desmolaiminae G. Schneider 1926

Terschellingiinae Andrassy 1976, syn. Lorenzen 1981 : 255
Anticyathus Cobb 1920
Desmolaimus de Man 1880
Linhomoella Cobb 1920
Megadesmolaimus Wieser 1954
Metalinhomoeus de Man 1907
   Monhysteriella Kreis 1929, syn. Wieser 1956 : 43
Prosphaerolaimus Filipjev 1918
Sarsonia Gerlach 1967
Terschellingia de Man 1888
Terschellingioides Timm 1967

Discussion: Andrassy (1976 : 103) includes *Desmolaimus* among other genera in the "Terschellingiinae". This genus is the type of the Desmolaiminae, so both families are synonymous with each other. The Desmolaiminae were not listed by Andrassy.

Eleutherolaiminae Gerlach & Riemann 1973

Eleutherolaimus Filipjev 1922
Eumorpholaimus Schulz 1932
Metalaimus Kreis 1928

Linhomoeinae Filipjev 1922

Anticyclus Cobb 1920
Didelta Cobb 1920
Disconema Filipjev 1918
Halinema Cobb 1920
   Zanema Cobb 1920, syn. Wieser 1956 : 40
Linhomoeus Bastian 1865
   Eulinhomoeus de Man 1907, syn. Wieser 1956 : 50
Monhysteroides Timm 1961
Paralinhomoeus de Man 1907
   Crystallonema Cobb 1920, syn. Gerlach 1963 : 628
   Paradesmolaimus Schulz 1932, syn. Cobb 1935 : 480
Perilinhomoeus Stekhoven 1950
Prolinhomoeus Timm 1961
Sphaerocephalum Filipjev 1918

Dubious genera of Linhomoeidae:

Aponcholaimus Allgen 1957, dub. Gerlach 1963 : 651
Bathylaimella Allgen 1930, dub. Gerlach 1963 : 651
Chloronemella Allgen 1929, dub. Gerlach 1967 : 39
Cryptolaimus Cobb 1933, dub. Hope & Murphy 1972 : 48
Halicylindrolaimus Allgen 1959, sub. Hope & Murphy 1972 : 49
Nijhoffia Allgen 1935, dub. Lorenzen 1981 : 256
Paraegialoalaimus Allgen 1934, dub. Lorenzen 1981 : 251

Discussion: only the type species is known for the dubious genera created by Allgen. In every case the type species has been described so vaguely that one can only guess that they belong to the Linhomoeidae.

AXONOLAIMOIDEA Filipjev 1918

There is no known holapomorphy with which to establish the holophyly of the Axonolaimoidea. The most important characteristic is that the amphids are always spiral or bow-shaped (very seldom they may also be pore-shaped, see Axonolaimidae), but this feature also occurs within the Linhomoeidae.

Additional features: The cuticle is smooth or striated and in most Comesomatidae it is covered with punctations. Personal observations have shown that intact animals fixed in formalin always colour well in cotton blue. The labial sensilla are papilliform or not visible. In most cases the 6+4 cephalic sensilla are situated in two separate circles (exceptions among the Comesomatidae). The posterior 4 cephalic sensilla are longer than or, at the most, equal in length, to the anterior 6 cephalic sensilla. The buccal cavity may have 6 (Axonolaimidae) or 3 (Comesomatidae partim) tooth-like structures on its front edge. Pharyngeal tubes occur frequently within the Axonolaimidae and Comesomatidae. The pharynx never widens at the end to form a muscular bulb. The females have two outstretched ovaries (seldom only one posterior ovary, see the Axonolaimidae); the males possess two testes or only one anterior testis. Ventrally situated pre-anal papillae occur, above all, in the Comesomatidae and only seldom in the Axonolaimidae.

Discussion: in extent the Axonolaimoidea resemble most fully Chitwood & Chitwood's (1950 : 22) super-family of the same name, whereas de Coninck's (1965 : 609) Axonolaimoidea only contains the Axonolaimidae in their presently understood sense. The four families Axonolaimidae, Diplopeltidae, Coninckiidae and Comesomatidae are differentiated and discussed in the order listed above.

Axonolaimidae Filipjev 1918

There is no known holapomorphy with which one can establish the holophyly of the Axonolaimidae. The 6 tooth-like structures on the front edge of the buccal cavity are very characteristic of the family. This feature is not considered to be a holapomorphy of the family because it also occurs in a similar form in some genera of the Comesomatidae. The 6 tooth-like structures are only insignificant in *Axonolaimus, Ascolaimus* and *Apodontium*. In the remaining genera of the family however, they are rather large and can be moved (Fig. 13); they are referred to as odontia. According to Chitwood & Chitwood (1950 : 71), they are formations of the labial region, but they could just as well originate from the anterior region of the buccal cavity; an analysis is still needed.

Additional features: Connected with the 6 tooth-like structures, the buccal cavity has a markedly cuticularized, columnar or conical wall. Posterior to the buccal cavity in the pharynx there are one ventral and two subdorsal pharyngeal tubes, similar to those which also occur in the Comesomatidae, Plectidae, Rhabditoidea and in some Monhysteridae. The cuticle is smooth in most cases and seldom striated. The 6+4 cephalic sensilla, which stand in two separate circles, are always dominated by the posterior 4. The amphids are curved into a loop shape (inverted U- or O-shaped) and are pore-shaped only in *Apodontium*; in many species the ventral arm of the amphid is longer than the dorsal arm. In animals at juvenile stage I, the amphids only form a spiral with one loop (Fig. 13; Lorenzen 1972b and 1973a for *Axonolaimus* and *Odontophoroides*, and recent personal observations on *Odontophora*). There are two testes, the posterior one of which continues forwards at first and only folds over in the anterior part. In most cases there are two ovaries, seldom only one posterior ovary (*Synodontium, Odontophoroides*). In most species the anterior gonad lies to the left or the right of the intestine and the posterior gonad lies on the side opposite to it; in *Ascolaimus* however the length is completely variable. The tail is always conical. The caudal glands open through a common pore. The family is marine.

Discussion: Chitwood & Chitwood (1950), Wieser (1956) and Gerlach & Riemann (1973) saw the Axonolaimidae merely as a subfamily of a much more extensive family "Axonolaimidae", whereas de Coninck (1965) and Vitiello (1974) saw the family as having more or less the same range as in the current work. The present author favours the decision of the last named authors because the "Axonolaimidae" were very heterogenous and because, despite the absence of an established holophyly, the Axonolaimidae in their present sense are very well characterized.

OUTLINE OF A PHYLOGENETIC SYSTEM   209

The species *Parachromagasteriella annelifer* (lapsus: *annelifera*), which was originally part of the "Cylindrolaiminae", is made a new member of *Axonolaimus*. The species is now called: *Axonolaimus annelifer* (Wieser 1956). The species is known to the author from samples from Chile (the species was described from Chile); the buccal cavity is similar in structure to that of *Axonolaimus*.

Personal observations show that the genera *Apodontium* and *Synodontoides* are exactly the same as *S. procerus* and are thus considered to be synonyms of one another. The only *Synodontoides* species is therefore now called: *Apodontium procerum* (Gerlach 1957).

The genera of the Axonolaimidae:

Apodontium Cobb 1920
   Synodontoides Hopper 1963, syn. Lorenzen 1981 : 260
Ascolaimus Ditlevsen 1919
   for *Bathylaimus* Ditlevsen 1918, a homonym
   Oligomonohystera Micoletzky 1922, syn. Stekhoven & de Coninck 1932 : 154
Axonolaimus de Man 1889
Margonema Cobb 1920
Odontophora Bütschli 1874
   Conolaimus Filipjev 1918, syn. Allgen 1929 : 309
   Trigonolaimus Ditlevsen 1918, syn. Allgen 1929 : 307
Odontophoroides Boucher & Helléouet 1977
Parascolaimus Wieser 1959
Parodontophora Timm 1963
Pseudolella Cobb 1920
   Pseudolelloides Timm 1957, syn. Gerlach 1962 : 97
Synodontium Cobb 1920

Comesomatidae   Filipjev 1918

The holophyly of the Comesomatidae is established by the following holapomorphy: the amphids are spiral and have at least 2½ loops (discussion point 27); this feature is unique within the Monhysterida. Personal observations on *Sabatiera celtica* (from the sublittoral region of Helgoland) show that the amphids in juveniles of stage I hardly look any different from those of adult animals from the same species.

The following three very typical features cannot currently be used to establish the holophyly of the family because in each case there are exceptions which cannot yet be considered phylogenetic.

a) The cuticle is always striated, and in most species the body annules

are punctated. Such ornamentation is unique within the Monhysterida; personal research and data in the literature show that within the Comesomatidae it is, of course, absent in *Cervonema, Laimella, Paracomesoma,* as well as in *Sabatieria hilarula.*
b) The adult males of many species have slightly arched, ventrally situated pre-anal supplements of the sort that do not appear anywhere else in the Chromadoria.
c) Within species the anterior gonad often lies constantly to the left of the intestine and the posetrior gonad to the right of it; within the Chromadoria this feature occurs regularly in the Xyalidae and the Coninckiidae, a family with few species, and it also occurs sporadically in other taxa.

Additional features: The buccal cavity is often cup-shaped and in the anterior region in the Dorylaimopsinae it has 3 or 6 tooth-like structures which also occur in a similar form in the Axonolaimidae and which are probably not homologous with the teeth of the Chromadorina. According to Jensen (1979a : 84) and personal observations, at the end of the buccal cavity many species have a very small dorsal tooth, the phylogenetic interpretation of which is uncertain. In many species one ventral and two subdorsal pharyngeal tubes are atached to the buccal cavity. The 6 labial sensilla are short. The two cephalic sensilla circles are usually separate from one another; they are joined in *Metacomesoma* and *Laimella*. The 6 anterior cephalic sensilla are usually shorter than the 4 posterior cephalic setae; they are approximately the same length in *Cervonema* (Fig. 1), *Metacomesoma* and *Pierrickia.* The females have two outstretched ovaries and the males have two testes facing in opposite directions from one another. The tail has a cylindrical end section. In many *Sabatieria* species the cervical gland is sexually dimorphic in structure: in the males it consists of one ventral and two large, subventral cells which Riemann (1977b) was the first to describe, whereas in the females it consists of only one ventral cell. It is not known whether sexual dimorphism of the cervical gland also occurs in other genera of the Comesomatidae. The family is marine.

Discussion: the Comesomatidae were put into the "Monhysterata" by Filipjev (1934) because of the outstretched ovaries, into the "Monhysterina" by Chitwood & Chitwood (1950) because of the pharyngeal tubes, and into the "Chromadorida" by Wieser (1954), de Coninck (1965), Gerlach & Riemann (1973) and Andrassy (1976) because of punctated cuticle and the spiral amphids. In Lorenzen (1981), in agreement with Filipjev, the Comesomatidae were put into the Monhysterida, above all because of the outstretched ovaries (discussion point 51 and section 5.2.1.).

Recently, Platt (1985) has replaced the Comesomatidae into the Chromadorida because of the punctated cuticle, the multi-spiral amphids and the presence of precloacal supplements. Within the Chromadorida, the combination of these features occurs in the Chromadoroidea. Because of his decision, Platt concluded the absence of punctations in *Laimella, Cervonema, Setosabatieria* and some species of *Paracomesoma* to be a secondary one (i.e. punctations lost during evolution). I would argue to retain the Comesomatidae in the Monhysterida on the following reasons:

— The precloacal supplements of Comesomatidae differ from those of the Chromadorida, that is, probably they have evolved independently in Comesomatidae and different taxa of Chromadorida.
— Precloacal supplements are also known from other taxa of the Monhysterida.
— Punctations of the cuticle have evolved independently in different taxa of the Chromadorida and Rhabditida; recently, Riemann (1986) found this character in the new genus *Nicasolaimus* (Axonolaimidae, Monhysterida). Therefore, punctations of the cuticle do not necessarily reflect a closer relationship between Chromadoroidea and Comesomatidae.
— A multi-spiral amphid may rather easily evolve from a loop-shaped one (or vice versa); this may be found frequently within Desmodoroidea (Chromadorida), but I have also seen an abnormal specimen of *Comesoma* sp. with a multi-spiral amphid on the left and a loop-shaped amphid on the right side (Fig. 28).
— The buccal cavity of certain comsomatid species resembles strongly that of certain axonolaimid species (compare, for example, that of *Paramesonchium*/Comesomatidae with that of *Ascolaimus*/Axonolaimidae).

Gerlach & Riemann (1973) have put the "Acantholaiminae" into the Comesomatidae. In the present study this sub-family is put into the Chromadoridae (see the Chromadoridae for reasons). The following classification of the Comesomatidae is in line with Jensen (1979a):

Comesomatinae   Filipjev 1918

Comesoma Bastian 1865
Comesomoides Gourbault 1980
Metacomesoma Wieser 1954
Paracomesoma Home & Murphy 1972

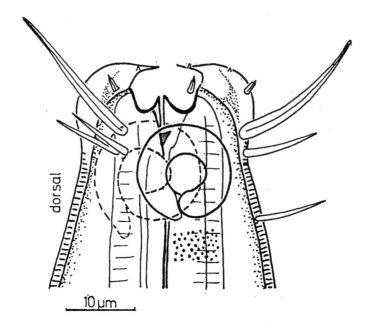

Fig. 28: Anterior end of an abnormal male of *Comesoma minimum* sensu Wieser 1954; on the left side, the amphid is multispiral as usual; on the right side, it is loop-shaped (abnormal condition). The male comes from southern Chile.

Sabatieriinae   Filipjev 1934

Cervonema Wieser 1954
Laimella Cobb 1920
Pierrickia Vitiello 1970
Sabatieria Rouville 1903
   Actarjana Hopper 1967, syn. Jensen 1979a : 89
   Parasabatieria de Man 1907, syn. Filipjev 1922 : 205
Scholpaniella Sergeeva 1973

Dorylaimopsinae   de Coninck 1965

Dorylaimopsis Ditlevsen 1918
   Mesonchium Cobb 1920, syn. Jensen 1979a : 91
   Pepsonema Cobb, syn. Cobb 1935 : 466
   Xinema Cobb 1920, syn. Filipjev 1934 : 27
Hopperia Vitiello 1969
Metasabatieria Timm 1961
Paramesonchium Hopper 1967
Vasostoma Wieser 1954

Dubious genera of the Comesomatidae:

Dolichosomatum Allgen 1951, dub. Hope & Murphy 1972 : 49
Grahamius Gerlach & Riemann 1973 (for *Grahamia* Allgen 1959, a homonym;, dub. Hope & Murphy 1972 : 50
Notosabatieria Allgen 1959, dub. Hope & Murphy 1972 : 50
Ungulilaimus Allgen 1958, dub. Hope & Murphy 1972 : 51

Diplopeltidae Filipjev 1918

There is no holapomorphy with which to support the holophyly of the Diplopeltidae. The cuticle may be smooth or striated. The 4 cephalic setae of the third sensilla circle are always predominant, while the anterior 6+6 sensilla are very small or not visible at all. The amphids curve into a bow-shape and consist of a spiral with one loop. Ocelli are occasionally present (Diplopeltinae partim). The buccal cavity is narrow and tubular or very small; tooth-like formations are never present. In some cases the pharynx has one ventral and two subdorsal pharyngeal tubes (Diplopeltinae partim); a muscular end bulb does not occur. There are two outstretched ovaries (only one anterior ovary in *Cylindrolaimus monhystera*). There are usually two testes; personal research has shown that the posterior testis is either anteriorly directed and is reflexed in the anterior section, or is posteriorly directed completely; occasionally the posterior testis is absent. Pre-anal organs are absent. The tail is conical or has a cylindrical end section. The caudal glands open together or separately. The Diplopeltinae are marine, the Cylindrolaiminae are limno-terrestrial.

Discussion: The Diplopeltidae contain only the genera of the "Diplopeltinae", "Campylaiminae" and "Cylindrolaiminae" (all 'Axonolaimidae") of the Bremerhaven checklist of Gerlach & Riemann (1973/ 1974), with the following exceptions: *Alaimonema* is put in the Desmodoridae (Chromadorinae), *Aegialoalaimus* and *Diplopeltoides* are put in the Aegialoalaimidae, *Domorganus* in the Ohridiidae, *Polylaimium* in the Leptolaimidae, *Tarvaia* in the Tarvaiidae, *Chitwoodia* in the Tubolaimoididae (all Leptolaimina) and *Paratarvaia* in the Coninckiidae (Axonolaimoidea). With the exception of *Paratarvaia*, all the genera listed are distinguished by having antidromously reflexed ovaries.

The holophyly of the two sub-families of the Diplopeltidae cannot, at present, be established.

Diplopeltinae Filipjev 1918

Campylaiminae Chitwood 1937, syn. Lorenzen 1981 : 264

The cuticle is smooth or striated. The amphids are bow-shaped or have a spiral contour with one turn. The stoma lies terminally or is displaced to the dorsal side (*Pararaeolaimus nudus* after Lorenzen 1973a, *Diplopeltis* partim, *Campylaimus*).

Discussion: Both the former "Diplopeltinae" and the former "Campylaiminae" contain species with striated cuticles and dorsally displaced stoma. Since it is not possible to distinguish between these two subfamilies using the form of the tail (conical or with a cylindrical end section) or even using other features, they are seen as synonymous with one another.

It should be possible to differentiate between *Araeolaimoides* and *Araeolaimus* using the amphids, which in *Araeolaimoides* are longer than they are wide. This feature also holds true in the type species of *Araeolaimus*, *A. bioculatus*. The two genera are therefore seen as synonymous. The three *Araeolaimoides* species are thus now called: *Araeolaimus microphthalamus* de Man 1893, *A. ovalis* (Wieser 1956) and *A. paucisetosus* (Wieser 1956).

The cylindrical end section of the tail distinguishes *Parachromagasteriella* from *Southerniella* (conical tail). According to Vitiello (1970), in females of *S. conicauda* the tail is conical and in males it has a cylindrical end section. Both genera are thus seen as synonymous. The *Parachromagasteriella* species are therefore now called: *Southerniella arctica* (Allgen 1953), *S. cylindricauda* (Allgen 1933) and *S. zosterae* (Allgen 1933). *P. annelifer* has already been put in *Axonolaimus* earlier on in the text.

The following four species are removed from *Diplopeltula* and put into *Diplopeltis*; they are now called *Diplopeltis incisus* (Southern 1914), *D. indicus* (Gerlach 1962), *D. intermedius* (Gerlach 1954) and *D. onustus* Wieser 1956. All four species have an unstriated cuticle and are much more similar to the type species of *Diplopeltis* (*D. typicus*) than they are to the type species of *Diplopeltula* (*D. breviceps*).

The genera of the Diplopeltinae:

Araeolaimus de Man 1888
  Araeolaimoides de Man 1893, syn. Lorenzen 1981 : 265
  Coinonema Cobb 1920, syn. Gerlach 1953a : 26
  Parachromagaster Allgen 1929, syn. de Coninck & Stekhoven 1933 : 97
Campylaimus Cobb 1920
Diplopeltis Cobb in Stiles & Hassal 1905
  for *Dipeltis* Cobb 1891, a homonym

OUTLINE OF A PHYLOGENETIC SYSTEM 215

Discophora Villot 1875 (a homonym), syn. Stiles & Hassal 1905 : 100
Diplopeltula Gerlach 1950
Metaraeolaimoides de Conick 1936
Pararaeolaimus Timm 1961
Pseudaraeolaimus Chitwood 1951
Southerniella Allgen 1932
Parachromagasteriella Allgen 1933, syn. Lorenzen 1981 : 266
Striatodora Timm 1961

Cylindrolaiminae   Micoletzky 1922

The cuticle is striated. The amphids form a spiral with one turn. The stoma is always terminal. Males are unknown in the majority of species.

Type and only genus: *Cylindrolaimus* de Man 1880

Coninckiidae   Lorenzen 1981

No known holapomorphy can be cited with which to establish the holophyly of the Coninckiidae. The combination of the following features is typical of the family and is unique within the Monhysterida:
a) The amphids are curved into an O-shape.
b) The 6+4 cephalic setae, which are arranged in two separate circles, are equal in length.
c) The buccal cavity is tiny.

Additional features: The cuticle is smooth or striated. The pharynx thickens slightly at the end. There are two outstretched ovaries and two testes which face in opposite directions; personal observations have shown that the anterior gonad lies to the left and the posterior gonad to the right of the intesine (Section 9.2.1.).

Type and only genus: *Coninckia* Gerlach 1956
Paratarvaia Wieser & Hopper 1967, syn. Lorenzen 1981, 266

Discussion: the family of the Coninckiidae has been created because of the unique combination of features within the Monhysterida mentioned above. Previously *Coninckia* belonged to the "Linhomoeidae" and *Paratarvaia* to the "Axonolaimidae". The two genera are seen as synonymous for the following reason: the author has 5 adults and several juveniles of the type species of *Coninckia* , *C. circularis*, from the sub-littoral of the North Sea, near Helgoland. In the individual animals the body annules and the cuticular plate on which the amphids are situated are differentially defined and, in some cases, hardly recognizable. As a result, the only two criteria for distinguishing between *Coninckia* and *Paratarvaia* are invalid.

5.3 Outline of a phylogenetic system for the Enoplia

ENOPLIA Pearse 1942

The holophyly of the Enoplia is established by the holapomorphy that the amphids are probably non-spiral at a primary level (discussion point 20). In addition, the fovea, in contrast to the aperture, is enlarged like a pocket. In instances where spiral amphids do occur within the Enoplia (Enchelidiidae, Tripyloididae, Trefusiidae), they are considered a secondary and not a primary condition. At the moment, no other holapomorphies can be cited for the establishment of the Enoplia.

Additional features: In most species the cuticle is not striated. Of the 6+4 cephalic sensilla, the 6 of the second sensilla circle are almost always longer than or, at the most, equal in length to the 4 of the third sensilla circle. Within the head region, the pharynx may just as often be attached to the wall of the buccal cavity as to the body wall. In many species the pharynx swells slightly in the posterior section, but, with the exception of *Syringolaimus* (Ironidae), it never forms a markedly muscular bulb, as is typical in many Chromadoria. The ovaries are always antidromously reflexed, except in *Cytolaimium exile* (Trefusiidae, outstretched ovaries). In cases where caudal glands are present, they always open, as far as is known, through a common pore. The Enoplia contain very many more larger species (over 3mm in length) than the Chromadoria.

Recently, only the Enoplida, Trefusiida and Dorylaimida have been retained in the Enoplia; the Mermithoidea, Trichuroidea and Dioctophymatoidea have been removed from the Enoplia and placed in the Secernentea (Lorenzen 1982).

5.3.1. Classification of the freeliving Enoplia into orders
Just as the holophyly of the freeliving Enoplia has been doubted so little since Filipjev (1918), so has their classification into taxa of the next lower rank and the substantiation of such a classification proved

OUTLINE OF A PHYLOGENETIC SYSTEM 217

equally difficult so far. The difficulties are reflected in the individual authors' differing gross systematic classifications of the freeliving Enoplia (Table 9): the freeliving Enoplia are divided, in part, into two and, in part, into three equal-ranking taxa at the next level down, and the systematic position of the Tripyloididae, Mononchoidea, Alaimidae, Ironidae and some smaller taxa is, in part, assessed very differently.

Table 9: Systematic classification of the freeliving Enoplia by different authors. The names and positioning of the taxa are those used by the authors concerned.

Filipjev (1934)	Chitwood and Chitwood (1950)	Clark (1961)

Enoplidae
  9 sub-families
 	
Trilobidae
  Trilobinae
  Mononchinae
  Tripyloidinae
 	
Dorylaimidae
  Alaiminae
  Ironinae
  Tylencholaiminae
  Dorylaiminae
	Enoplina
	  Enoploidea
	    Enoplidae
	    Oncholaimidae
	  Tripyloidea
	    Tripylidae
	    Mononchidae
	    Alaimidae
	    Ironidae
	
	Dorylaimina
	  Dorylaimoidea
		Enoplina
		  Enoploidea
		    Enoplidae
		    Lauratonematidae
		    Oncholaimidae
		  Tripyloidea
		    Tripylidae
		    Ironidae
		
		Alaimina
		  Alaimidae
		
		Dorylaimina
		  Mononchoidea
		  Dorylaimoidea
		  Diphtherophoroidea

de Coninck (1965)	Andrassy (1976)	The present study

Enoplida
  Enoplina
    Tripyloidea
    Enoploidea
    Oncholaimina
      Oncholaimoidea
 	
Dorylaimida
  Dorylaimina
    Mononchoidea
    Dorylaimoidea
    Diphthero-
      phoroidea
  Alaimina
	Enoplida
	  Enoplina
	  Oncholaimina
	  Tripylina
	
	Dorylaimida
	  Mononchina
	  Dorylaimina
	  Diphtherophorina
	  Mermithina
		Enoplida
		  Enoplina
		    Enoplacea
		    Oncholaimacea
		  Tripyloidina
		
		Trefusiida
		
		Dorylaimida
		  Dorylaimina
		  Mononchina
		
		Bathyodontina

The only previous criterion for the division of the freeliving Enoplia into two orders was produced in the 1930s by Chitwood & Chitwood (comprehensive description 1950):

a) In the Enoplida ("Enoplina" in Chitwood & Chitwood), at least one pharyngeal gland opens in each of the three pharyngeal sectors in or directly posterior to the buccal cavity; in the dorsal pharyngeal sector there is one pharyngeal gland and in each of the two subventral pharyngeal sectors there is a total of one or two pharyngeal glands.
b) In the Dorylaimida ("Dorylaimina" in Chitwood & Chitwood), all the pharyngeal glands (as a rule, a total of 5: 1 dorsal, 4 subventral) open posterior to the nerve ring.

Chitwood & Chitwood (1950 : 24) have used the criterion inconsistently: they put the "Mononchidae" in the "Enoplina" and not in the "Dorylaimina", although they knew that all pharyngeal glands in the "Monochidae" open posterior to the nerve ring. Clark (1961 : 127) has raised this inconsistency and he was the first to put the Mononchoidea in the Dorylaimida. However, in 1962 Clark also introduced an inconsistency: he proved that, contrary to earlier assumptions, all the pharyngeal glands in the Alaimidae open posterior to the nerve ring; nevertheless, he did not put the Alaimidae in the "Dorylaimina", but listed them next to the "Enoplina" and "Dorylaimina" (Table 9) as a sub-order of equal rank. De Coninck raises this inconsistency in 1965. The inconsistency of Chitwood & Chitwood, as well as by Clark, lay in the fact that, on the one hand they recognized and stressed the phylogenetic value of the patterns according to which the pharyngeal glands open, but on the other hand, through their systematic classification, they failed to realize the significance that the dorylaimoid pattern in particular (all pharyngeal glands open posterior to the nerve ring) could have evolved independently within the "Enoplina" and "Dorylaimina" (Chitwood & Chitwood) and within the "Dorylaimina" and "Alaimina" (Clark) respectively. In other words, the phylogenetic value of a feature was recognized and emphasized on the one hand, but on the other hand was made to look dubious when the system was constructed.

For a long time the patterns according to which the pharyngeal glands open in the Enoplida and Dorylaimida were considered unique within the nematodes. Chitwood & Chitwood had found (see their comprehensive work of 1950) that in the Chromadoria and Secernentea the subventral pharyngeal glands always open well posterior to the buccal cavity and the dorsal pharyngeal gland almost always opens in or near the buccal cavity (an exception in the freeliving Secernentea: in the Aphelenchoidea all the pharyngeal glands open well posterior to the buccal cavity but anterior to the nerve ring, thus differing from the

Dorylaimida in the second aspect). Only recently has it been shown (Riemann 1977a) that, in representatives of the Chromadoria too, pharyngeal glands may open not only dorsally, but also subventrally in or near the buccal cavity. Riemann found this feature in species of the Siphonolaimoidea (Monhysterida); in addition he had also rediscovered a similar observation in zur Strassen (1904) and points out that Decraemer (1976, 1978) has described the feature for species of the Desmoscolecoidea (Monhysterida). In the current work the feature was also found in Tubolaimoides (Leptolaimina, Tubolaimoididae) (Fig. 24). On account of the recent observations there is now no reason to establish the holophyly of the Enoplida by using the feature of the way in which the pharyngeal glands open. Consequently the only hitherto possible establishment of the holophyly of the Enoplida is now invalid.

98. However, the pattern according to which the pharyngeal glands open in the Dorylaimida is, as far as is known, unique within the nematodes and occurs in all species of the Dorylaimida, with the result that is considered a holapomorphic feature with which the holophyly of this order can be established.

During the course of the present work a new feature was found for the establishment of the holophyly of the Enoplida and was developed into a systematic criterion: the Enoplida have metanemes which are interpreted as stretch receptors and which have not been found outside the Enoplida. Their existence is thus considered a holapomorphy with which the holophyly of this order is established (discussion point 32).

The Enoplida, in its newly understood sense, contains most of those families which were also formerly part of this order. Some of the families and sub-families listed under the "Enoplida' by Gerlach & Riemann (1974) have, however, been moved to other orders: the Monochromadorinae, *Tobrilia* ("Tobrilinae"), the Prismatolaiminae and Odontolaiminae have been put in the Chromadorida, the Alaimidae and Cryptonchidae in the Dorylaimida, the Onchulinae, Trefusiidae and Lauratonematidae in the Trefusiida, and the Rhaptothyreidae, in agreement with Hope (1977), near the Mermithoidea. The Trefusiida form a paraphyletic group because they represent the nonholophyletic remains when the holophyletic Enoplia are divided into holophyletic taxa (Enoplida, Dorylaimida) of the highest possible rank. The paraphyly of the Trefusiida indicates that the inter-relationships within the Enoplia are not yet fully understood.

The inter-relationships discussed within the freeliving Enoplia are illustrated in Fig.29.

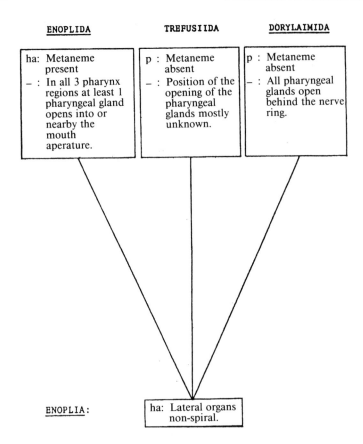

Fig. 29: The phylogenetic relationships within the freeliving Enoplia. The Trefusiida are a paraphyletic group. The names of the holophyletic taxa have been underlined.
ha: holapomorphy, which serves to demonstrate the holophyly.
p: plesiomorphy within the holophyletic taxon directly above.
–: a phylogenetic assessment of the feature is not available.

5.3.2. Classification of the Enoplida

ENOPLIDA Filipjev 1929

The holophyly of the Enoplida is established by the holapomorphy that metanemes are present (discussion point 32). As far as is currently known, metanemes do not occur outside the Enoplida. At present no other holapomorphies are known for the order.

Additional features: The cuticle is smooth or no more than weakly striated. The labial sensilla are papilliform or short and setiform. The cephalic sensilla are almost always setiform and, in a few families, they are jointed. The labial region is often divided into three lips. At least one pharyngeal gland always opens in each of the three pharyngeal sectors in or near the buccal cavity; as was established in section 5.3.1., this feature cannot be utilised, at least for the moment, to establish the holophyly of the Enoplida. The ovaries are always antidromously reflexed. There are usually two ovaries and two testes which faces in opposite directions to one another. The majority of the taxa are marine; only a few are freshwater.

Discussion: The extent of the Enoplida is determined in discussion point 32. Within the Enoplida a distinction is made between the extensive, holophyletic sub-order Enoplina and the relatively small, paraphyletic sub-order Tripyloidina. The paraphyly of the Tripyloidina indicates that the inter-relationships within the Enoplida are not yet fully understood.

ENOPLINA Chitwood & Chitwood 1937

The holophyly of the Enoplina is established by the fact that in most of its species the caudal glands penetrate far into the pre-caudal body region (discussion point 58); as far as is known at present, this feature never occurs outside the Enoplina. As a result of this extension into the pre-caudal body region, the volume of the caudal glands is increased as is also probably the productive power. Current data shows that within the Enoplina, the caudal glands lie completely in the tail only in *Enoplus* (Enoplidae), and Anticomidae, the Halalaiminae (Oxystominidae) and in part of the Ironidae. These taxa are included in the Enoplina on the basis of other features.

Discussion: Since it is unlikely that an apomorphy which was acquired relatively late should degenerate several times completely into the plesiomorphic condition, the fact that the caudal glands lie completely in the tail in the three named taxa of the Enoplina is probably a primary condition. As a result, only the genetic potential for the development of this feature – that the caudal glands extend into the pre-caudal region – can be drawn upon as a holapomorphy for the establishment of the holophyly of the Enoplina.

The Enoplina, in their current form, correspond extensively with the "Enoplidae" of Filipjev (1934) and the "Enoploidea" of Chitwood & Chitwood (1950). The authors saw it as characteristic for the taxon that a cephalic capsule is developed, at least at a primary level. This feature is commented upon in the discussion on the Enoplacea. The feature of the extension of the caudal glands into the pre-caudal region has hitherto not played a role as a systematic criterion. Within the Enoplina a distinction is made between the Enoplacea and Oncholaimacea, the holophyly of which can be established in both cases and the positions of which both lie between that of sub-order and that of super-family.

ENOPLACEA

The holophyly of the Enoplacea is established by the following two holapomorphies:

99. a) The musculature of the pharynx inserts in the buccal cavity region on the body wall and, in most cases, the inner layer of the body cuticle is differentiated to form a cephalic capsule in the region of the insertion point (this differentiation is absent in the Oxystominidae and Ironidae); as far as is known, this complex of characteristics does not occur outside the Enoplacea.
b) In cases where a cervical gland occurs, it lies in the pharyngeal body section except in part of the Ironidae (discussion point 62); this feature is not known in any other Adenophorea except in *Linhystera* (Xyalidae) and has hitherto not been of importance as a systematic criterion.

Discussion: the first holapomorphy (discussion point 99) was used as a systematic criterion for the first time by Filipjev (1918 : 23). Filipjev used this criterion, on the one hand in 1918 and 1934 to characterize the entire Enoplia and, on the other hand in 1934 (in the key on page 8) to characterize only the "Enoplidae", which correspond more or less to the present Enoplina. He interpreted the absence of this feature in all the taxa classed as Oncholaimacea and Tripyloidina in the present work as a secondary condition, because, it seems, remains of a primitive cephalic capsule may be present in at least some instances. Inglis (1964) has devoted a great deal of attention to the construction of the head in the Enoplida and (on page 298) confirms Filipjev's view at least for the Enoplida: "I am sure that originally such an attachment existed in all root stocks of the Enoplida and its absence in some forms is almost certainly due to its later loss not it never having existed in an ancestral form." Inglis did not substantiate his view. In his research on representatives from all families of the Onchoilaimacea and Tripyloidina, the author was not able to find any remains of a cephalic shell and there is no circumstancial evidence to support the assumption that the cuticularization, found almost directly posterior to the labial sensilla in representatives of the Oncholaimidae (Fig. 4a) and Enchelidiidae (Fig. 33) represents such remains.

Since the inner layer of the body cuticle in the cephalic region does not differentiate into a cephalic capsule in the Trefusiida and Chromadoria either, the author considers the absence of this feature in the Oncholaimacea and Tripyloidina as a primitive condition.

Within the Enoplina a distinction is made between the holophyletic super-family Enoploidea and the paraphyletic sub-family Ironoidea.

ENOPLOIDEA Dujardin 1845

The holophyly of the Enoploidea is established by the following two holapomorphies:

a) Personal research has shown that the anterior and posterior gonads are always found to the left of the intestine (discussion point 53); this feature occurs outside the Enoploidea at the most in isolated species.
b) The adult males posses a single ventrally situated pre-anal tubule (discussion point 55); this feature is unique within the Enoplida. In a very few species the tubule is absent or 9 tubules are present (*Epacanthion multipapillatum*). The deviations from the norm are seen as derived because, on the basis of other features, the taxa concerned clearly belong to the Enoploidea.

Additional features: All three types of metanemes are found; in most cases the metanemes only lie dorsolaterally, less frequently they may also be ventrolateral (Enoplidae, Phanodermatidae). The wall of the pharynx, at least in the posterior section, undulates in the Thoracostomopsidae and Phanodermatidae and it is flat in the Enoplidae, Anoplostomatidae and Anticomidae; the undulating contour is brought about by the closely packed sequence in which muscular and glandular areas alternate. So far it has not been possible to locate a cervical gland in the Thoracostomopsidae and Anoplostomatidae; a cervical gland is present in the Enoplidae, Phanodermatidae and Anticomidae and is limited to the pharyngeal body section. The Enoploidea are marine, only isolated species of the Enoplolaiminae penetrate up river as far as freshwaters.

Discussion: The undulating contour of the pharynx wall which is found in the Thoracostomopsidae and Phanodermatidae is unique within the freeliving nematodes, at least in so strongly expressed a form; it formerly held good as a family characteristic essentially only for the Phanodermatidae.

Within the Enoploidea a distinction is made between the Enoplidae, Thoracostomopsidae, Anoplostomatidae, Phanodermatidae and Anticomidae.

Enoplidae Dujardin 1845

The holophyly of the Enoplidae is established by the following two holapomorphies:

a) In cases where metanemes are present, adult animals have dorsolateral and ventrolateral loxometanemes of type II, and juveniles only have dorsolateral loxometanemes of type I (Lorenzen 1981a; discussion point 37); this combination of features is unique within the Enoplida. A caudal filament is sometimes present. No metanemes were found in *Enoplus schulzi* from New Zealand (= *Ruamowhitia orae*); they have probably been reduced, for the animals studied were only about 1.2mm long.
100. b) The buccal cavity only contains three mandibles and no onchia; this feature is also unique within the Enoplida.

The following feature may also represent a holapomorphy of the Enoplida:

c) According to Chitwood & Chitwood (1950 : 52) and personal observations, muscle bundles travel obliquely forwards from the posterior edge of the pharynx to the body wall; according to Chitwood & Chitwood this feature is otherwise known within the Enoplida only in *Metoncholaimus*.

Additional features: Personal observations have shown that the cuticle has oblique rows of fine dots, at least in *Enoplus brevis* and *E. michaelsensi*. The outlets of the very large epidermal glands can be of any shape. The lips are low and the labial sensilla are papilliform. The amphids either lie in the region of the well-developed cephalic capsule and are not connected to the post-capsule cuticle (by an incision of the suture), or they lie directly posterior to the cephalic capsule. The cephalic organs (cephalic slits) lie ventro-frontally to the lateral cephalic setae and each one looks like an "introverted cirrus" according to Inglis (1964 : 273). Pharyngeal musculature inserts onto the cephalic capsule. The pharyngeal glands open dorsally and subventrally at the end of the buccal cavity. The outer wall of the pharynx is smooth. According to personal observations the cervical gland lies constantly to the left of the pharynx. The females have two antidromously reflexed ovaries and the males usually have two testes, facing in opposite directions to one another, and only seldom have a single, anterior testis (*Enoplus schulzi*). The anterior end of the very large pre-anal tubule lies constantly to the right of the dorsal plane of the body. In juveniles and adult females the caudal glands lie completely in the tail, whereas in the males they penetrate a short way into the pre-caudal body region (personal observation).

Only genus of the Enoplidae:
Enoplus Dujardin 1845
Enoplostoma Marion 1870, syn. Marion 1875 : 500
Ruamowhitia Yeates 1967, syn. Lorenzen 1981 :279

Discussion: The extent of the present Enoplidae is drastically reduced in comparison with that of the former "Enoplidae". In general, the Thoracostomopsidae and Chaetonematinae were previously also put in the "Enoplidae"; only *Thoracostomopsis* was separated from the other "Enoplidae"; and put in a family of its own by some authors (among them Filipjev 1927). The former "Enoplidae" (with *Thoracostomopsis* and excluding *Chaetonema*) probably form a holophyletic group for the following reason:

101. In contrast with all other Enoplida they are singled out by the presence of mandibles and also, presumably, of onchia at a primary level (onchia are absent in *Enoplus*).

The Thoracostomopsidae, in their newly understood sense and the Chaetonematidae are removed from the "Enoplidae" for the following reasons:

a) The Thoracostomopsidae differ from *Enoplus* in the structure of the metanemes, the extension of the caudal glands well into the precaudal region, the differentiation of the outlets of the epidermal glands and the setiform condition of the labial sensilla; in the first three features the Thoracostomopsidae resemble the Phanodermatidae extensively.

b) The Chaetonematidae differ from *Enoplus* in the form and position of the amphids and metanemes; they differ from *Enoplus* and the Thoracostomopsidae in that the musculature of the pharynx has a completely different form of insertion in the buccal cavity region.

After studying two paratypes (2♂♂) of the type species of *Ruamowhitia*, *R. orae* Yeates 1967, the author has come to the conclusion that this species, as well as *R. halophila* Guirado 1975, are synonymous with *Enoplus schulzi* Gerlach 1952. The amphids of *R. orae* were erroneously described as spiral with one loop; personal observations have shown that they look exactly like those which Gerlach (1953 : 7) described in examples of *Enoplus schulzi* from Chile. Neither Yeates (1967) or Guirado (1975) have compared their species with *Enoplus schulzi*. *E. brachyuris* Ditlevsen 1932 is also very similar. Since no criterion has been found to support a distinction between the genera *Enoplus* and *Ruamowhitia*, the two genera are judged to be synonymous.

Thoracostomopsidae Filipjev 1927

The holophyly of the Thoracostomopsidae is established by the following holapomorphies:

a) Only dorsolateral orthometanemes with a robust scapulus but no caudal filament are present (Lorenzen 1981a, discussion point 40); this feature is unique within the Enoplina. The presence exclusively of dorsolateral orthometanemes occurs within the Enoplina otherwise only in the Oxystomininae, where, however, the scapulus is less well-developed and has a small caudal filament.

The following two features are also unique within the Enoploidea:

b) In addition to three mandibles (1 dorsal and 2 subventral), there are also three onchia in the spacious buccal cavity. This complex of features cannot be called upon to establish the holophyly of the Thoracostomopsidae, because mandibles also occur in the Enoplidae and, according to Inglis (1964 : 293), onchia are absent probably at a secondary level.

c) The labial sensilla are robust and setiform in structure (papilliform only in *Fenestrolaimus*); since setiform labial sensilla are seen as plesiomorphic within the Adenophorea (discussion point 5), this feature cannot be used to establish the holophyly of the Thoracostomopsidae.

The following feature was previously completely unknown:

d) The epidermal glands each have a particularly well differentiated outlet (Fig. 19). The feature is very common within the Thoracostomopsidae. Since it also occurs outside this family in representatives of the Phanodermatidae, it cannot be introduced to establish the holophyly of the Thoracostomopsidae.

Additional features: The inner layer of the cuticle in the cephalic region is differentiated to form a cephalic capsule onto which pharyngeal musculature is attached. Cephalic organs (cephalic slits) are often present; they vary a great deal in form and are situated frontally through ventrofrontally to the lateral cephalic bristles. The amphids are either small and situated directly posterior to the cephalic shell (Figs 30 and 31), or they are not recognizable at all and are, therefore, probably absent. As far as is curently known, there are no muscles connecting the posterior end of the pharynx to the body wall. No cervical gland or cervical pore has, as yet, been found, which suggests that they are probably absent. The males have two testes which face in opposite directions to one another, and the females have two antidromously reflexed ovaries (a single, posterior ovary in *Mesacan-*

OUTLINE OF A PHYLOGENETIC SYSTEM

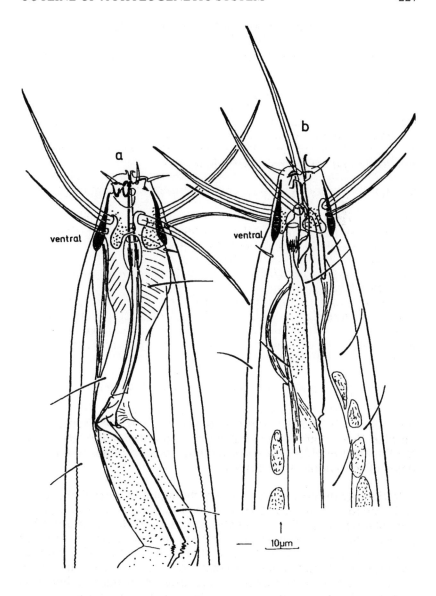

Fig. 30: *Thoracostomopsis barbata* (Thoracostomopsidae) with the spear a) withdrawn and b) everted. a: adult female from sublittoral region of the German Bay, 1st July 1973; b: adult female from a harbour wall in Helgoland (German Bay), 22nd August 1975. The preserved specimens originate from Dr Riemann's collection in Bremerhaven.

thion monhystera alone). My own observations have shown that the caudal glands always penetrate into the pre-caudal region; data in the literature to the contrary are dubious (Section 4.6.1.).

Thoracostomopsinae   Filipjev 1027

The holophyly of the Thoracostomopsinae is established by the following holapomorphy:

102. In the oral cavity there is a long spear which is unique among the freeliving nematodes. According to Inglis (1964 : 281), it is made up of the elements of all three buccal cavity sectors. Inglis doubts whether the spear can justify its name because no author has reported having seen it everted. This doubt can now be removed (Fig. 30): the spear really can be everted out of the buccal cavity. Personal observations have shown that the pharyngeal glands open dorsally and subventrally on the jointed part of the spear.

The author was unable to identify with any certainty the outstretched mandibles reported by Inglis (1964 : 281).

Only genus: *Thoracostomopsis* Ditlevsen 1918

Trileptiinae   Gerlach & Riemann 1974

The holophyly of the Trileptiinae is established by the following holapomorphy:

103. The three onchia which are equal in size are situated well forwards in the spacious buccal cavity (Fig. 31); this position of the onchia is unique within the Enoploidea. Personal observations have shown that the pharyngeal glands open through the onchia. According to Inglis (1964 : 280) the mandibles are small; the author was unable to find mandibles in the material available to him.

The author found only two caudal glands, both of which penetrate into the pre-caudal region.

Only genus: *Trileptium* Cobb 1933
for *Trilepta* Cobb 1920, a so-called homonym (hitherto unconfirmed)

Enoplolaiminae   de Coninck 1965

Enoploidinae de Coninck 1965, syn. Lorenzen 1981 : 285
Oxyonchinae de Coninck 1965, syn. Lorenzen 1981 : 285

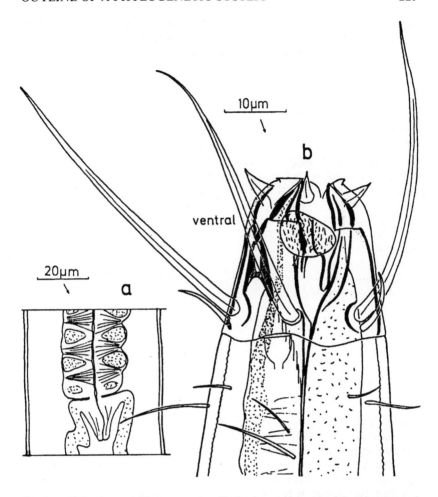

Fig. 31: *Trileptium* sp. (Thoracostomopsidae). a) posterior end of the pharynx and the cardia; b) anterior end: the onchia are situated very far forward, and the cephalic organ, which is oval in outline, is situated between the lateral labial and cephalic sensilla. Adult female, sandy beach on the Island of Chiloé (south Chile), 1st September 1973.

There is no known holopomorphy with which the holophyly of the Enoplolaiminae can be established. The following complex of features is very characteristic: In the oral cavity there are always three mandibles and three onchia at a primary level. One mandible and one onchium together form a unit which can be moved back and forth by specialized pharyngeal muscles, whereby the frontal section of the unit is moved in line with the body axis (Inglis 1964 : 285). A pharyngeal gland opens through each onchium. In part, all three onchia are the same length, and in part, the dorsal onchium is distinctly smaller than the two subventral onchia. The two subventral onchia are always mirror-images of each other; one of the two is never larger than the other, as is the case in most Oncholaimoidea.

Discussion: The "Enoploidinae" and "Oxyonchinae" are seen as synonymous with the Enoplolaiminae for the following reasons:
a) The three sub-families are distinguished from one another by de Coninck (1965) on the basis of the mandibles and the onchia. The criteria orginate from Wieser (1953), among others. On analysis of the head in the "Enoplidae" Inglis (1964 : 275, 292) established that, in part, the criteria were based on misunderstandings. Thus, in their present form at least, the criteria are not suitable for establishing subfamilies.
b) According to Fig. 468 of de Coninck (1965 : 657) the three sub-families only represent three stages, from which follows, in all probability their polyphyly.

Personal observations on *Fenestrolaimus* sp. from the Kattegatt have shown that *Fenestrolaimus* and *Trichenoplus* resemble one another in every respect; they are thus seen as synonymous. The type and only species of *Trichenoplus*, *T. antarcticus*, is therefore now called *Fenestrolaimus antarcticus* (Mawson 1956). Strange to say, no males have as yet been described for *Fenestrolaimus* species.

The genera of the Enoplaiminae:

Africanthion Inglis 1964
Cryptenoplus Riemann 1966
Enoploides Ssaweljev 1912
Enoplolaimus de Man 1893
Epacanthion Wieser 1953
   Hyalacanthion Wieser 1959, syn. Inglis 1966 : 89
Fenestrolaimus Filipjev 1027
   Trichenoplus Mawson 1956, syn. Lorenzen 1981 : 287
Filipjevia Kreis 1928
   Enoplonema Kreis 1934, an invalid substitution
Hyptiolaimus Cobb 1930
Mesacanthion Filipjev 1927
Mesacanthoides Wieser 1953
Metenoploides Wieser 1953
Oxyonchus Filipjev 1927
Paramesacanthion Wieser 1953

Parasaveljevia Wieser 1953
Parenoplus Filipjev 1927
Saveljevia Filipjev 1927

Anoplostomatidae   Gerlach & Riemann 1974

The following features are considered to be holapomorphies with which the holophyly of the Anoplostomatidae is established:

104.   a) The buccal cavity is spacious, toothless and surrounded by pharyngeal tissue only in the posterior section; this complex of features is unique within the Enoploidea.

105.   b) There is a cephalic capsule, but it is not an insertion point for musculature; this feature is also unique within the Enoploidea.

Additional features: The amphids are always situated well behind the cephalic capsule and have a small aperture and a relatively large fovea, in the adult females and juveniles at least; this difference in size is not so marked in other representatives of the Enoploidea. Only dorsolateral loxometanemes are present; they belong predominantly to type I and only a small number are of type II (Lorenzen 1981a). They all have a caudal filament. The labial region consists of three lips, each with two very short labial sensilla. Personal observations show that the pharyngeal glands open dorsally and ventrally directly posterior to the buccal cavity. The outline of the pharynx is smooth. It has not, as yet, been possible to establish the existence of a cervical gland; only a cervical pore has been observed (Wieser 1953 for *Anoplostoma camus*, *Chaetonema amphora* and *C. captator*). The males have two testes, the posterior one of which proceeds forwards a little and is then reflexed (personal observations). The females have two antidromously reflexed ovaries. Personal research indicates that the three caudal glands always penetrate into the pre-caudal region in *Anoplostoma vivipara* and *Chaetonema riemanni*; data in the literature to the contrary are dubious (Section 4.6.1.).

Discussion: the Anoplostomatinae and – this is new – the Chaetonematinae are placed in the Anoplostomatidae. Formerly, the Anoplostomatinae were seen as closely related to the Oncholaimidae and the Chaetonematinae as members of the "Enoplidae". The constant position of the gonads to the left of the intestine and the agreement in the remaining features listed justify the unification of the two subfamilies to form a single family of their own.

Jensen (1976) put *Pandolaimus* in the Anoplostomatidae; in the present study, the family Pandolaimidae is introduced within the Tripyloidina for this genus.

Anoplostomatinae   Gerlach & Riemann 1974

The holophyly of the Anoplostomatinae is established by the following holapomorphy:

106. Adult males have a copulatory bursa; this feature is unique within the Enoploidea. A pre-anal tubule is never present. The amphids are the same in both sexes.

Only genus: *Anoplostoma* Bütschli 1874

Chaetonematinae   Gerlach & Riemann 1974

The holophyly of the Chaetonematinae is established by the following holapomorphy: the amphids show extreme sexual dimorphism in their structure (discussion point 30). This sexual dimorphism was unknown previously and is illustrated in Fig. 11; it is unique within the freeliving nematodes. The sexual dimorphism remained undiscovered because the amphids in the males were mistaken for Steiner's organ and the amphids in adult females and juveniles were not recognized.

The Steiner's organ lies in the vicinity of the amphids (Fig. 11). In many species the 4 cephalic setae of the third sensilla circle are situated anterior to the 6 cephalic of the second sensilla circle. The adult males have a pre-anal tubule in only a few species.

Only genus: *Chaetonema* Filipjev 1927
  Dubious genus of the Anoplostomatidae:
  Donsinema Allgen 1949, dub. Hope & Murphy 1972 : 49

Phanodermatidae   Filipjev 1927

There is no known holapomorphy with which to establish the holophyly of the Phanodermatidae. Dorsolateral and ventrolateral orthometanemes are present, as well as a few loxometanemes of type I in *Phanoderma*. None of the metanemes has a caudal filament (Lorenzen 1981a). The amphids lie in the region of the cephalic capsule or directly posterior to it. The labial sensilla are always papilliform. Pharyngeal musculature inserts into the cephalic capsule. Scattered data in the literature and personal research on *Phanoderma* and *Crenopharynx* show that pharyngeal glands open dorsally and subventrally immediately posterior to the buccal cavity. The outline of the pharynx, at least in the posterior part, is undulating; this undulation is caused by the closely packed sequence of glandular and muscular areas. A cervical gland is usually present; it always lies completely within the pharyngeal body section. The males have two testes which face in opposite directions to one another, and the females have two antidromously reflexed ovaries. The adult males of most species have a pre-anal tubule; this is absent in *Crenopharynx* and *Phanodermopsis*. Data in the literature and personal research on *Phanoderma campbelli* and *P. necta* show that caudal glands always penetrate into the pre-caudal body region, within *Phanoderma* at least. It is uncertain whether this feature also holds good in the remaining genera.

Discussion: For a number of species of the Phanodermatidae outside the genus *Phanoderma* it has been reported that the caudal glands lie completely within the tail. The author was not able to detect the position of the caudal glands in the two male *Crenopharynx marioni* available to him. Since, among other things within the Thoracostomopsidae, the position of the caudal glands was often recorded incorrectly (Section 4.6.1.), this feature needs to be re-examined.

Phanodermatinae Filipjev 1927

The holophyly of the Phanodermatinae is established by the following feature which is considered to be a holapomorphy:

107. The buccal cavity contains one smaller dorsal and two larger subventral teeth which all point forwards. Personal observations on *Phanoderma campbelli* and *P. necta* show that pharyngeal glands do not open through the teeth, so that the latter cannot be seen as homologous with the onchia of the Thoracostomopsidae. The structure of the teeth is unique within the Enoplida and is thus considered to be a holopomorphy of the Phanodermatinae. It is difficult to determine with any certainty whether the teeth protrude freely into buccal cavity or whether they only represent thickenings of the wall of the buccal cavity; Inglis (1964 : 271) explicitly stresses the validity of the first interpretation.

Discussion: Personal observations show that *Phanodermopsis necta* fulfills the genus diagnosis of *Phanoderma* and is thus put in this genus. The species is therefore now called *Phanoderma necta* (Gerlach 1957).

Only genus of the Phanodermatinae:

Phanoderma Bastian 1865
  Heterocephalus Marion 1870, syn Marion 1975 : 500
  Cophonchus Cobb 1920, syn Filipjev in Kreis 1926 : 157
  Phanodermina Allgen 1939, syn Gerlach 1962 : 86–87
    Phanodermatina Allgen 1942, an invalid substitution for *Phanodermina*

Subgen *Alyncoides* Wieser 1953
Subgen *Phanoderma* Bastian 1865

Crenopharynginae Platonova 1976

Crenopharyngidae Platonova 1976
Dayellidae Platonova 1976, syn. Lorenzen 1981 : 291

There is no known holapomorphy with which to establish the holophyly of the Crenopharynginae. The buccal cavity is always small and does not contain teeth typical of those of *Phanoderma*. The cephalic capsule is no more than a narrow ring.

Discussion: Platonova (1976) put only the genera *Crenopharynx* and *Nasinema* in the Crenopharynginae. The taxon is not acceptable in this form becuse its holophyly lacks foundation just as much as that of the remaining "Phanodermatinae". Indeed the holophyly of the Crenopharynginae with its extended range, likewise cannot be established, though it can be established for the Phanodermatinae with its new, reduced range. For this reason both sub-families are accepted. The "Dayellidae" was introduced by Platonova (1976 : 142) as a new family only in the key.

The author considers this family to be synonymous with the Crenopharynginae because the holophyly cannot currently be established.

The genera of the Crenopharynginae:

Crenopharynx Filipjev 1934
Dayellus Inglis 1964
Klugea Filipjev 1927
  Subgen *Klugea* Filipjev 1927
  Subgen *Nasinema* Filipjev 1927
    Gullmarnia Allgen 1929, syn. Filipjev 1934 : 9
Micoletzkyia Ditlevsen 1926
Paraphanoderma Inglis 1971
Phanodermella Kreis 1928
Phanodermopsis Ditlevsen 1926
  Galeonema Filipjev 1927, syn. Filipjev 1927 : 202

Anticomidae   Filipjev 1918

There is no known holapomorphy with which to establish the holophyly of the Anticomidae. In this family the metanemes are very variable in form (Lorenzen 1981a): in *Anticoma acuminata* there are only dorsolateral loxometanemes which are predominantly of type II and only in a few cases of type I; in *A. trichura* both dorsolateral and ventrolateral loxometanemes of type I occur; and finally, in *Paranticoma tubuliphora* there are only dorsolateral metanemes, two of which are orthometanemes and about ten of which are loxometanemes of type I. The buccal cavity is small and is surrounded by pharyngeal tissue. The cephalic capsule is very narrow; pharyngeal musculature inserts into it. The pharyngeal glands open dorsally and subventrally immediately posterior to the buccal cavity. The outline of the pharynx is smooth. In most cases the cervical gland is very extensive and lies on the left of the pharynx in *Anticoma* and dorsal to the pharynx in *Paranticoma* (personal observations). The cervical pore is situated on a spine-like projection in *Paranticoma*; such a pore is absent in the remaining genera of the Anticomidae. The males have two testes or a single, anterior testis and the females have two antidromously reflexed ovaries. The adult males of most species have a pre-anal tubule; it is absent in *Anticoma allgeni*, *A. strandi*, *Anticomopsis typicus*, *Antopus serialis* and *Paranticoma caledoniensis*. Data in the literature and personal research show that the caudal glands lie completely in the tail.

Discussion: The Anticomidae first received family status from Hope & Murphy (1972 : 6). Before that they were regarded as a sub-family of the Leptosomatidae on account of the smooth outline of the pharynx wall and despite the existence of a pre-anal tubule in adult males; the latter is found nowhere else in the Leptosomatidae but is wide-spread in the Enoploidea. The constant position of the gonads to the left of the intestine and the existence of the pre-anal tubule in adult males justifies the separation of the Anticomidae from the Leptosomatidae and their classification in the Enoploidea. Here they are probably particularly closely related to the likewise non-holophyletic Phanodermatidae on account of the development of the cephalic region.

Hope & Murphy (1972 : 6) removed *Barbonema*, *Parabarbonema* and *Tubolaimella* from the Leptosomatidae and placed them in the Anticomidae, and Platonova (1976 : 112) considered the Barbonematinae, Platycominae and the Parabarbonematinae, which are not more closely defined, to be sub-families of the Anticomidae. The genera named have already been put back into the Leptosomatidae by Gerlach & Riemann (1974), and the sub-families named are dealt with in the same way in the current work. The reason is this, that the males do not have a pre-anal tubule in the genera and sub-families concerned and the position of the gonads relative to the intestine shows intraspecific variation in *Platycoma* (= *Parabarbonema*).

The genera of the Anticomidae:
Anticoma Bastian 1865
  Stenolaimus Marion 1870, syn. Cobb 1891 : 773
Anticomopsis Micoletzky 1930
Antopus Cobb 1933
Cephalanticoma Platonova 1976
Odontanticoma Platonova 1976
Paranticoma Micoletzky 1930

IRONOIDEA de Man 1876

There is no known holapomorphy with which to establish the holophyly of the Ironoidea. The Ironidae, Leptosomatidae and Oxystominidae are classed in this super-family.

Ironoidae de Man 1876

The holophyly of the Ironidae is established by the following complex of features which is considered to be holapomorphic:

108. On the anterior edge of the buccal cavity there are three or four movable teeth; pharyngeal glands do not open through these teeth (the pharyngeal glands open further back in the buccal cavity, see Chitwood 1960 for *Trissonchulus reversus*, van der Heiden 1975 for *Ironus tenuicaudatus, Thalassironus jungi, Trissonchulus oceanus* and *Dolicholaimus* sp.; personal observations on *Ironus ignavus, Parironus bicuspis* and *Trissonchulus acutus*). Where three teeth are present, one is situated dorsally and two subventrally, and all three teeth are more or less equal in length; where four teeth are present, two smaller teeth are situated dorsally and two larger teeth subventrally. In juvenile animals the teeth for the following stages are in pharyngeal pouches behind the functioning teeth. The ability to move the teeth in a radial way and the disposition of new teeth behind the functioning teeth in juvenile animals is unique within the Enoplida, so that this complex of features is considered to be a holapomorphy of the Ironidae. Van der Heiden (1975) has analysed the construction and mode of function of the buccal cavity structure in *Ironus, Thalassironus, Trissonchulus* and *Dolicholaimus*.

The following feature is also characteristic of the Ironidae: of the 6 + 4 cephalic sensilla the 4 of the third sensilla circle are usually predominant, only in *Thalassironus* are the 6 cephalic sensilla of the second sensilla circle dominant. Since all cephalic sensilla are papilliform in *Dolicholaimus, Pheronus, Trissonchulus* and *Syringolaimus*, a phylogenetic assessment of the feature of the frequent dominance of the posterior 4 cephalic sensilla is rejected.

The following features are strikingly different within the Ironidae:
a) In *Ironus* there are dorsolateral and ventrolateral orthometanemes which are very delicately built; in *Parironus* there are only dorsolateral loxometanemes of type II which are very robust and in *Trissonchulus* and *Syringolaimus* no metanemes were found at all (Lorenzen 1981a).
b) Personal observations show that within species the position of the gonads relative to the intestine is completely variable in *Ironus* and *Syringolaimus* and completely constant (anterior and posterior gonad to the right of the intestine) within species in *Parironus* (Section 8.2.1.).
c) The pharyngeal gland extends right into the post-pharyngeal region of the body in *Syringolaimus* and, according to van der Heiden (1975 : 430) in *Ironus, Thalassironus, Trissonchulus* and *Dolicholaimus* too, whereas in *Parironus bicuspis* (personal observations) it is limited to the pharyngeal region of the body and is found constantly to the right of the pharynx. In *Syringolaimus* the cervical pore lies slightly frontal to the nerve ring, whereas in *Ironus, Thalassironus, Trissonchulus, Dolicholaimus* (according to der Heiden 1975 : 430) and *Parironus bicuspis* (personal observation) it is found at the same level as the cephalic sensilla.
d) In *Dolicholaimus, Parironus* partim and *Thalassironus* the caudal glands extend into the pre-caudal body region, and in *Ironus* and *Parironus bicuspis* (Table 6) they are situated completely in the tail.
e) *Ironus* is freshwater, whereas all the other genera are marine or, at the most, only penetrate into brackish waters.

Additional features of the Ironidae: The pharynx inserts, at least in some genera (e.g. *Thalassironus*), onto the body cuticle in the buccal cavity region. The females have two antidromously reflexed ovaries, seldom only one posterior ovary (*Trissonchulus oceanus* and *T. raskii*). The males either have two testes facing in opposite directions or a single anterior testis (see Section 8.2.2.).

Discussion: On account of the considerable differences within the Ironidae, a classification into sub-families is necessary. However, at present, only the holophyly of the Ironinae can be established.

Chitwood (1951 : 646), Riemann (1970a : 382), Gerlach & Riemann (1973 : 25) and Andrassy put *Syringolaimus* in the Rhabdolaimidae ("Araeolaimida") or else in the Ironidae. *Ironus* cannot be put in the Rhabdolaimidae, because the structure of the dorsal tooth is completely different in the Rhabdolaimidae and no subventral teeth are present. Indeed, in *Syringolaimus* the pharynx does thicken at the end into a bulb-like formation and, in this region at least, is lined on the inside by a thickened cuticle, as is found frequently within the

Rhabdolaimidae and Leptolaimina but never within the Enoplida; on the other hand, a buccal cavity is never present within the Leptolaimina, but is characteristic of the Ironidae. Since, at present, the buccal cavity of the Ironidae but not the pharynx of the Leptolaimina can be considered holapomorphic, *Syringolaimus* is returned to the Ironidae and here to the non-holophyletic Thalassironidae.

Ironinae   de Man 1876

The holophyly of the Ironinae is established by the following two holapomorphies:

a) The delicately built, dorsolateral and ventrolateral orthometanemes occur in a strictly alternating sequence (Lorenzen 1981a, discussion point 42); this feature is otherwise known only in the Tripylidae.
109.   b) The Ironinae are limnotic, whereas all other Ironidae are marine.

Discussion: In contrast to Andrassy (1976 : 200) and Coomans & van der Heiden (1979 6), only *Ironus* is put in the Ironinae because otherwise the holophyly of no other sub-family of the Ironidae can be established. Andrassy's characteristic feature for the "Ironinae" – that they have only 4 instead of 6 + 4 cephalic sensilla – is based on an error: all Ironidae have the full compliment of 6 + 4 cephalic sensilla; 6 are often overlooked.

Only genus of the Ironinae:

Ironus Bastian 1865
   Nanonema Cobb in Stiles & Hassal 1905 (for *Cephalonema* Cobb 1893, a homonym), syn. Micoletzky 1922a : 323–324

Thalassironinae   Andrassy 1976

Conliinae Coomans & van der Heiden 1979, syn. Lorenzen 1981 : 289

Discussion: The Coniliinae are seen as synonymous with the Thalassironinae because holophyly cannot be established for either of the two sub-families and because those genera which Andrassy (1976 : 200) and Coomans & van der Heiden (1979 : 6) included in the "Ironinae" are put into the newly understood Thalassironinae excluding *Syringolaimus*.

OUTLINE OF A PHYLOGENETIC SYSTEM 239

The genera of the Thalassironinae:
Conilia Gerlach 1956
Dolicholaimus de Man 1888
Ironella Cobb 1920
Parironus Micoletzky 1930
Pheronus Inglis 1966
Syringolaimus de Man 1888
Thalassironus de Man 1889
Trissonchulus Cobb 1920

Leptosomatidae Filipjev 1916

There is no known holapomorphy with which to establish the holophyly of the Leptosomatidae. The body size varies between about 2mm (*Leptosomella phaustra*) and about 50mm (*Thoracostoma* partim); the largest, freeliving, marine nematodes therefore belong to the Leptosomatidae. The labial sensilla are mostly papilliform; they are setiform only in *Barbonema flagrum* and *B. horridum*. The 6 longer and 4 shorter cephalic sensilla form a single circle; often the cephalic sensilla are only very short. The amphids are pocket-shaped. Personal observations on *Platycoma sudafricana, Deontostoma arcticum* and *Synonchus longisetosus* (see Lorenzen 1981a) show that they have large numbers of metanemes (up to 230), made up of orthometanemes and loxometanemes of type I. They are situated dorsolaterally and ventrolaterally (*Platycoma, Deontostoma*) or only dorsolaterally (*Synonchus*). A caudal filament is usually present. Many species have ocelli. The buccal cavity is narrow and may have tooth-like structures. Within species the cephalic capsule is very variable in form; pharynx muscles insert onto it. The pharyngeal glands open in the buccal cavity. The pharynx is always smooth in outline. A cervical gland and a cervical pore are known only in a few species (*Leptosomatum* partim, *Leptosomella, Platycomopsis, Syringonomus*); the cervical gland is always situated in the pharyngeal body section in these species. The females have two antidromously reflexed ovaries and the males, as far as is known, have two testes facing in opposite directions. Personal observations have shown that the position of the gonads relative to the intestine is variable in *Deontostoma* and *Platycoma*. Adult males never have pre-anal tubules, but often have sub-ventral and/or ventral pre-anal papillae. The majority of species have caudal glands. Where these are present, they always extend, as far as is known, into the pre-caudal body region. The family is marine.

Discussion: There have always been difficulties in distinguishing between the three famiies Leptosomatidae, Phanodermatidae and

Anticomidae, because in all three families there are species in which the adult males do not have pre-anal tubules, the cephalic capsule is weakly developed and the buccal cavity shows hardly any differentiation. Since Filipjev (1927) such species have been put in the Leptosomatidae, if the outline of the pharynx is undulating. For a long time the Anticomidae have also been put in the Leptosomatidae for the same reason. In the current work the Anticomidae are put in the Enoploidea on account of the constant position of the gonads to the left of the intestine and the existence of a pre-anal tubule in the males of many species. Possibly in the future, when further research has been carried out, it will be necessary to move individual species or genera within the three, hitherto non-holophyletic families Phanodermatidae, Anticomidae and Leptosomatidae.

The following division of the Leptosomatidae into sub-families originates from de Coninck (1965) and Platonova (1970 and 1976). It is unclear whether the sub-families are holophyletic or not.

Barbonematinae   Platonova 1976

Barbonema Filipjev 1927

Platycominae   Platonova 1976

Parabarbonematinae Platonova 1976, syn. Lorenzen 1981 : 300
Pilosinema Platonova 1976
Platycoma Cobb 1894
Platycomopsis Ditlevsen 1926
  Dactylonema Filipjev 1927, syn. Filipjev 1927 : 201
  Parabarbonema Inglis 1964, syn. Coles 1977 : 9
Proplatycoma Platonova 1976

Discussion: In Platonova (1976 : 113) the Parabarbonematidae only appear as a taxon in the diagnostic key. The synonymization of the Parabarbonematinae and the Platycominae follows logically from the synonymization of *Parabarbonema* and *Platycomopsis*.

Leptosomatinae   Filipjev 1916

Leptosomatides Filipjev 1918
Leptosomatum Bastian 1865
Leptosomella Filipjev 1927
Paraleptosomatides Mawson 1956
Syringonomus Hope & Murphy 1969
Tubolaimella Cobb 1933

Synonchinae Platonova 1970

Anivanema Platonova 1976
Corythostoma Hope & Murphy 1972
Eusynonchus Platonova 1970
Macroncus Inglis 1964
Paratuerkiana Platonova 1970
Sadkonavis Platonova 1979
Synonchoides Wieser 1956
Synonchus Cobb 1894
  Subgen *Fiacra* Southern 1914
  Subgen *Jaegerskioeldia* Filipjev 1916
  Subgen *Synonchus* Cobb 1894
Triaulolaimus Platonova 1979
Tuerkiana Platonova 1970

Thoracostomatinae de Coninck 1965

Deontostoma Filipjev 1916
Pseudocella Filipjev 1927
Thoracostoma Marion 1870
Triceratonema Platonova 1976

Cylicolaiminae Platonova 1970

Cylicolaimus de Man 1889
  Nudolaimus Allgen 1929, syn. Allgen 1935 : 15
Metacylicolaimus Stekhoven 1946
Paracylicolaimus Paranova 1970

Dubious genera of the Leptosomatidae:

Leptosomatina Allgen 1951, dub. Hope & Murphy 1972 : 49
Ritenbenkia Allgen 1957, dub. Hope & Murphy 1972 : 51

Oxystominidae Chitwood 1935

Oxystomatini Filipjev 1918; the name is invalid because the type genus *Oxystoma* Bütschli 1874 is a homonym for *Oxystoma* Dumeril 1806.

There is no known holapomorphy with which to establish the holophyly of the Oxystominidae. The combination of the following three feature complexes is characteristic of the family and is unique within the Enoplida:

a) The outstretched body is very thin at the anterior end (approximately 4–15 μm).
b) The narrow, tubular (Oxystomininae, Halalaiminae) or funnel-shaped buccal cavity (Paroxystomininae) does not contain teeth.
c) The 6 + 4 cephalic sensilla are very slender and are always in two separate circles; two separate circles is considered plesiomorphic within the nematodes (discussion point 4).

Additional features: Setiform labial sensilla are found in several genera (Section 8.1). Within species the amphids are unusually polymorphic. Only orthometanemes occur; these are situated either dorsolaterally and ventrolaterally (Halalaiminae) or only dorsolaterally (Oxystomininae) (Lorenzen 1981a). They have a very short caudal filament. The cephalic capsule is not particularly well developed; nevertheless the pharynx inserts onto the body cuticle in the region of the buccal cavity. Personal observations on *Oxystomina* and *Thalassolaimus* show that the pharyngeal glands open at some distance from the nerve ring. In the posterior half the pharynx shows glandular thickening and has an irregular, slightly undulating outline. A cervical gland has hitherto only been found in representatives of the Oxystomininae and Halalaiminae, but not in representatives of the Paroxystomininae; the cervical gland always lie completely in the pharyngeal body section. The males have two testes which face in opposite directions to one another, or only one anterior testis; the females have two antidromously reflexed ovaries or (Oxystomininae) only one posterior ovary. The position of the gonads relative to the intestine is variable (personal observation) (Section 8.2.1.). The caudal glands in the Halalaiminae lie completely in the tail and in the Oxystomininae and Paroxystomininae they extend into the pre-caudal body region.

Oxystomininae Chitwood 1935

The holophyly of the Oxystomininae is established by the following holapomorphy: only the posterior ovary is present (discussion point 47); this feature is otherwise, at the most, sporadic within the Enoplida (Section 8.2.2.).

It is possible that the following feature can also be used to establish the holophyly of the Oxystomininae: only dorsolateral orthometanemes are found. For the time being, a phylogenetic interpretation of the feature is rejected because members of the Paroxystomininae have not yet been studied for the presence of metanemes.

OUTLINE OF A PHYLOGENETIC SYSTEM

Discussion: Several changes are made with reference to Gerlach & Riemann's (1973/1974) checklist:

a) *Adorus* was formerly put in the Alaimidae on account of the supposedly alaimid-like appearance of its anterior end. According to Jensen (1979c : 86) and personal observations, *Adorus* fully resembles the genus *Oxystomina* in the pharyngeal position of the cervical gland, in the extension of the caudal glands into the precaudal region, in the structure of the buccal cavity and in the position of the 6 + 4 cephalic sensilla. Within this genus there are also species with papilliform cephalic sensilla (e.g. *O. unguiculata*), with the result that the genera *Oxystomina* and *Adorus* are considered to be synonymous. The former *Adorus* species are therefore now called *Oxystomina tenuis* (Cobb, in Thorne 1939) and *O. astridae* (Jensen 1979).

b) *Alaimonemella* was hitherto put in the Trefusiidae. Its type and only species, *A. simplex* Allgen 1935, is probably identical to *Litinium bananum* Gerlach 1956 above all because of the similarity in the peculiar structure of the tail. For this reason the author regards *Litinium* and *Alaimonemella* as synonymous so that *A. simplex* is now called *Litinium simplex* (Allgen 1935). On account of the inadequate and, as far as the amphids are concerned, misleading description of *L. simplex*, this species is defined as a *species inquirenda* by the author.

c) *Nemacoma* is seen as synonymous with *Halalaimus*.

d) *Porocoma* is placed in the Xenellidae (Trefusiida).

e) *Thalassoalaimus aquaedulcis* is placed in *Aulolaimus* (Chromadorida, Aulolaimidae).

f) *Angustinema* is defined as a dubious genus of uncertain affinities.

The genera of the Oxystomininae:

Litinium Cobb 1920
 Alaimonemella Allgen 1935, syn. Lorenzen 1981 : 304
Nemanema Cobb 1920
Oxystomina Filipjev 1921
 for *Oxystoma* Bütschli 1874, a homonym
 Oxystomatina Stekhoven 1935 is an invalid emendation of *Oxystomina*
 Adorus Cobb in Thorne 1939, syn. Lorenzen 1981 : 304
 Asymmetrica Kreis 1929, syn. Hope & Murphy 1972 : 9
 Nemanemella Filipjev 1927, syn. Wieser 1953 : 33
 Oxystomella Filipjev 1946, syn. Hope & Murphy 1972 : 9
 Schistodera Cobb 1920, syn. Filipjev in Kreis 1926 : 157, *nomen oblitum*
Thalassoalaimus de Man 1893
Wieseria Gerlach 1956

Halalaiminae de Coninck 1965

The holophyly of the Halalaiminae is established by the following holapomorphy: The aperture of the amphids is represented by a groove which runs longitudinally, and a pocket is practically non-existent (discussion point 31); the occurrence of this complex of features, both in juveniles and in adults, is unique within the Enoplida. It otherwise only occurs in adult males of *Chaetonema* (Anoplostomatidae).

Discussion: The following alterations are made *vis-a-vis* Gerlach & Riemann (1974) and Juario (1974):

a) *Nemacoma* was previously put in the Oxystomininae, but it resembles *Halalaimus* in the possession of two ovaries and other characteristics. The two genera are thus seen as synonymous. The type and only species of *Nemacoma*, *N. borealis*, is therefore now called *Halalaimus borealis* (Steiner 1916). The caudal end in both *H. borealis* and *H. similis* Allgen 1930 swells in a characteristic way to form a ball. Since the two species resemble each other in other features, *H. similis* is thus declared a synonym of *H. borealis*.

b) Recently *Nuada* and *Pachydora* have been regarded as sub-genera of *Halalaimus*. These two sub-genera and the sub-genus *Halalaimus* differ from one another only in differing proportions, e.g. whether the 6 + 4 cephalic setae are situated close or less close together, whether the amphids are situated immediately or not quite immediately posterior to the head end and whether the amphids are slightly thicker or slightly thinner. Such a division of the genera creates an artificial impression and is therefore not accepted.

c) *Nualaimus* was introduced as a sub-genus of *Halalaimus* and is thought to differ from the other *Halalaimus*-species because of its setiform labial sensilla. Since the type species of *Halalaimus*, *H. gracilis*, also has setiform labial sensilla according to the original description, *Nualaimus* must, therefore, be a synonym of *Halalaimus*.

Only genus of the Halalaiminae:
   Halalaimus de Man 1888
      Halalaimoides Cobb 1933, syn. Hope & Murphy 1972 : 9
      Nemacoma Hope & Murphy 1972 (for *Acoma* Steiner 1916, a homonym), syn. Lorenzen 1981 : 306
      Nuada Southern 1914, syn. Stekhoven 1935 : 21
         Tycnodora Cobb 1920, syn. Gerlach & Riemann 1974 : 474
      Nualaimus Juario 1974, syn. Lorenzen 1981 : 306
      Pachydora Wieser 1953, syn. Lorenzen 1981 : 306

OUTLINE OF A PHYLOGENETIC SYSTEM

Paroxystomininae de Coninck 1965

There is no known holapomorphy with which to establish the holophyly of the Paroxystomininae. The occurence in adult males of preanal supplements arranged in two subventral rows is very characteristic.

Discussion: Subventrally situated pre-anal supplements are also common in the Leptosomatidae. In particular, the structurally complex pre-anal supplements of *Paroxystomina* resemble those of *Thoracostoma trachygaster* (Leptosomatidae) in structure and arrangement; those of many Eurystomininae (Enchelidiidae) are also similar, though they are arranged in a single, ventral row.

The genera of the Paroxystomininae:
Maldivea Gerlach 1962
Paroxystomina Micoletzky 1924

ONCHOLAIMACEA

As is the case in most Enoplacea, the caudal glands penetrate into the pre-caudal region of the body, whereas the cervical glands, unlike those of the Enoplacea, extend into the post-caudal region of the body. The second feature is considered plesiomorphic within the freeliving Adenophorea (discussion point 62).

Since the Oncholaimacea only consist of the holophyletic super-family of the Oncholaimoidea, they are a holophyletic group.

ONCHOLAIMOIDEA Filipjev 1916

The holophyly of the Oncholaimoidea is established by the following three features which are considered to be holapomorphies:
a) The anterior and posterior gonads always lie on the righthand side of the intestine (discussion point 53); this feature occurs outside the Oncholaimoidea no more than sporadically in the freeliving nematodes.
b) In species of all the larger sub-families of the Oncholaimoidea there are always dorsolateral and ventrolateral orthometanemes which have a pronounced caudal filament and which are, irrespective of body size, delicately built (Lorenzen 1981a, discussion point 38); this complex of features is unique within the Enoplida.
c) Personal observations have shown that, in all the species listed in Section 8.2.1, the cervical gland lies to the right of the intestine (discussion point 63); this feature otherwise only occurs sporadically

in the Enoplina (personal observation). The cervical gland is only seldom absent. In cases where the cervical gland and the anterior gonad are found together, the former is displaced to the ventral side.

The following complex of features is also unique within the Enoplida:

d) The buccal cavity is spacious and is surrounded in pharyngeal tissue only in the posterior section, the pharynx inserts onto the cephalic region and not onto the body wall, and the pharyngeal glands open dorsally and subventrally through onchia which are replaced by flat, cuticular structures in the Pelagonematinae and possibly also in the Krampiinae. No attempt is made at a phylogenetic assessment of this complex of features, because an analysis is still necessary to determine whether the rudiments of pharyngeal tissue at the posterior end of a spacious buccal cavity should generally be considered plesiomorphic or apomorphic within the freeliving Adenophorea.

Additional features of the Oncholaimoidea: The cuticle is smooth (only in *Oncholaimoides* is it striated). The labial sensilla are usually papilliform; they are setiform in *Pontonema ardens* and *Ditlevsenella* partim. The 6 longer and the 4 shorter cephalic sensilla are usually short and setiform, they are never jointed and, in adult animals at least, they are always arranged in a single circle. In many cases the amphids are non-spiral and pouch-shaped; they are dorsally spiral in many Enchelidiidae. The lips are usually merged into one another. Of the three onchia, one of the two subventral onchia is usually the largest; less often the two sub-ventral onchia are equal in size and are bigger than the dorsal onchium (*Meyersia, Metaparoncholaimus, Wiesoncholaimus,* Pontonematinae); only extremely seldom are all three onchia equal in size (*Oncholaimus keiensis, O. leptos*). In some taxa of the Enchelidiidae the pharynx has an undulating outline. There are two ovaries or a single, anterior ovary; extremely rarely is only a posterior ovary present (*Calyptronema sabulicola*). The males generally have two testes, seldom only one posterior testis (*Adoncholaimus thalassophygas*). The adult males of many species possess one ventral or two subventral rows of pre-anal supplements; a copulatory bursa is seldom present (*Oncholaimelloides, Oncholaimellus*). The Oncholaimoidea are marine.

Discussion: In extent, the Oncholaimoidea correspond most fully to the "Oncholaimidae" of Chitwood & Chitwood (1950) and to the "Oncholaimina" of de Coninck (1965). *Anoplostoma* (Anoplostomatidae) and the Dioncholaiminae (formerly Oncholaimidae) are no longer included in the super-family. In the current study, the Anoplostomatidae are put in the Enoploidea because of the constant position of the gonads to the left of the intestine, and the Dioncholaiminae have been a synonym of the Mononchinae (Dorylaimida, Mononchidae) ever since Andrassy (1976 : 203) declared *Dioncholaimus* to be a synonym of *Mononchus*.

The Oncholaimoidea consist of two families, the Oncholaimidae and the Enchelidiidae. Inglis (1974 : 299) doubted that they were more closely related to one another and regarded their similarity as "results of massive convergence". The three holapomorphies listed above, all of which are based on completely new characteristics, strengthen the circumstantial evidence for a close relationship between the Oncholaimidae and Enchelidiidae.

The author has not succeeded in establishing the holophyly of the Oncholaimidae or the Enchelidiidae. The reasons are listed as the following:

a) The ability or inability to move a subventral onchium as a systematic criterion: according to Filipjev (1934 :), the ability to move a subventral onchium is typical of the Enchelidiidae, whereas the inability to move all three onchia is typical of the Oncholaimidae. Within the Enchelidiidae one subventral onchium can, in fact, be moved (Fig. 32), but neither the present author, nor other authors have been able to establish the ability to move in a subventral onchium within the Eurystomininae. Consequently de Coninck (1965 : 662) states that, in the Enchelidiidae (erroneously referred to as Eurystominidae) the dominant, subventral onchium is very strongly developed and is only sometimes moveable. However, one of the two subventral onchia are also very strongly developed in the Oncholaimidae, so that in this form the feature cannot be used to differentiate between the two families.

b) Dominance of one of the two subventral onchia as a systematic criterion: it has already been mentioned that one of the two subventral onchia is larger than the remaining onchia in many Oncholaimidae and all Enchelidiidae. Scattered references in the literature and personal observations suggest that there probably exist inherent laws at the genus level and higher, as to whether the right or the left subventral onchium is the larger of the two. This feature has hitherto played no role as a criterion in the systematics of the Oncholaimoidea. In the course of the present work a study was made of the onchia of 16 species of the following 10 genera of the Oncholaimoidea in which one of the two subventral onchia is larger than the other one: *Oncholaimelloides, Viscosia* (both Oncholaimellinae), *Adoncholaimus* (Adoncholaiminae), *Oncholaimus* (Oncholaiminae), *Belbolla, Ditlevsenella, Eurystomina, Pareurystomina* (all Eurystomininae), *Calyptronema* and *Enchelidium* (both Enchelidiinae). From the literature and the authors' own results, the following picture is presented: where a

subventral onchium is found to dominate within the buccal cavity, the left subventral onchium always dominates in the Oncholaiminae and the right subventral onchium always dominates in the Oncholaimellinae, Adoncholaiminae, Eurystomininae, and Enchelidiinae. Only one exception was found during the author's investigations: in *Calyptronema maxweberi* either the right subventral or the left subventral onchium is the largest. According to Ditlevsen (1918 : 204), the left subventral onchium also seems to be the biggest in *Ditlevsenella danica* (described as *Oncholaimus demani*). The author only studied 40 individuals from 16 species, so only after further investigations will it be possible to establish whether the inherent laws suggested above are a reliable systematic criterion or not.
c) The outline of the pharynx as a systematic criterion within the Oncholaimoidea: Chitwood & Chitwood (1950 : 23) and de Coninck (1965 : 650) both separate the Oncholaimidae from the Enchelidiidae by the criterion that the outline of the pharynx is smooth in the former and it is usually undulating in the latter. This criterion allows for the existence of Enchelidiidae with a smooth pharyngeal outline; and the pharynx of many Eurystomininae do in fact have a smooth outline. Thus the criterion of the pharyngeal structure is not suitable for establishing the holophyly of one of the two families of the Oncholaimoidea.
d) The demanian organ is a genito-intestinal connection found in adult females and within the freeliving nematodes it only occurs among the Oncholaimidae. Rachor (1969) analysed the organ in detail. Within the Oncholaimidae the structure of this organ varies greatly; so far it has been found in representatives of the Oncholaimellinae, Adoncholaiminae and Oncholaiminae, but not in representatives of the Pelagonematinae, Krampiinae, Pontonematinae and Octonchinae. There is no analysis available to indicate whether the absence of the demanian organ within the last four sub-families named is probably primary or probably secondary. As a result the feature cannot, at least for the moment, be used as a systematic criterion to establish the holophyly of the Oncholaimidae, but at most, could be used to establish the holophyly of an indeterminate part of the Oncholaimidae.

Since the holophyly of the Oncholaimidae and Enchelidiidae cannot yet be established, one must not ignore the possibility that the Oncholaimoidea should be divided into more than two families. This is what Andrassy (1976) has done. He has divided the "Oncholaimina", which are identical with the Oncholaimoidea in the current work, into the "Pelagonematoidea" (only family: "Pelagonematidae"), "Enchelidioidea" (three families: "Eurystominidae", "Enchelidiidae", "Belbollidae") and "Oncholaimoidea" (two families: "Mononcholaimidae", "Oncholaimidae"). The "Pelagonematoidea" are said to

OUTLINE OF A PHYLOGENETIC SYSTEM 249

differ from the "Oncholaimoidea" due to sexual dimorphism in the formation of the buccal cavity of the former: the buccal cavity in the females is said to be wide and in the males reduced. This is an error. The buccal cavity in the "Pelagonematoidea" is the same in both sexes. This destroys the most important difference between the two taxa. The second difference put forward by Andrassy is connected with the onchia, which are small or absent altogether in the "Pelagonematoidea" and are said to be large in the "Oncholaimoidea". This gradual size difference cannot currently be assessed from a phylogenetic point of view and does not justify putting the two taxa in the same rank. Therefore Andrassy's division of the "Oncholaimina" into three sub-families does not solve the problem of dividing the Oncholaimoidea into phylogenetic sub-taxa of the highest possible rank. In the following, the "Pelagonematoidea" are once again seen as a sub-family of the Oncholaimidae, as was commonly accepted before.

In the following, the families, sub-families and genera of the Oncholaimoidea are listed, to a large extent, as they are in Gerlach & Riemann's (1973/1974) checklist. At the beginning of the discussion there are already indications of reductions in the extent of the taxon.

Oncholaimidae Filipjev 1916

There is no known holapomorphy with which to establish the holophyly of the Oncholaimidae.

Discussion: De Coninck (1965) was the first to divide the family with its many species into sub-families. Of all the seven sub-families thus produced, the "Oncholaiminae" remained large and heterogenous. Gerlach and Riemann (1974), using above all the results of Rachor (1969), have divided the large number of species, which de Coninck attributed to the "Oncholaiminae", "Oncholaimellinae" and "Mononcholaiminae", into sub-families in a completely new way. Andrassy (1976) has not taken up this classification, but, apart from shifts in rank, has adhered to de Coninck's concept of the sub-families; in particular, all the genera listed by de Coninck and some genera introduced later have been put in the "Oncholaiminae". As thus understood the "Oncholaiminae" were divided by Belogurov and Belogurova (1978) into tribe and sub-tribe. They based their work to a large extent on the results of Rachor (1969), as did Gerlach & Riemann (1974), but, unlike the latter, once again isolated *Viscosia* ("Oncholaiminae") from *Mononcholaimus* and *Oncholaimellus* (outside the "Oncholaiminae"), without giving any reasons. Gerlach & Riemann (1974 : 561) have expressly pointed out the close relationship between the three genera named, with the result that this is one of the two reasons for not taking Belogurov and Belogurova's systematic division into consideration in the following. The second reason: the systematic division of a sub-family into tribe and sub-tribe is not common practice in nematology at the moment, and can only be justified when the holophyly of the sub-families concerned has already been established. This has not yet been achieved.

Pelagonematinae de Coninck 1965

Curvolaiminae Adrassy 1976, syn. Lorenzen 1981 : 313

Onchia are small or absent altogether. The females usually have two ovaries, only seldom a single, anterior ovary (*Curvolaimus*). A demanian organ has not yet been found and is therefore probably absent.

Discussion: According to Andrassy (1976 : 190) the "Curvolaiminae" (only genus: *Curvolaimus*) are characterized by having a single posterior ovary. This is an error, thus making the foundation of the sub-family invalid.

Andrassy (1973b and 1976) put *Thalassogenus* in the Pelagonematinae. According to Jensen (1976 : 262) the genus belongs to the "Tripyloidea or even Mononchida". In the current work, *Thalassogenus* is put in the Monochidae because of the structure of the buccal cavity, the pharynx and the tail.

Belogurov (1977) considers *Bradybucca* to be closely related to *Pontonema* (Pontonematinae) and creates the tribe "Bradybuccini". Since no onchia have been described for the type species of *Bradybucca*, *B. rhopolurus*, the author includes the genus in the Pelagonematinae.

The genera of the Pelagonematinae:
  Anoncholaimus Cobb 1920
  Bradybucca Stekhoven 1956
    for *Anoncholaimus* Stekhoven 1950, a homonym
  Curvolaimus Wieser 1953
  Pelagonema Cobb 1894
    Cavilaimus Wieser 1951, syn. Wieser 1953 : 102
  Pelagonemella Kreis 1932
  Phaenoncholaimus Kreis 1934
  Pseudopelagonema Kreis 1932
  Vasculonema Kreis 1928

Krampiinae de Coninck 1965

Onchia are described for some species but not for others (among them the type specimen). Only an anterior ovary is present. A demanian organ has not yet been found.

Only genus:
 Krampia Ditlevsen 1921
  Conolaimella Allgen 1930, syn. Stekhoven in Kreis 1934 : 105

Oncholaimellinae de Coninck 1965

Mononcholaiminae de Coninck 1965, syn. Gerlach & Riemann 1974 : 561

As far as is known, the right subventral onchium is always bigger than the remaining onchia, if the others occur at all. The females usually have two ovaries, seldom only one anterior ovary (*Oncholaimelloides*). A simple demanian organ is known in *Oncholaimellus* and *Viscosia*; the simplicity of the organ is described as primitive by Rachor (1969 : 159).

Discussion: In the following, *Mononcholaimus* and *Meroviscosia* are described as synonyms of *Viscosia*. Reason: according to Riemann (1966 : 203), the type species of *Mononcholaimus*, *M. elegans*, is very similar to the type species of *Viscosia*, *V. viscosa*. Riemann shows convincingly that, in *V. viscosa*, the dorsal and left subventral onchia are often difficult or even impossible to detect. This gives one the impression that one is in fact dealing with a *Mononcholaimus* species. The difference between the type species of *Viscosia* and the type species of *Mononcholaimus* is therefore so slight, that the division of the two genera is not acceptable. According to Kreis (1934 : 192), the main difference between *Meroviscosia* and *Viscosia* is that, in the former, the buccal cavity extends posteriorly only half as far dorsally as it does ventrally, with the result that dorsal onchium lines the posterior half of the dorsal wall of the buccal cavity. The buccal cavity of *Viscosia* can be interpreted in exactly the same way, with the result that there is no difference between the genera.

The former *Mononcholaimus* species are therefore now called: *Viscosia bandaensilis* Lorenzen 1981 – for *Viscosia bandaensis* (Kreis 1932), a subjective homonym of *Viscosia bandaensis* Kreis 1932, *V. brevidentata* (Vitiello 1967), *V. conicaudata* (Kreis 1932), *V. diodon* (Wieser 1951) (described as *Oncholaimellus diodon*), *V. elegans* (Kreis 1924), *V. filiformis* (Kreis 1932), *V. gabriolae* (Allgen 1951), *V. glaberoides*

(Allgen 1932), *V. keiensilis* Lorenzen 1981 – for *Viscosia keiensis* (Kreis 1932), a subjective homonym of *Viscosia keiensis* Kreis 1932, *V. klatti* (Allgen 1941), *V. longidentata* (Stekhoven & Adam 1931) (described as *Oncholaimus longidentatus*), *V. norvegica* (Allgen 1946), *V. papillata* (Kreis 1932), *V. parasetosa* (Kreis 1932), *V. profunda* (Vitiello 1970), *V. rustica* (Kreis 1929) (already put in *Viscosia* by Lorenzen 1974a: 319), *V. separabilis* (Wieser 1953), *V. setosa* (Kreis 1932), *V. tasmaniensis* (Allgen 1927), *V. viscosula* Lorenzen 1981 – for *Viscosia viscosa* (Allgen? 1930), a subjective homonym of *Viscosia viscosa* (Bastian 1865).

The only *Meroviscosia* species is now called *Viscosia longicaudata* (Kreis 1932).

As a result of *Viscosia* and *Mononcholaimus* being declared synonymous, *Viscosia papillata* Chitwood 1951 becomes a subjective homonym of *Viscosia papillata* (Kreis 1932) (described as *Mononcholaimus papillatus*) and is therefore now called *Viscosia papillatula* Lorenzen 1981.

Cacolaimus is taken from the Oncholaiminae and put in the Oncholaimellinae, because, according to Kreis (1934: Fig 122a), the right subventral onchium is larger than the left subventral onchium. The type species of *Cacolaimus*, *C. papillatus*, is described from a juvenile animal, with the result that the number of ovaries is unknown.

The genera of the Oncholaimellinae:

Cacolaimus Kreis 1932
Oncholaimelloides Timm 1969
Oncholaimellus de Man 1890
Oncholaimoides Chitwood 1937
Viscosia de Man 1890
  Meroviscosia Kreis 1932, syn. Lorenzen 1981 : 317
  Mononcholaimus Kreis 1924, syn. Lorenzen 1981 : 317
  Steineriella Allgen 1932 (for *Steineria* Ditlevsen 1928, a homonym), syn. Hope & Murphy 1972 : 15)
  Steineriella Kreis 1934, a later and therefore invalid substitution for *Steineria* Ditlevsen 1928

Adoncholaiminae Gerlach & Riemann 1974

Either the two subventral onchia are more or less the same size (*Meyersia*) or the right subventral onchium is larger than the other onchia (the remaining genera). The females always have two ovaries. The demanian organ is always formed and, according to Rachor

(1969 : 157), it reaches the highest level of development in this subfamily.

The genera of the Adoncholaiminae[1]:

Adoncholaimus Filipjev 1918
Kreisoncholaimus Rachor 1969
Metoncholaimoides Wieser 1953
Meyersia Hopper 1967

Oncholaiminae   Filipjev 1916

As far as is known, the left subventral onchium is almost always larger than the other onchia: very rarely are all three onchia the same size (*Oncholaimus keiensis, O. leptos*). The females always have a single, anterior ovary (contradictory data in the literature are dubious, see Section 8.2.2). A demanian organ is only present in some of the species and, according to Rachor (1969), is absent altogether in other species.

The genera of the Oncholaiminae:

Fotolaimus Belogurova and Belogurov 1974
Metaparoncholaimus de Coninck & Stekhoven 1933
Metoncholaimus Filipjev 1918
Oncholaimus Dujardin 1845
  Oncholaimium Cobb 1930, syn. Rachor 1969 : 137
Prooncholaimus Micoletzky 1924
Pseudoncholaimus Kreis 1932
Wiesoncholaimus Inglis 1966

Pontonematinae   Gerlach & Riemann 1974

Both subventral onchia are approximately the same size and are larger than the dorsal onchium. The females always have two ovaries. Neither Rachor (1969) nor other authors have been able to find a demanian organ.

Discussion: Personal observations on *Pontonema ditlevseni* indicate that the *Convexolaimus* species are nothing more than early juvenile

---

[1] Belogurov & Belogurova (1978 : 29) also mention the genus *Admirandus* Belogurov & Belogurova 1978. The genus is a nomen nudum, because the pertinent study (Belogurov & Belogurova 1978a) does not constitute a publication.

stages of *Pontonema* (Fig 4a-b). This proves the synonymy of the two genera. The two *Convexolaimus* species are therefore now called *Pontonema filicaudatum* (Kreis 1928) and *P. teissieri* (Vitiello 1967) and are defined as *species inquirendae* because they were only described from very early juvenile stages.

The genera of the Pontonematinae:
Filoncholaimus Filipjev 1927
Pseudoparoncholaimus Kreis 1932, syn. Kreis 1934 : 161
Pontonema Leidy 1855
Convexolaimus Kreis 1928, syn. Lorenzen 1981 : 319
Paroncholaimus Filipjev 1916, syn. Cobb & Steiner 1934 : 57

Octonchinae   de Coninck 1965

There are many subventral teeth. The only species known so far was described from what appears to be a young female, with the result that data on the number of ovaries and the presence of a demanian organ are unavailable.

Only genus:
Octonchus Clark 1961
for *Polydontus* Schulz 1932, a homonym

Enchelidiidae   Filipjev 1918

Enchelidiinae Filipjev 1918
Symplocostomatinae Filipjev 1918, syn. Chitwood 1960 : 373
Belbollinae Andrassy 1976, syn. Lorenzen 1918 : 319
Eurystomininae Chitwood 1935, syn. Lorenzen 1981 : 319
Pareurystomininae Andrassy 1976, syn. Lorenzen 1981 : 319
Thoonchinae Gerlach & Riemann 1974, syn. Lorenzen 1981 : 319

There is no known holapomorphy with which to establish the holophyly of the Enchelidiidae. The amphids are, in part, non-spiral (at least within the genus *Calyptronema*) and, in part, dorsally-spiral (at least within the genera *Belbolla, Ditlevsenella, Eurystomina, Megeurystomina, Pareurystomina* and *Polygastrophora* – data in the literature and personal observations). As far as is known, the right subventral onchium is usually larger than the remaining onchia in the buccal cavity, only seldom is the left subventral onchium larger (known with certainty only in the species *Calyptronema maxweberi*, in which either the right or the left subventral onchium is the biggest – personal observations). It appears that the dominant onchium can be extended

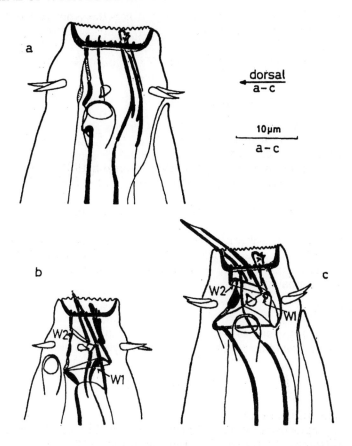

Fig. 32: *Calyptronema maxweberi* (Enchelidiidae). The main onchium can be everted from the buccal cavity, but the other subventral onchium, which is probably used for catching hold of prey, cannot. The amphids are non-spiral. a and c are drawn from adult females, and b from a juvenile. From Lorenzen (1969: 235).

out of the buccal cavity at least within the genera *Calyptronema* (Fig. 32), *Polygastrophora* and *Symplocostoma*, but probably not in *Ditlevsenella* (Fig. 33), *Eurystomina*, *Megeurystomina*, *Pareurystomina* and *Thoonchus*; the feature still needs analysis. In *Calyptronema*, *Symplocostoma* and *Polygastrophora* partim, the buccal cavity and the pharynx are sexually dimorphic, in that the organs concerned are reduced in the males and no longer perform their original function. The females almost always have two ovaries, very rarely only the posterior ovary (*Calyptronema sabulicola*). A demanian organ is absent according to all former observations. The adult males have two ventrally situated pre-anal supplements with a very characteristic structure in *Belbolla* partim, *Eurystomina, Ledovitia, Pareurystomina* and probably also in *Megeurystomina* (males hitherto unknown).

Discussion: Ever since Filipjev (1934), the group of species which are included in the Enchelidiidae have generally been divided into the Eurystomininae and Enchelidiinae. Gerlach & Riemann (1974) and Andrassy (1976) are the first to have taken on a more extensive classification. The author has been unable to establish the holophyly of any of the resulting sub-families. Therefore, the synonyms listed above are adopted instead. More detailed reasoning is given in the following section.

The systematic status of *Belbolla* and *Polygastrophora* has always caused a great deal of trouble: in both genera the posterior section of the pharynx consists of a row of bulb-like, muscular swellings. The latter are unique and can, therefore, be interpreted as circumstantial evidence for the close relationship of the two genera. However, *Belbolla* is put in the Eurystomininae because of the absence of sexual dimorphism in the formation of the buccal cavity and the pharynx and because of the presence of the two, characteristic, pre-anal supplements, whereas *Polygastrophora* is put in the Enchelidiinae because of the sexual dimorphism in the structure of the buccal cavity and the pharynx in one of several known species (in *P. quinquebulba*) and because of the absence of the two characteristic pre-anal supplements. If the sub-division of the Enchelidiidae is to be retained in the way shown – into the Eurystomininae and Enchelidiinae – and the two sub-families are to be regarded as holophyletic, then one must draw the following conclusions which, according to the present state of knowledge, cannot be shown to be likely:

a) The pharyngeal bulbs have developed independently within the two sub-families.
b) Since sexual dimorphism in the structure of the buccal cavity only appears in some of the *Polygastrophora* species (it is definitely not present in *P. omercooperi*), it must have evolved independently at least twice within the Enchelidiinae, if the holophyly of the *Polygastrophora* is not to be questioned.
c) The dorsally-spiral amphids must have evolved independently within the two sub-families, although the feature is unique within the Enoplida.

The problems that arise are not solved by the introduction of the Bolbollinae.

OUTLINE OF A PHYLOGENETIC SYSTEM 257

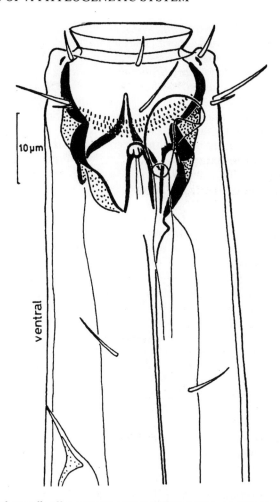

Fig. 33: *Ditlevsenella* aff. *murmanica* (Enchelidiidae). The amphids are dorsally spiral and slightly displaced towards the dorsal side. It was previously unknown that the labial region forms a rounded edge. The buccal cavity is dominated by the right, subventral onchium. Adult female, sublittoral sand, Kattegat, 29th July 1976.

Thoonchus, Ditlevsenella (Fig. 33; both "Thoonchinae") and the "Pareurystomininae" are so similar to *Eurystomina* in the structure of the labial region, the buccal cavity and the amphids, that a division of the "Thoonchinae" and "Pareurystomininae" from the "Eurystomininae" is not justifiable.

Within the genus *Calyptronema*, Wieser (1953b : 145) makes a distinction between the sub-genera *Calyptronema* (=*Catalaimus*) and *Dilaimus*, whereas Gerlach & Riemann (1974 : 613) distinguish between the sub-genera *Calyptronema* (=*Dilaimus*) and *Catalaimus*. Both divisions are rejected because Wieser (1953 : 165) regards the type species of the three genera, *Calpytronema paradoxum*, *Catalaimus acuminatus* and *Dilaimus pauli* to be synonymous with one another. Subsequently the species has been called *Calyptronema acuminatum*. The synonymy of the three accompanying sub-genera follows automatically from the synonymy of the three type species.

The genera of the Enchelidiidae:
Bathyeurystomina Lambshead & Platt 1979
Belbolla Andrassy 1973 for *Bolbella* Cobb 1920, a homonym
  Bolbellia Gerlach & Riemann 1974, a later and therefore invalid substitution for *Bolbella* Cobb 1920
Calyptronema Marion 1870
  Bradystoma Stekhoven 1943, syn. Wieser 1953 : 164
  Catalaimus Cobb 1920, syn. Wieser 1953 : 145
  Dilaimus Filipjev in Kreis 1926, syn. Gerlach & Riemann 1974 : 613
  Rhinoplostoma Allgen 1929, syn. Wieser 1953 : 145
Ditlevsenella Filipjev 1927
Enchelidium Ehrenberg 1836, dub. Wieser 1953 : 158
  (the genus is not listed below under "Dubious genera of the Oncholaimoidea", because it is the type genus of the Enchelidiidae)
Eurystomina Filipjev 1921
  for *Eurystoma* Marion 1870, a homonym
  Marionella Cobb 1922, a later and therefore invalid substitution for *Eurystomina* Marion 1870
  Gerlachystomina Inglis 1926, syn. Warwick 1969 : 396
Ledovitia Filipjev 1927
Lyranema Timm 1961
Megeurystomina Luc & de Coninck 1959
Pareurystomina Micoletzky 1930
Polygastrophora de Man 1922
Symplocostoma Bastian 1865
  Amphistenus Marion 1870, syn. Marion 1875 : 500
  Isonemella Cobb 1920, syn. Filipjev 1927 : 184
  Lasiomitus Marion 1870, syn. Wieser 1953 : 164
  Parasymplocostoma Schulz 1931, syn. Stekhoven 1935 : 59
Symplocostomella Micoletzky 1930
Thoonchus Cobb 1920

Dubious genera of the Oncholaimoidea:
Asymmetrella Cobb 1920, dub. Hope & Morphy 1972 : 48
Doryonchus Kreis 1932, dub. Chitwood 1960 : 356
Enchelidiella Allgen 1954, dub. Hope & Murphy 1972 : 49
Fimbrilla Cobb in Stiles & Hassal 1905 (for *Fimbria* Cobb 1894, a homonym), dub. Hope & Murphy 1972 : 49
Illium Cobb 1920, dub. Hope & Murphy 1972 : 49

TRIPYLOIDINA de Coninck 1965

There is no known holapomorphy with which to establish the holophyly of the Tripyloidina. The accompanying families, in contrast to the Enoplina, are singled out by the following plesiomorphic features: where a cervical gland occurs, it extends into the post-pharyngeal region of the body; the caudal glands are always situated completely in the tail; the pharynx inserts in the cephalic region and never in the body cuticle, so that, in particular, there is no cephalic capsule.

Discussion: Since the Tripyloidina represent the non-holophyletic remains after the separation of the holophyletic Enoplida, they are paraphyletic. De Coninck (1965 : 615) created the Tripyloidina as a sub-order of the "Araeolaimida" and only put the Tripyloididae in it.

In the following, the families Tripyloididae, Tobrilidae, Tripylidae, Triodontolaimidae, Rhabdodemaniidae and Pandolaimidae are put in the Tripyloidina.

TRIPYLOIDIDAE Filipjev 1928

The holophyly of the Tripyloididae is established by the following two features which are considered to be holophylies:

a) Where metanemes are present, they are almost exclusively ventrolateral loxometanemes of type II, while dorsolateral loxometanemes occur only rarely or are absent altogether (Lorenzen 1981a, discussion point 39); the predominance of ventrolateral metanemes is unique within the Enoplida.

b) Only the anterior testis is present (discussion point 44); this feature is constant within families otherwise only in the Rhabdodemaniidae and Pandolaimidae which on account of other features (structure of the buccal cavity and the amphids etc.) are not clusely related to the Tripyloididae.

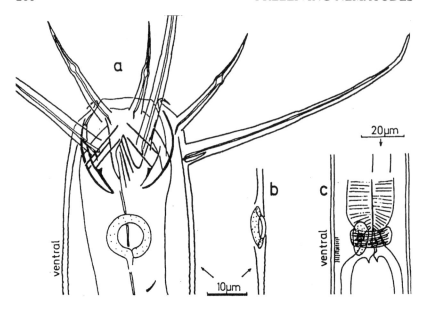

Fig. 34: *Gairleanema angremilae* Warwick & Platt, 1973 (Tripyloididae). a) anterior end of an adult male; b) amphid of another male, seen from the ventral side; c) posterior end of the pharynx and cardia of the first male. The drawings were made using the syntypes. Original.
Additional comment on the original description: the amphids are dorsally spiral and clearly lie posterior to the lateral cephalic setae; Warwick & Platt mistook a structure directly posterior to the lateral cephalic setae for amphids. Metanemes could not be detected. Only the anterior testis is present; it lies subventrally right or subventrally left of the intestine. The cardia is surrounded by a sphincter.

Additional features: The 6, often setiform labial sensilla and the 6 longer and the 4 shorter cephalic sensilla may be jointed; the 6+4 cephalic sensilla are always situated at the same level. The amphids are usually ventrally-spiral and seldom dorsally-spiral (*Gairleanema*). The labial region is divided into three lips. The buccal cavity often has three to four portions which lie one behind the other; there is seldom only one portion (*Gairleanema, Ingenia*). Teeth-like projections are common in the buccal cavity; a dorsal tooth is dominant in *Tripyloides amazonicus, T. acherusius, Ingenia* and *Gairleanema* (Fig 34). Personal observations on *Bathylaimus, Gairleanema* and *Tripyloides* show that a cervical gland is not developed; there is no data in the literature to suggest the contrary. The females possess two antidromously reflexed ovaries. The gonads lie ventral to the intestine.

OUTLINE OF A PHYLOGENETIC SYSTEM   261

The three caudal glands lie completely in the tail. The family is marine; only a very few species penetrate into freshwater (e.g. *Tripyloides acherusius*).

Discussion: Until now the systematic status of the Tripyloididae has remained unsettled. Filipjev (1918 : 180, 1934 : 14), Gerlach (1966 : 31, 32) and Gerlach & Riemann (1974 : 443) put the family in the "Enoplida" despite the spiral amphids and because of the similarity in head structure to the Tobrilidae. On the other hand, de Coninck & Stekhoven (1933 : 95), de Coninck (1965 : 615) and Andrassy (1976 : 119) put the family in the "Araeolaimida" because of the spiral amphids, whereas Chitwood & Chitwood (1937, 1950 : 22), who rejected the order "Araoelaimida" put the family in the "Chromadorina". Now the Tripyloididae belong definitively to the Enoplida because of the existence of metanemes.

Spiral amphids occur within the Enoplida otherwise only in the Enchelidiidae, in which, however, the fovea is never round. The Tripyloididae are very similar to the Tobrilidae in the structure of the oral cavity. *Gairleanema* (Fig. 34) in no way contradicts the family diagnosis of the Tripyloididae and is therefore removed from the "Enoplidae" and put in the Tripyloididae. This arrangement specifically implies the assertion that, in *Gairleanema*, metanemes are absent at a secondary and not at a primary level.

The genera of the Tripyloididae:

Arenasoma Yeates 1967
Bathylaimus Cobb 1894
  Bathylaimoides Allgen 1947, syn. Wieser 1956 : 30
  Cothonolaimus Ditlevsen 1919 (for *Macrolaimus* Ditlevsen 1918, a homonym) syn. de Man 1922 : 119
  Parabathylaimus de Coninck & Stekhoven 1933, syn. Gerlach 1951 : 206
Gairleanema Warwick & Platt 1973
Ingenia Gerlach 1957
Paratripyloides Stekhoven 1950
Tripyloides de Man 1886
  Nannonchus Cobb 1913, syn. Wieser 1956 : 53

<u>Tobrilidae</u>   de Coninck 1965

Trilobinae Filipjev 1918; the name is invalid because the type genus, *Trilobus* Bastian 1865 is a homonym
Baicalobrilinae Tsalolikhin 1976, syn. Lorenzen 1981 : 329
Kurikaniinae Tsalolikhin 1976, syn. Lorenzen 1981 : 329

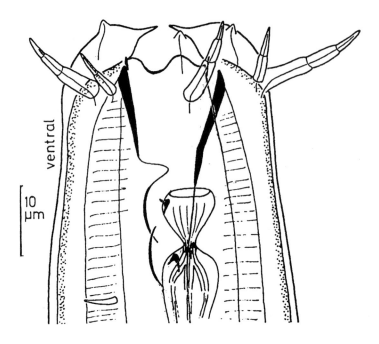

Fig. 35: Anterior end of *Tobrilus grandipapillatus* (Tobrilidae). It was previously unknown that jointed cephalic setae may occur within the family. In the buccal cavity there are two subventral teeth, each in a subventral pouch; there is no record of a dorsal tooth in the entire family. Adult male, Lake Selenter near Kiel (freshwater), 12th October 1975.

Quasibrilinae Tsalolikhin 1976, syn. Lorenzan 1981 : 329
Tobriloidinae Tsalolikhin 1976, syn. Lorenzen 1981 : 329

The holophyly of the Tobrilidae is established by the following two features which are considered to be holapomorphies:

110. a) The adult males have blister-like pre-anal supplements, which are unique within the Enoplida and which otherwise only occur in a similar form in *Echinotheristus* (Monhysterida, Xyalidae).
b) Where metanemes are present, dorsolateral loxometanemes of type II are found (Lorenzen 1981a, discussion point 41); this feature occurs within the Tripyloidina otherwise only in the Triodontolaimidae which, however, on account of other features, are not closely related to the Tobrilidae.

Additional features: The cuticle is smooth or striated. The 6 labial sensilla are always papilliform. The 6 longer and 4 shorter cephalic sensilla are usually situated in a single circle and seldom in two separate circles (*Tobriloides*); personal observations have shown that all 6+4 cephalic setae may be jointed (Fig. 35) which was hitherto unknown. The aperture of the non-spiral amphids is always considerably smaller than the fovea. The spacious buccal cavity possesses a large, anterior chamber without teeth and, posterior to that, a right subventral and a left subventral chamber each with a small subventral tooth which is only very rarely absent (*Quasibrilus kurikania*); a dorsal tooth is never present. The cardia consists of three spherical lobed cells which are generally interpreted as cardial glands. Personal observation and scanty references in the literature suggest that a cervical gland does not occur. The caudal glands lie completely in the tail. The number of caudal glands is given as two or three in the literature and it was difficult to establish the exact number during personal investigations. The females have two antidromously reflexed ovaries and the males have two testes which face in opposite directions to one another. Personal research has shown that the gonads lie somewhere between subventral and ventral to the intestine. The family is predomintly limnotic, only a few species are terrestrial.

Discussion: Until now the Tobrilidae have almost always been put together with the Tripylidae in a single family ("Tripylidae"). The two families do indeed show similarities in the pocket-shaped amphids, the presence of a cardia with many cell nuclei (up to 100 nuclei, according to Chitwood & Chitwood 1950 : 90), the ventral to subventral position of the gonads and the freshwater habitat. However, none of the features listed can, as far as is known, be used to establish the holophyly of the Tripylidae and Tobrilidae as a single unit; in particular there has only been very sporadic research carried out on a few nematodes into the number of cell nuclei in the cardia, and an abundance of nuclei in this organ is also known to exist in representatives of the Mononchidae (Dorylaimida) (Chitwood & Chitwood 1950 : 90). The differences between the Tobrilidae and the Tripylidae are not insignificant: the Tobrilidae never have a dorsal tooth, whereas most Tripylidae do; the Tobrilidae only have dorsolateral metanemes of type II, whereas the Tripylidae have both dorsolateral and ventrolateral orthometanemes; and finally, the structure of the pre-anal papillae and the structure of the cuticle are completely different in the two families. Due to these differences it seems unlikely that the two families are closely related and, in agreement with Tsalolikhin (1976 : 351), they are therefore both given the rank of family.

In the present study, *Tobrilia* is removed from the Tobrilidae and put in the Rhabdolaimidae (Leptolaimina) (reasons to be found in the new systematic location). Tsalolikhin (1976) has described five species of Tobrilidae from Lake Baikal. These are all new species, and four new genera and three new sub-families have been created for them. Two additional sub-families have been created for *Tobrilus* and *Tobriloides*. The differences between the five sub-families are so small that they can, at the most, be used to distinguish between genera. The author therefore adopts the synonyms listed above. The tails of the species from Lake Baikal have unusually slender end sections in comparison with the body width in the anal region which raises the suspicion that the animals may have been badly squashed. It is possible that in the case of Tsalolikhin's genera, the characteristics, which above all concern the buccal cavity, are based on damaged specimens.

The genera of the Tobrilidae:

Baicalobrilus Tsalolikhin 1976
Kurikania Tsalolikhin 1976
Paratrilobus Microletzky 1922
Quasibrilus Tsalolikhin 1976
   Lamuania Tsalolikhin 1976, syn. Lorenzen 1981 : 332
Tobriloides Loof 1973
Tobrilus Andrassy 1959
   for *Trilobus* Bastian 1865, a homonym

<u>Tripylidae</u>   de Man 1876

The holophyly of the Tripylidae is established by the following two features which are considered to be holapomorphies:

a) Where metanemes are present they occur in an alternating sequence of dorsolateral and ventrolateral orthometanemes which are very delicately built and only possess a short caudal filament (Lorenzen 1981a, discussion point 42); this feature is unique within the Tripyloidina and occurs in a similar form otherwise only in *Ironus* (Ironidae).

111.   b) At the end of the narrow buccal cavity there is a small dorsal tooth, through or near which the dorsal pharyngeal gland opens (Fig 36); a tooth of this kind is unique within the Tripyloidina. It is only absent in *Paratripyla*; this absence is interpreted as a secondary condition because the genus belongs to the Tripylidae on the basis of other features and because the opening of the dorsal pharyngeal gland is still recognizable in the corresponding place. In addition to the dorsal tooth, one or two tiny subventral teeth also occur in the buccal cavity. According to Chitwood & Chitwood (1937 : 518) pharyngeal glands also open subventrally into the buccal cavity; however, these openings have not been recorded by other authors, nor have they been found during personal observations.

OUTLINE OF A PHYLOGENETIC SYSTEM

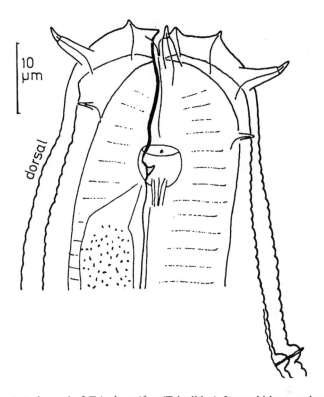

Fig. 36: Anterior end of *Tripyla setifera* (Tripylidae). It was hitherto unknown that the 6 cephalic setae may be jointed. In the narrow buccal cavity there is a dominant dorsal tooth and a tiny subventral tooth (the latter can be seen in the drawing through the aperture of the amphid). On the ventral side the anterior-most of 8 cervical papillae can be recognised. Adult male, Lake Selenter near Kiel (freshwater), 12th October 1975.

Additional features: Within the genus *Tripyla* at least, the outer layer of the cuticle surrounds the body of juvenile and adult animals like a cuticle that has not been shed. The labial sensilla are short and papilliform. The 6 longer and 4 shorter cephalic sensilla are usually arranged in two separate circles; they form a single circle in *Paratripyla* and *Trischistoma*. It was hitherto unknown that the 6 longer cephalic sensilla may be jointed, as long as they are setiform (Fig. 36). The non-spiral amphids are pocket-shaped. The wide cardia has many cell nuclei. A cervical gland has not yet been observed and is therefore probably absent. The females have two ovaries or only one anterior ovary and the males have two opposite facing testes or a single anterior testis. The ovaries are antidromously reflexed. Personal observations showed that the gonads lie subventral to ventral to the intestine (Section 8.2.1). The three caudal glands, which are only recognizable in good preparations, lie completely in the tail. The family is predominantly fresh water; only a few species are terrestrial.

Discussion: Formerly the Tripylinae, Tobrilinae and Monochromadorinae were usually put in the Tripylidae (see Gerlach & Riemann 1974). Judging from the present state of knowledge, it is very unlikely that the holophyly of these wide reaching "Tripylidae" or of only a joint group of two or three of the sub-families listed, could ever be justified. As a result the extent of the Tripylidae, as they are now understood, is limited to the former Tripylinae. The former Tobrilinae receive the rank of family, and the Monochromadorinae are put in the Rhabdolaimidae (Leptolaimina); these decisions are discussed in more detail in the new systematic locations of the respective groups.

The genera of the Tripylidae:

Abunema Khera 1971
Andrassya Brzeski 1960
Multidens Muchina 1978
Paratripyla Brzeski 1964
Tripyla Bastian 1865
    Promononchus Micoletzky 1923, syn. Micoletzky 1925 : 128
Trischistoma Cobb 1913
    Tripylina Brzeski 1963, syn. Brzeski 1965 : 449

Triodontolaimidae   de Coninck 1965

The holophyly of the Triodontolaimidae is established by the two following features which are considered to be holapomorphies:

112.   a) The buccal cavity contains three characteristic teeth which are equal in size; a pharyngeal gland opens at the base of each tooth.

b) Only dorsolateral loxometanemes of type II are present (Lorenzen 1981, discussion point 41); this feature occurs within the Tripyloidina otherwise only in the Tobrilidae which, because of other features, are, however, not more closely related with the Triodontolaimidae.

Additional features: There are 6 labial papillae and 6+4 cephalic setae which occur in a single circle and which are more or less equal in length. The non-spiral amphids are outstretched and pocket-shaped. The inner layer of the body cuticle is not differentiated in the cephalic region thus forming a cephalic capsule and the pharynx is never attached to the body cuticle in the cephalic region. The pharynx has an undulating outline in the posterior section. The cervical gland extends into the post-pharyngeal region of the body. The females possess two antidromously reflexed ovaries and the males have two testes that face in opposite directions. Pre-anal supplements are absent. The gonads lie ventral to the intestine. The three caudal glands lie completely in the tail. The family is marine.

Only genus: *Triodontolaimus* de Man 1893

Discussion: The only species of the Triodontolaimidae, *Triodontolaimus acutus*, was found and redescribed most recently by Lorenzen (1978b); prior to that it was last described by de Man in 1893. The recent redescription clearly indicated that, contrary to earlier assumptions (Filipjev 1934 : 11, de Conick 1965 : 651), the Triodontolaimidae are not particularly closely related to the Enoploidea or the Leptosomatidae.

Rhabdodemaniidae Filipjev 1934

The holophyly of the Rhabdodemaniidae is established by the following two features which are considered to be holapomorphic:

a) The amphids do not have an aperture or a fovea which is visible using a light microscope; instead, on either side a gently sinusoidal, fibrous band is faintly visible which runs subcutaneously as far as the nerve ring and which ceases to be sinusoidal in the posterior half (Fig. 10, discussion point 29). This structure was first found in the course of the current work and, as far as is known, it is unique within the free-living nematodes.

113. b) There are 6 teeth on the anterior edge of the buccal cavity (2 per sector of the buccal cavity) and 3 teeth in the middle region of the buccal cavity (Fig. 10). In combination with the absence of a cephalic capsule, this feature is unique within the Enoplida and is therefore considered to be a holapomorphy of the Rhabdodemaniidae. Within the Enoplida a similar dentation is otherwise only known in species of the Leptosomatidae (see e.g. *Deontostoma washingtonense*), but there is always a cephalic capsule present in these species. In the Rhabdodemaniidae the three teeth of the middle region of the buccal cavity are either equal in size or the dorsal tooth is larger than the two subventral teeth. Personal observations on *Rhabdodemania minor* failed to show any instance of pharyngeal glands opening through the teeth; on the contrary they open directly posterior to the buccal cavity, in fact dorsally slightly further posterior than subventrally.

The following three features occur both in the Rhabdodemaniidae and in the Pandolaimidae, so that a phylogenetic interpretation is rejected for the time being.

c) The lateral epidermal borders are unusually broad; in addition, personal observations have shown that, in the Rhabdodemaniidae, the epidermis cells in the lateral borders are bubble-like and turgescent in structure to an unusually marked extent which is unknown to the author in other representatives of the Enoplida.

d) Only a single, anterior testis is present.

e) There are only two caudal glands, both of which lie completely within the tail; contradictory literature is dubious.

Additional features: The cuticle is very thick. The labial sensilla are papilliform. The 6 longer and the 4 shorter cephalic setae are situated at more or less the same level; in some species the 4 cephalic setae (third sensilla circle) may even be situated anterior to the 6 sensilla of the second circle. There are dorsolateral and ventrolateral orthometanemes, both types have a caudal filament (Lorenzen 1981a). Personal observations showed that the cervical gland extend into the postpharyngeal body region. The females have two antidromously relfexed ovaries. The position of the gonads relative to the intestine is completely variable (personal observations) (section 8.2.1.). The family with its numerous species, is marine.

Discussion: Filipjev (1934 : 12) regarded the Rhabdodemaniidae as a family of a very uncertain systematic position. Wieser (1959 : 8), de Coninck (1965 : 658) and Platonova (1974), on the other hand, considered the family to be closely related to the "Enoplidae" (Enoplidae and Thoracostomopsidae). This view is untenable for the following reasons: unlike the Rhabdodemaniidae, no cephalic capsule is found in the "Enoplidae", the lateral epidermis borders are very narrow, the amphids and the metanemes are of a completely different structure and the latter are also arranged differently. As there is no circumstantial evidence to suggest that the Rhabdodemaniidae are descended from ancestors with cephalic capsules, they are put in the Tripyloidina. The similarities to the Pandolaimidae possibly indicate a close relationship between the two families.

Conistomella corresponds completely to *Rhabdodemania* in the structure of the head and the tail, so that both genera are now regarded as synonymous. Stekhoven (1942 : 236) only distinguished *Conistomella* from *Symplocostoma* (Enchelidiidae). The only *Conistomella* species is now called *Rhabdodemania brevicaudata* (Stekhoven 1942). It is herewith declared *species inquirenda* because it was only described from a juvenile and could be clearly recognized again.

Only genus of the Rhabdodemaniidae:

Rhabdodemania Baylis & Daubney 1926
for *Demania* Southern 1914, a homonym
Conistomella Stekhoven 1942, syn. Lorenzen 1981 : 339
Pendulumia Allgen 1954, syn. Hope & Murphy 1972 : 13

Pandolaimidae Belogurov 1980

Pandolaimidae Lorenzen 1981b, syn. nov.

There is no known holapomorphy with which to establish the holophyly of the Pandolaimidae. The combination of the following three complexes of features is unique within the Enoplida.

a) There are dorsolateral and ventrolateral orthometanemes, each with a very short caudal filament.

b) The 6+4 cephalic setae are more or less equal in length. They are each divided into two segments (personal observations) and they are arranged in two clearly separated circles.

c) The buccal cavity is spacious, toothless and only surrounded by pharyngeal tissue in the posterior section.

Additional features: The lateral epidermal borders are very broad. Amphids were not detected either by Jensen (1976) or by the author. There is no cephalic capsule, and the pharynx never inserts onto the body cuticle. The pharyngeal glands open dorsally and subventrally directly posterior to the buccal cavity. No cervical gland could be found and is therefore probably absent. The females have two antidromously reflexed ovaries and the males have a single, anterior testis. Personal observations showed that the gonads lie subventrally to ventrally to the intestine. There are only two caudal glands which according to personal observations, lie completely in the tail (Jensen's statement, 1976 : 258, that the caudal glands extend into the precaudal body region, is based on an error). This family has very few species and is marine.

Type and only genus:

Pandolaimus Allgen 1929
Allgenia Strand 1934 (for *Fimbriella* Allgen 1929, a homonym), syn. Jensen 1976 : 261
Filipjeviella Allgen 1935, syn. Jensen 1976 : 261
Filipjevinema Allgen 1953, syn. Gerlach 1963 : 650
Metapelagonema Sergeeva 1972, syn. Jensen 1976 : 261

Discussion: For a long time *Pandolaimus* was considered to be a member of the "Linhomoeidae". Only Jensen (1976) put the genus in the Enoplida, in the Anoplostomatidae. Personal research shows that *Pandolaimus* cannot be put in the Anoplostomatidae or in any of the other existing families of the Enoplida without thereby disturbing the holophyly of the family or super-family concerned. Therefore, a new family is created for the genus. Similarities with the Rhabdodemaniidae are noted in the discussion of this family. There are striking similarities with the Halanonchinae (Trefusiida, Trefusiidae) in the structure of the buccal cavity and in the form and distribution of the cephalic setae; however, the author was not able to find metanemes in any of the Trefusiidae, so that, for this reason, *Pandolaimus* is not put in the Halanonchinae.

5.3.3. Classification of the Trefusiida
TREFUSIIDA Lorenzen 1981

The holophyly of the Trefusiida cannot be established. The amphids are usually non-spiral; they are spiral in only part of the Trefusiidae, but in these cases the canalis lies laterally on the body. The 6+4 cephalic sensilla are setiform and arranged in two well separated circles except in *Trefusialaimus* (Trefusiidae) and the Lauratonematidae. Metanemes are absent in all the species so far studied for this feature (see Table 1). In most species it is unknown where the pharyngeal glands open; in *Onchulus* the dorsal pharyngeal glands open through the dorsal tooth, and in *Trefusialaimus* the dorsal gland and the subventral glands open in the vicinity of the buccal cavity. Where caudal glands are present they either lie completely in the tail, or their position is unknown.

Discussion: The Trefusiida belong to the Enoplida because the amphids are usually non-spiral and because even in spiral amphids the canalis lies on the lateral line of the body (discussion point 20). The members of the Trefusiidae are not put in the Enoplida nor in the Dorylaimida for the following two reasons:

a) The species studied of the Trefusiida do not possess either the metanemes typical of the Enoplida, or a dorsal pharyngeal gland which opens into the posterior half of the pharynx which is typical of the Dorylaimida. In addition the Trefusiida differ from the Dorylaimida in the form and position of the 6+4 cephalic sensilla and in the structure of the buccal cavity.

b) No subtaxon of the Trefusiida on the basis of any other feature forms a holophyletic group with subtaxa of the Enoplida or Dorylaimida.

The creation of the order Trefusiida is a provisional measure. The non-holophyletic status of the order indicates that the inter-relationships between the large taxa of the freeliving Enoplia is not yet fully understood.

Simpliconematidae   Blome & Schrage 1985

Within the Trefusiida, Blome & Schrage (1985) erected the new family Simpliconematidae with the single new genus *Simpliconema*. The description of the single species is based on a single male. It is characterized by the following characters: there are three lips, each provided with two labial sensilla; the pharyngeal glands seem to open in the frontal part of the pharynx; the cervical gland is located in the pharyngeal body region; metanemes are absent; the amphids are circular; the single testis is located left to the intestine. Because of the first four characters, the new genus is not placed in the Monhysterida, but in a new family in the Trefusiida. However, the authors do not mention a strong resemblance between *Linhystera* (Xyalidae, Monhysterida) and *Simpliconema*. Both genera have the same arrangement of cephalic sensilla, the amphids are circular, the cervical gland is situated in the pharyngeal body region, and the anterior testis is located left to the intestine. The finding of females of *Simpliconema* could clarify the systematic position of this genus.

Trefusiidae   Gerlach 1966

There is no known holapomorphy with which to establish the holophyly of the Trefusiidae. The cuticle is smooth. The labial sensilla are usually papilliform and only seldom setiform (*Rhabdocoma* partim). The 6+4 cephalic setae are usually arranged in two well separated circles; they are at the same level only in *Trefusialaimus*. The 6 anterior cephalic setae are jointed (Figs 2 and 3a–b). In some species, lateral setae which may be homologous to the dereids (discussion point 2) are situated posterior to the amphids. The amphids are either non-spiral (*Trefusia* partim, *Trefusialaimus*, *Rhabdocoma* partim, *Halanonchus*), ventrally-spiral (personal observations on *Rhabdocoma* sp. from the North Sea) or dorsally-spiral (personal observations on *Cytolaimium exile* and *Rhabdocoma americana*, both from the sublittoral of Chile). There are no metanemes in species of *Trefusia*, *Rhabdocoma* and *Halanonchus*, not even in animals 7mm in length (Lorenzen 1981a). The labial region is generally divided into three lips. The buccal cavity is conical (Trefusiinae) or barrel-shaped (Halanonchinae) and, as far as is known, always toothless. In *Trefusialaimus*, pharyngeal glands open dorsally and subventrally posterior to the buccal cavity; the position of the openings is unkown in the remaining taxa. A cervical pore may occasionally be found; proof of the accompanying cervical gland is missing so far. The males usually have two testes which face in opposite directions to one another, and the females usually have two ovaries; occasionally there is only an anterior testis (*Trefusialaimus*) and only a posterior ovary (*Rhabdocoma* partim, Halanonchinae). In most cases the ovaries are antidromously reflexed, they are outstretched only in *Cytolaimium exile*. The gonads lie subventral to ventral to the intestine. In adult males, ventrally situated (*Trefusia, Halanonchus*) or subventrally situated (*Cytolaimium, Trefusialaimus*) pre-anal and post-anal supplements are found (the latter only in a sub ventral position). It is unkown whether the caudal glands lie completely in the tail or not. The familly is marine.

Discussion: In the current work, *Alaimonemella* is removed from the Trefusiidae and put in *Litinium* (Oxystominidae). The Trefusiidae are characterized above all by plesiomorphies. Otherwise the family is rather heterogenous, particularly as regards the position of the 6+4 cephalic setae, the structure of the buccal cavity and the arrangement of the male supplementary copulation organs (papillae).

Trefusiinae   Gerlach 1966

Cytolaimium Cobb 1920
Rhabdocoma Cobb 1920
Trefusia de Man 1893
    Bognenia Allgen 1932, syn. de Coninck & Stekhoven 1933 : 44
Trefusialaimus Riemann 1974

Halanonchinae   Wieser & Hopper 1967

Halanonchus Cobb 1920
    Latilaimus Allgen 1933, syn. Gerlach 1964 : 23

Onchulidae   Andrassy 1963

The holophyly of the Onchulidae is established by the following feature which is considered to be holapomorphic:

114. a) The pharynx has an undulating outline, which is produced by the close succession of glandular and muscular regions; this feature, which is constant within families, only occurs within the Enoploidea. Since the two taxa are not more closely related to one another, the feature must have developed independently.

The following two complexes of features are also very characteristic:

b) The 6 longer and 4 shorter cephalic setae are arranged in two well separated circles and, in many species at least, the 6 anterior cephalic setae are jointed; both features are considered to be plesiomorphic within the Adenophorea (discussion points 4 and 5).

c) The very spacious buccal cavity contains either three teeth equal in size (1 dorsal, 2 subventral) or one large dorsal tooth and small subventral teeth. A phylogenetic assessment of the feature is currently rejected because very similar buccal cavities are found in the Diplogasteridae (Secernentea) (Fig. 12). As far as the author knows, no mention has previously been made of the, in part, very great similarities. Personal observations on *Onchulus nolli* have shown that the dorsal tooth does not develop before the shedding of the skin in juvenile animals. As least the dorsal pharyngeal gland probably opens through the accompanying dorsal tooth; it is not known where the subventral pharyngeal glands open.

Additional features: The labial sensilla are often short and setiform. The amphids are pocket-shaped and non-spiral, and they lie posterior to the posterior 4 cephalic setae. Metanemes are absent in *Onchulus* and *Stenonchulus* (personal observations). A cervical gland and a cervical pore have not yet been recorded and are, therefore, probably absent. The females have two antidromously reflexed ovaries and the males have two testes which face in opposite directions to one another. Personal observations have shown that the position of the gonads relative to the intestine is completely variable (Section 8.2.1). Adult males have ventrally situated pre-anal supplements which may extend into the neck region. Caudal glands have not yet been found and are, therefore, probably absent. The family is limnotic/terrestrial.

Discussion: The Onchulidae, together with the Prismatolaimidae, have until now been members of the family "Prismatolaimidae" (Filipjev 1934, Andrassy 1963, de Coninck 1965, Gerlach & Riemann 1974), without this ever being substantiated. There are only fleeting similarities between the two families; these similarities are restricted to the spaciousness of the buccal cavity, the amphids which lie well posterior, and the limnotic habits of the two families. However, the differences between the two families are quite considerable: in the Prismatolaimidae, in the newly understood sense, the 6+4 cephalic setae are situated in a single circle, the amphids are dorsally-spiral, at the most there is a small dorsal tooth at the end of the buccal cavity, the pharynx has a smooth outline and, as far as is currently known, the position of the gonads relative to the intestine is constant within species. What is more, according to Riemann (1977c : 224) the spicula in the Onchulidae become outstretched under hydrostatic pressure, and this is interpreted as primitive within the nematodes, and is otherwise only known in the Tripylidae (Enoplida), and in the Diphtherophoridae and Trichodoridae (Dorylaimida). In the current work, the Prismatolaimidae are put in the Leptolaimina (Chromadorida).

The genera of the Onchulidae:
Cyathonchus Cobb 1933
Kinonchulus Riemann 1972
Limonchulus Andrassy 1963
Onchulus Cobb 1920
Paronchulus Altherr 1972
Pseudonchulus Altherr 1972
Stenonchulus W. Schneider 1940

Laurathonematidae   Gerlach 1953

The holoyphyly of the Lauratonematidae is established by the following three features which are considered to be holapomorphies:

115. a) The vulva lies very close to the anus (*Lauratonemoides*) or together they form a cloaca (*Lauratonema*); both conditions are unique within the Enoplida. A cloaca in females is otherwise unknown in freeliving nematodes, but in parasitic nematodes it is found in Rondonia (Atractidae). The presence of a single, anterior ovary which is antidromously reflexed in the Lauratonematidae, is correlated with the position of the vulva, pushed well towards the posterior.

b) Personal observations show that the males have a single posterior testis (Fig. 16a, discussion point 45); this feature was hitherto unknown in the Lauratonematidae and, within the Enoplia, had otherwise only been observed in *Adoncholaimus thalassophygas* (Oncholaimidae, see Section 8.2.1).

c) The ovary (= anterior gonad) constantly lies to the left and the testis (= posterior gonad) constantly lies to the right of the intestine (discussion point 53); this feature is unique within the Enoplia.

Additional features: The cuticle is finely, but very clearly striated. The 6 labial sensilla are papilliform. The 6 longer and 4 shorter cephalic setae are arranged in a single circle. The amphids are non-spiral and pocket-shaped. Personal observations on *Lauratonema* failed to indicate the existence of metanemes. The buccal cavity is conical and toothless. The pharynx is very probably attached to the body cuticle in the region of the buccal cavity (personal observations);

there is, however, no cephalic capsule. It is not known where the pharyngeal glands open; the author was also unable to locate the openings. A cervical gland has also not been found and is, therefore, probably absent. The adult males do not possess pre-anal supplements. The three caudal glands lie completely in the tail. The family, which has few species, is marine.

Discussion: Within the Enoplia, the insertion of the foremost section of the pharynx onto the cuticle of the body is typical of the Enoplacea. Since the feature also occurs outside the Enoplida (e.g. in *Glochinema*: Epsilonematidae, see Lorenzen 1974b), it is not used as an argument for putting the Lauratonematidae in the Enoplacea. A fine, clear annulation is rare within the Enoplia; it occurs in a similar form in the Xenellidae and some representatives of the Halalaiminae (Oxystominidae).

The genera of the Lauratonematidae:

Lauratonema Gerlach 1953
Lauratonemoides de Coninck 1965

Xenellidae   de Conick 1965

116. The holophyly of the Xenellidae is established by the following feature which is considered to be holapomorphic: the finely and clearly striated cuticle has several longitudinal rows of cuticular elevations; longitudinal rows of decorations are very rare within the Enoplia; they are known in *Halalaimus longistriatus* (Oxystominidae), *Dolicholaimus marioni* (Ironidae) and *Encholaimus* (Dorylaimina). The feature occurs more frequently outside the Enoplia.

Additional features: There are 6 short labial sensilla and there are 6 + 4 cephalic setae which are in two separate circles. All cephalic setae are more or less equal in length. The amphids are non-spiral and are wide and pocket-shaped. Personal research on *Xenella* failed to reveal metanemes. The buccal cavity is very small. The pharynx expands glandularly at the end. It is not known where the pharyngeal glands open. In *Porocoma* the cervical pore lies on a setiform elevation. In *Xenella* there is only one anterior testis and, according to personal investigations, a single, posterior, antidromously reflexed ovary; in *Porocoma* there are two posterior ovaries, one of which however, extends forwards immediately from the vulva outwards, and then folds over posteriorly. As far as is currently known, the three caudal glands lie completely in the tail. The family is marine.

Discussion: The Xenellidae were previously put in the Chromadoria

– in the "Desmodorida". There is no argument in favour of such a classification. The Xenellidae are put in the Enoplia because of their non-spiral, wide, pocket-shaped amphids, and they are put in the Trefusiida because of the absence of metanemes, the setiform nature of the cephalic sensilla and their purely marine occurence. Reference is made to the similarity to the Oxystomininae as regards the tiny buccal cavity, the form and position of the 6 + 4 cephalic setae, the outstretched body and the presence of a single, posterior ovary: these similarities cannot currently be used as arguments for putting the Xenellidae in the Oxystominidae or somewhere close to them because, as yet, the holophyly of the Oxystominidae has never once been established (not even in the current work). *Porocoma* used to be put in the Oxystominidae; however, due to the similarities to *Xenella*, the genus very probably belongs to the Xenellidae. Within the freeliving nematodes, the location of the cervical pore on a setiform elevation otherwise only occurs in *Paranticoma* (Enoplida, Anticomidae) and *Punctodora exochopora* (Chromadorida, Chromadoridae).

The genera of the Xenellidae:

Porocoma Cobb 1920
Xenella Cobb 1920

5.3.4. Remarks on the Dorylaimida

Next to the Enoplida, the Dorylaimida are the second holophyletic order of the freeliving Enoplia. The holophyly of the Dorylaimida is established by the following feature which is considered to be holapomorphic: all pharyngeal glands open posterior to the nerve ring (discussion point 98). This feature is unknown in nematodes outside the Dorylaimida.

Additional features: The labial and cephalic sensilla are almost always papilliform; the few exceptions are listed in Section 8.1. The 6 + 4 cephalic sensilla are always situated at more or less the same level. The amphids are always non-spiral. Personal observations show that metanemes are absent in representatives of the Dorylaimina, Mononchidae, Alaimidae and Cryptonchidae (see Section 4.3.2.). A cervical gland has not yet been observed; only a cervical pore has been found within a few taxa. Caudal glands are present in the Mononchina, Bathyodontidae, Cryptonchidae and Mononchulidae; in each case they are situated completely in the tail. The males have two testes or only one anterior testis, and the females have two ovaries or only one anterior ovary or only one posterior ovary (see Section 8.2.2). The ovaries are always antidromously reflexed. The order is limnotic/

terrestrial and is only represented at a secondary level by some species in brackish surroundings.

Discussion: Contradictory views still predominate as to the extent of the Dorylaimida. Above all, arguments persist over the systematic position of the Alaimidae, Cryptonchidae and Isolaimiidae which have been put into the Dorylaimida by some authors and into taxa outside the Dorylaimida by other authors (see discussion in the families named). In the current work, the three families are put in the Dorylaimida because, for them as well as for the other taxa of the Dorylaimida, it has been proved that all pharyngeal glands open posterior to the nerve ring.

Within the Dorylaimida, the Dorylaimina is the sub-order with the most species. The sub-order is characterized by the fact that a subventral tooth has developed into a moveable spear and the pharynx consists of a narrow anterior section and a wide posterior section. The sub-order Mononchina possesses the second largest number of species within the Dorylaimida. Its membership of the Dorylaimida has not been contested since Clark (1961). The species of the Mononchina are characterized by a spacious, barrel-shaped buccal cavity in which there is at least one dorsal tooth which is never smaller than any subventral teeth that may be present. After the division of the Dorylaimida into the Dorylaimina and Mononchina, there remains a heterogenous residue of seven families with few species. These families have hitherto been divided into three sub-orders and one order in the following way:

Alaimina Clark 1971: Alaimidae
Bathyodontina Coomans & Loof 1970: Bathyodontidae, Cryptonchidae, Mononchulidae
Diphtherophorina Coomans & Loof 1970: Diphtherophoridae, Trichodoridae
Isolaimida Timm 1969: Isolaimiidae

The ranking of the four taxa into sub-orders and orders shows that the inter-relationships of the high-ranking taxa of the Dorylaimida are not yet fully understood. The author expresses the existing phylogenetic uncertainty in the following way: the seven families listed are united to form a single, non-holophyletic sub-order, for which the name Bathyodontina is chosen from the available names. According to Riemann (1977c : 224), the Diphtherophoridae and Trichodoridae are particularly primitive because of the structure of the body cuticle and the spicular apparatus (see also discussion on the Onchulidae).

BATHYODONTINA   Coomans & Leaf 1970

Bathyodontidae   Clark 1961

Bathyodontus Fielding 1950
  Mirolaimus Andrassy 1956, syn. Andrassy in Hopper & Cairns 1959 : 126

Discussion: Coomans & Loof (1970 : 194) also put *Cryptonchus* in the Bathyodontidae. If this were to be accepted then the Bathyodontidae would be a synonym of the Cryptonchidae because the latter family has already been introduced in 1937.

Cryptonchidae   Chitwood 1937

Cryptonchus Cobb 1913
  Cryptonchulus Sukul 1969, syn. Andrassy 1976 : 199
  Ditlevsenia Micoletzky 1929, syn. Filipjev 1936 : 105–106
  Gymnolaimus Cobb 1913, syn. Andrassy 1976 : 199

Discussion: *Gymnolaimus* corresponds exactly to *Cryptonchus* in the structure of the buccal cavity and the pharynx, and in the possession of a single, anterior ovary; the two genera differ only in the slightly different structure of the cardia. This difference, however, does not justify a generic division.

Coomans & Loof (1970 : 194) were the first to put *Cryptonchus* into the Dorylaimida, and here into the Bathyodontidae, on account of the structure of buccal cavity. Prior to this, the genus was put in the Enoplida, close to the Ironidae, by Chitwood & Chitwood (1950 : 24), Goodey (1963 : 391), de Coninck (1965 : 649) and Gerlach & Riemann (1974 : 441), and into the Tobrilidae by Filipjev (1934 : 13, listed under "Trilobinae"). In the current study the Cryptonchidae is put in the Dorylaimida because Coomans & Loof (1970 : 189) have proved that all pharyngeal glands open posterior to the nerve ring in *Cryptonchus*.

Mononchulidae   de Coninck 1965

Mononchulus Cobb 1918
Oionchus Cobb 1913
  Enoplocheilus Kreis 1932, syn. Thorne 1935
Rahmium Andrassy 1973
  for *Stephanium* Rahm 1938, a homonym

OUTLINE OF A PHYLOGENETIC SYSTEM 279

Diphtherophoridae   Micoletzky 1922

Diphtherophora de Man 1880
Archionchus Cobb 1913, syn. Micoletzky 1922
Chaolaimus Cobb 1893, syn. Micoletzky 1922
Longibulbophora Yeates 1967
Tylolaimophorus de Man 1880
Triplonchium Cobb 1920, syn. Goodey 1963 : 514

Trichodoridae   Thorne 1935

Monotrichodorus Andrassy 1976
Paratrichodorus Siddiqi 1973
  Subgen *Atlantadorus* Siddiqi 1973
  Subgen *Nanidorus* Siddiqi 1973
  Subgen *Paratrichodorus* Siddiqi 1973
Trichodorus Cobb 1913

Isolaimiidae   Timm 1969

Isolaimium Cobb 1920

Discussion: *Isolaimium* was put in the "Plectidae" (Chromadorida) by Filipjev (1934 : 19), in the "Araeolaimida" by Goodey (1963 : 325) and by Gerlach & Riemann (1973 : 28), in the Dorylaimida by de Coninck (1965 : 669) and, finally, in the Mermithina by Andrassy. In the present study the Isolaimiidae are included in the Dorylaimida because, according to Coomans & Loof (1970 : 189), the pharyngeal glands open without doubt subventrally and probably dorsally in the posterior section of the pharynx, and because the structure of the buccal cavity, the pharynx and the male copulatory apparatus justifies such a classification.

Alaimidae   Micoletzky 1922

Alaimus de Man 1880
Amphidelus Thorne 1939
Etamphidelus Andrassy 1977
Paramphidelus Andrassy 1977

Discussion: For a long time *Adorus* was also put in the Alaimidae. In agreement with Andrassy (1976 : 195) and Jensen (1979 : 86), this genus is now put in the Oxystominidae.

The Alaimidae were put in the Dorylaimida by Filipjev (1934 : 15),

Thorne (1939 : 162), Goodey (1963 : 508) and de Coninck (1965 : 680), and in the Enoplida by Chitwood & Chitwood (1950 : 24), Gerlach (1966 : 34), Gerlach & Riemann (1974 : 455) and Andrassy (1976 : 196), whereas Clark (1961 : 134) ranked the "Alaimina" equal next to the "Enoplina" and "Dorylaimina". In the present study the family is included in the Dorylaimida because Clark (1962 : 119) has proved that all pharyngeal glands open posterior to the nerve ring.

MONONCHINA Jairajpuri 1969

The following two remarks should be made on the Mononchidae:

a) Andrassy (1976 : 203) has made *Dioncholaimus* Kreis 1932 synonymous with *Mononchus* Bastian 1865. *Dioncholaimus* was the type of the Dioncholaiminae de Coninck 1965. Therefore this sub-family is now a synonym of the Mononchidae Filipjev 1934.

b) In the present work *Thalassogenus* Andrassy 1973 is removed from the Oncholaimidae and put in the Mononchidae (see discussion on the Oncholaimidae).

5.4. Genera *incertae sedis* of the freeliving Adenophorea (excluding Dorylaimida)

The genera listed below cannot currently be classified as belonging to one of the existing families of the freeliving Adenophorea due to inadequate description of the type species. For this reason they must be regarded as dubious. The fact that they should belong to the Adenophorea is, in general, clearly recognizable. *Pseudorhabdolaimus*, on the other hand, probably belongs to the Rhabditida.

Acmaeolaimus Filipjev 1918, dub. Lorenzen 1981 : 354
Amblyura Ehrenberg 1828, dub. Filipjev 1918 : 13
Angustinema Cobb 1933, dub. Lorenzen 1981 : 354
Colpurella Cobb 1920, dub. Goodey 1963 : 326
Diplobathylaimus Allgen 1959, dub. Hope & Murphy 1972 : 48
Diplohystera de Cillis 1917, dub. Lorenzen 1981 : 354
Frostina Gerlach & Riemann 1974 (for *Frostia* Allgen 1952, a homonym), dub. Lorenzen 1981 : 354
Iotalaimus Cobb 1920, dub. Goodey 1963 : 521
Kreisia Allgen 1929, dub. Lorenzen 1981 : 354
Lineolia Gerlach & Riemann 1974 (for *Lineola* Kölliker 1845, a homonym), dub. Hope & Murphy 1972 : 49

Linolaimus Cobb 1933, dub. Hope & Murphy 1972 : 50
Litonema Cobb 1920, dub. Goodey 1963 : 521
Litotes Cobb 1920, dub. Hope & Murphy 1972 : 50
Longilaimus Allgen 1958, dub. Hope & Murphy 1972 : 50
Longitubopharynx Allgen 1959, dub. Hope & Murphy 1972 : 50
Macfadyenia Allgen 1953, dub. Goodey 1963 : 521
Nema Leidy 1856, dub. Bastian 1865 : 175
Notocamacolaimus Allgen 1959, dub. Hope & Murphy 1972 : 50
Notosouthernia Allgen 1959, dub. Hope & Murphy 1972 : 50
Nuadella Allgen 1928, dub. Hope & Murphy 1972 : 50
Phanoglene Nordmann 1840, dub. Bastian 1865 : 172
Potamonema Leidy 1856, dub. Bastian 1865 : 174
Pseudodilaimus Kreis 1928, dub. Lorenzen 1981 : 355
Pseudorhabdolaimus Soos 1937, dub. Lorenzen 1981 : 355
Rhabdotoderma Marion 1870, dub. Allgen 1942 : 13
Rhinonema Allgen 1928, dub. Hope & Murphy 1972 : 51

5.5. Remarks on the gross systematic classification of the freeliving Secernentea

The holophyly of the Secernentea is established by the following two features which are considered to be holapomorphic:

a) Phasmids are present (discussion point 59).
b) Caudal glands are absent (discussion point 61).

The remaining features which Chitwood & Chitwood (1950 : 12) and de Coninck (1965 : 591) list as characteristic of the Secernentea cannot currently be referred to as holapomorphies of this group; this applies in particular to the feature of the dereids (discussion point 2). In contrast to the Adenophorea, the Secernentea are in all probability primarily limnotic/terrestrial, for in cases where freeliving Secernentea live in marine surroundings they are found in the coastal region and have close relatives in freshwaters or in the earth, whereas conversely many of the freeliving Adenophorea are not represented in limnotic or terrestrial surroundings.

Filipjev (1934 : 29) distinguished between the "Anguillulidae" and "Tylenchidae" within the multitude of freeliving Secernentea. Technically this division is still retained; the two groups are called Rhabditida and Tylenchida. However, the extent of the two orders has been altered considerably since Filipjev. In particular, the Diplogasteroidea are no longer included in the Tylenchida, but have been put in the Rhabditida. In the light of known criteria and those acquired during the current work, the previous arguments for the gross systematic classification of the freeliving Secernentea are examined. The criteria acquired during the current work are as follows:

a) Formation of four additional cephalic setae or papillae in adult males as a criterion for the holophyly of the Diplogasteroidea (Fig. 12, discussion point 13). The feature was known.
b) Homodromously reflexed ovaries as a criterion for the holophyly of the Rhabditoidea. Previously no distinction was made between antidromously and homodromously reflexed ovaries (Fig. 15c–d, discussion point 50).
c) Antidromously reflexed ovaries as a plesiomorphy within the nematodes and therefore particularly within the Diplogasteroidea (discussion point 49).
d) Outstretched condition of the ovaries as a criterion for the holophyly of the Tylenchida (discussion point 51). The feature was unknown.
e) Anterior and posterior gonads on the same side of the intestine (both to the left or both to the right of the intestine) as a criterion for the holophyly of the Tylenchida (discussion point 53). The feature was almost unknown.
f) It is considered an apomorphy of the Rhabditida that, in most cases, the anterior gonad constantly lies to the right and the posterior gonad constantly lies to the left of the intestine within a given species (discussion 52). Personal observations have shown that in the Diplogasteridae, the testis lies ventral to the intestine, whereas the anterior ovary lies to the right and the posterior ovary lies to the left of the intestine, and in the Bunonematidae the position relative to the intestine is variable (see Section 8.2.1). The feature of the gonad position has been almost unknown for the Rhabditida too.

The holophyly of the Tylenchida is substantiated by an additional holapomorphy:

117. g) In the buccal cavity there is a characteristic spine, which is called a stomatostyle; all three sectors of the buccal cavity contribute to its formation.

From this list of criteria it emerges that the holophyly of the Rhabditoidea, Diplogasteroidea and Tylenchida can be established. Together, all three taxa represent almost all freeliving Secernentea: only for the Brevibuccidae, Chambersiellidae and Cylindrocorporidae are there no known criteria to suggest their classification in the Rhabditoidea or Diplogasteroidea. Character analyses have not produced any reliable criterion for joining two of the three groups Rhabditoidea, Diplogasteroidea and Tylenchida to form a holophyletic order. Only the circumstancial evidence of the gonad position was found that together the first two taxa probably form a holophyletic order which is named the Rhabditida: within the Rhabditoidea and Diplogasteroidea, as far as is currently known, the anterior gonad usually lies to the right and the posterior gonad to the left of the intestine, whereas, in the Tylenchida, as far as is currently known, the anterior and the posterior gonad always lie on the same side of the intestine. On account of this argument the Rhabditoidea and Diplogasteroidea are joined together to form the Rhabditida in the following, and only as a result of this decision do the Rhabditida and Tylenchida receive the rank of order. Before the gross systematic classification of the freeliving Secernentea is written down, the previous arguments which have been used for combining two of the three groups Rhabditoidea, Diplogasteroidea and Tylenchida, should be discussed.

Filipjev (1934 : 33) combined the Diplogasteroidea and the Tylenchida to form a single taxon because in both groups the pharynx has a pronounced bulb in the middle section and in the adjacent, posterior section it is glandular and only slightly muscular. In contrast the pharynx of the Rhabditoidea is very muscular in the posterior section and it has a pronounced end bulb with a valvular apparatus. Chitwood & Chitwood (1933 : 130 and 1950 : 80) considered the pharynx of the Rhabitoidea to be plesiomorphic and that of the Diplogasteroidea and Tylenchida to be apomorphic within the Secernentea. This provides support for Filipjev's decision.

It is common at the moment to combine Rhabditoidea and the Diplogasteroidea to form one order without giving any reasons for doing so. Probably some part is played in this by the arguments which, according to Ritter & Theodorides (1965 : 766), among other things support a close relationship between the Rhabditidae and Diplogasteridae; the species of both families show similarities in habitat, way of life and in the structure of the buccal cavity. Global similarities in habitat have no value in systematic argument. The similarities in the way of life refer to a widely scattered distribution in the earth, in fresh water and in rotting substances; there is no evidence for a family-typical specialization, so that this argument too cannot be used like a holapomorphy to support the union of the Rhabitoidea and Diplogasteroidea to form a single order. The similarities in the structure of the buccal cavity of the Rhabditidae and Diplogasteridae were worked out for *Rhabditis* and *Diplogaster* respectively by Osche (1952) and Weingärtner (1955). If the remaining families of the Rhabditida are included in the comparison, then the similarities are reduced to nothing more than the ability to homologize individual sections of the buccal cavity, and this can be used and has, in fact, been used for buccal cavities of taxa outside the Rhabditida. Therefore the similarities between the Rhabditoidea and Diplogasteroidea in the structure of the buccal cavity are very probably based only on plesiomorphy and thus cannot be used to establish the holophyly of the Rhabditida. Osche (1952 : 204) reports similarities between many *Rhabditis* and *Diplogaster* species in the number of bursal papillae in adult males. As far as the author is aware, the size of the taxon, the holophyly of which is possibly established by this feature, has not yet been analysed; the mere existence of bursal papillae is in any case probably only a plesiomorphic feature within the nematodes (discussion point 54). During studies using electron microscopes, Ward *et al.* (1975 : 327) have found 4 additional nerve endings, each with a sensory cilia, posterior to the 4 cephalic sensilla of the third sensilla circle in adult males of *Rhabditis*. These structures are very probably homologous with the 4 additional cephalic sensilla which occur regularly in adult males of the Diplogasteroidea (discussion point 12). As long as the feature is thought not to occur outside the Rhabditida, it can probably be used as a holapomorphy for establishing the holophyly of the order. Of course, assessment of the holapomorphy is still not possible because no attention has been paid to this feature in species outside the Rhabditida.

In conclusion, it appears that the holophyly of the Tylenchida is indeed well established, whereas the holophyly of the Rhabditida is not.

The gross systematic classification of the freeliving Secernentea:

RHABDITIDA

Rhabditoidea Örley 1880
  Rhabditidae Örley 1880
  Bunonematidae Micoletzky 1922
  Panagrolaimidae Thorne 1937
  Myolaimidae J Goodey 1963
  Cephalobidae Filipjev 1934
Diplogasteroidea Micoletzky 1922
  Diplogasterididae Micoletzky 1922
  Pseudodiplogasteroididae Körner 1954

Rhabditida *incertae sedis*:
  Brevibuccidae Paramonov 1956
  Chambersiellidae Thorne 1937
  Cylindrocorporidae T Goodey 1939

TYLENCHIDA Thorne 1949
Tylenchoidea Filipjev 1934
Aphelenchoidea Fuchs 1937

5.6 Summary of the outline for a phylogenetic system for freeliving nematodes

In the following, the outline, presented in this study, of a phylogenetic system for freeliving nematodes is represented graphically in figures 37–42. The names of the holophyletic taxa are underlined, in order to distinguish them from the non-holophyletic taxa. The numbers in the figures and in the captions refer to the discussion points, in which the holapomorphies for the establishment of the holophylies are to be found.

5.7 Concluding remarks: the attraction of an incomplete system

The system of freeliving nematodes, which is elaborated in the current work, is extremely incomplete for a scientific theory. In this respect it is similar to all previous systems of freeliving nematodes. In all probability, there will never be a scientifically complete system for the nematodes, because such a system would presuppose that the interrelationships between all nematode taxa were completely understood. This full understanding would be possible if, and only if;
   1) the system were to contain exlusively taxa, whose holophylies had been established, and if;
   2) all the branches within the phylogenetic tree were dichotomous.

The previous systems and that of the current work far from fulfill these two conditions. From figures 37–42 it is clear that, on the one hand, the holophyly of many taxa from the rank of sub-family to sub-class cannot be established (the names of these taxa are those that are not underlined in Figs 37–42), and on the other hand, many branches in the phylogenetic tree are not dichotomous but are in the form of clusters (the Leptolaimina in Fig. 39 are a particularly striking example of this).

286 FREELIVING NEMATODES

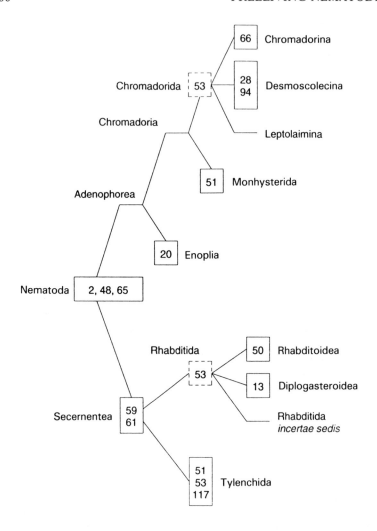

Fig. 37: The high level systematic inter-relationships within the freeliving nematodes.

2 (p.39): Nematoda: presence of 6+6+4 sensilla on the anterior end.
13 (p.46): Diplogasteroidea: development of 4 additional cephalic sensilla in adult males.
20 (p.64): Enoplia: amphids non-spiral.
28 (p.67): Desmoscolecina: amphids blistered.
48 (p.96): Nematoda: presence of a vulva.
50 (p.98): Rhabditoidea: ovaries homodromously reflexed.
51 (p.98): among others Monhysterida: ovaries outstretched.
51 (p.98): among others Tylenchida: ovaries outstretched.

OUTLINE OF A PHYLOGENETIC SYSTEM

In the current work, several new and known features have been developed as new systematic criteria, with the result that the system in the current work rests on more systematic criteria than has any other system of freeliving nematodes in the past. Despite the importance of this factor, a further difference from the previous systems seems, to the author, to have even greater importance: the incomplete nature of the system in the current work is elaborated explicitly, whereas in the previous system it was hardly recognisable for the two following reasons:

1) the creation of holphyletic taxa has, indeed, also been aspired to in earlier systems, but as yet serious attempts have only seldom been made to establish the respective holophylies as well;

2) in most cases in previous systems, no distinction was made between holophyletic taxa; they were just listed next to one another with undifferentiated claims to validity.

For these reasons, the former systems of freeliving nematodes are mainly of diagnostic value only, in that they make it easier to deal with the diversity of species; they hardly satisfy, however, the criteria of a scientific theory, because the corresponding line of argument is largely missing and because hardly any distinction is made between what has been substantiated and what had not been substantiated. One exception is Filipjev's system (1918/1921, 1934), which contains, as yet, the most extensive foundation for a scientific theory and which is corroborated in large parts in the present study.

The incomplete nature of a system does have some attraction, scientifically. The only prerequisite is that it is also recognisable as incomplete. For this reason, a purely diagnostic system of a group of organisms,

---

Fig. 37 *continued*

53 (p.101): among other Chromadorida and Rhabditida: within species the anterior gonad is often constantly to the right of the intestine and the posterior gonad to the left of it; this feature is only a frequent apomorphy within both taxa and cannot currently be considered a holapomorphy of the two taxa.
53 (p.101): among others Tylenchida: anterior and posterior gonad always on the same side of the intestine.
59 (p.113): phasmids present.
61 (p.113): Secernentea: caudal glands are absent.
65 (p.124): Nematoda: presence of a spicular apparatus; post-embryonic development over four stages.
66 (p.133): Chromadorina: vestibulum with 12 folds.
94 (p.160): Desmoscolecina: cephalic setae of the third sensilla circle on peduncles.
117 (p.281): Tylenchida: formation of a characteristic spear in the buccal cavity (stomatostyle).

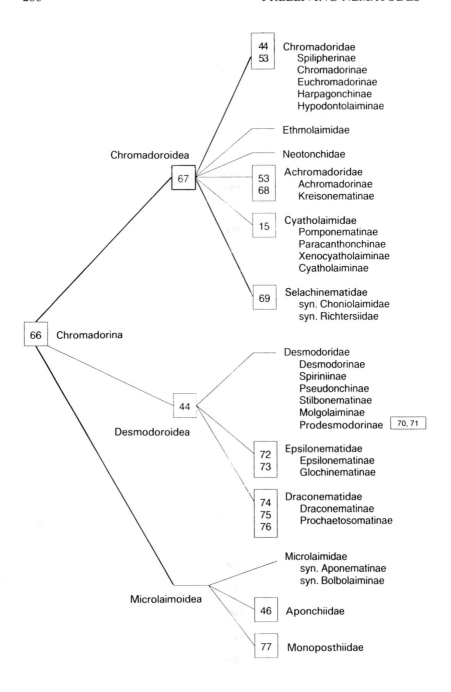

OUTLINE OF A PHYLOGENETIC SYSTEM

despite its incomplete nature, is just as unlikely to attract further research as is a complete phylogenetic system, for in the former, the lack of completion is not recognisable and in the latter, it is absent. Phylogenetic systematics are bound to remain incomplete for a long time yet. Much greater effort must be made in the future than has hitherto been made to ensure that this lack of completion is recognisable. The scientific attraction of a recognisably incomplete system lies in the starting points and stimulation it provides for further research. Only in this way can fresh life be breathed into the largely stagnating study of phylogenetics, and only in this way can we escape from the dilemma which was described in the introduction, the dilemma in which systematics, and not only the systematics of nematodes, has found itself for decades.

It is hoped that the present study will provide stimulation of this sort.

Fig. 38: The inter-relationships within the Chromadorina.

15 (p.46): Cyatholaimidae: both circles of cephalic sensilla are situated at the same level.
44 (p.95): among others Chromadoridae: only one anterior testis present.
44 (p.95): among others Desmodoroidae: only one anterior testis present.
46 (p.96): among others Aponchiidae: only one anterior ovary present.
53 (p.101): among others Chromadoridae: the anterior gonad lies consistently to the right and the posterior gonad to the left of the intestine.
53 (p.101): among others Achromadoridae: the anterior and the posterior gonad consistently lie on the same side of the intestine.
66 (p.133): Chromadorina: vestibulum with 12 folds.
67 (p.135): Chromadoroidea: punctated cuticle.
68 (p.142): Achromadoridae: parthenogenetic reproduction common.
69 (p.146): Selachinematidae: the buccal cavity is spacious and does not have teeth, which are homologous with those of the other Chromadorina.
70 (p.153): Prodesmodorinae: parthenogenetic reproduction common.
71 (p.153): Prodesmodorinae: purely freshwater habitat.
72 (p.153): Epsilonematidae: ovaries situated posterior to the dorsal bend of the body (i.e. in the posterior curve of the epsilon of S-shaped body).
73 (p.153): Epsilonematidae: the stilt (ambulatory) setae are situated anterior to or in the same region of the body that the females have ovaries.
74 (p.154): Draconematidae: ovaries situated anterior to the dorsal bend of the body (i.e. in the middle section of the Z or S-shaped body).
75 (p.155): Draconematidae: the subventral stilt setae are situated posterior to that region of body, in which the females have ovaries.
76 (p.155): Draconematidae: adhesion tubes on the dorsal side of the anterior end.
77 (p.159): Monoposthiidae: amphids non-spiral.

FREELIVING NEMATODES

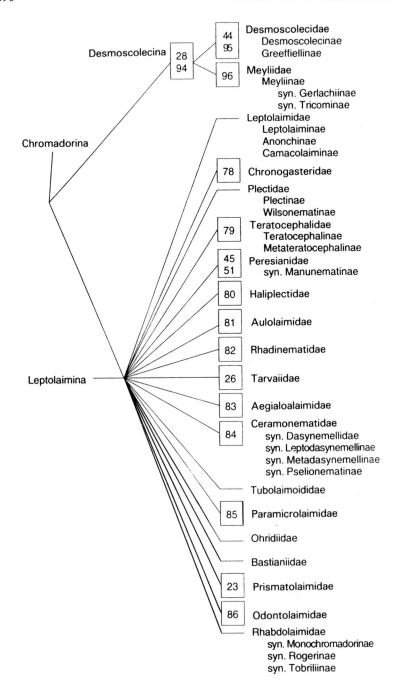

Fig. 39: The inter-relationships within the Leptolaimina and Desmoscolecina.

23 (p.65): Prismatolaimidae: amphids constantly dorsally-spiral (Fig. 9d on p.54).
26 (p.67): Tarvaiidae: loops of the amphids run from the inside towards the outside (Fig. 8b on p.52).
28 (p.67): Desmoscolecina: amphids with blistered corpus gelatum, which has a very strong outer skin.
44 (p.95): among other Desmoscolecoidea: only one anterior testis present.
45 (p.96): among others Peresianidae: only one posterior testis present.
51 (p.98): among others Monhysterida: ovaries outstretched.
78 (p.170): Chronogasteridae: only one anterior ovary present.
79 (p.173): Teratocephalidae: the labial region consists of 6 flap-like lips separated by deep grooves.
80 (p.176): Haliplectidae: the pharynx is weakly muscular in the anterior section; it has a slight bulb in the middle section and a very well-developed bulb at the end.
81 (p.177): Aulolaimidae: the pharynx consists of a cuticularized tube without surrounding muscle tissue in the anterior section and it is very muscular in the posterior section.
82 (p.178): Rhadinematidae: vestibulum is reinforced by 6 cuticularized, longitudinal rods; the buccal cavity section is reinforced by a cuticularized ring.
83 (p.179): Aegialoalaimidae: the pharynx, at least in the middle section and to some extent in the anterior section, has extremely thin walls and no muscle; it swells to form a bulb at the end.
84 (p.181): Ceramonematidae: body ridges decorated in a characteristic way.
85 (p.186): Paramicrolaimidae: characteristic buccal cavity teeth (Fig. 25 on p.186).
86 (p.192): Odontolaimidae: dorsal tooth can be extended like a spear (Fig. 27 on p.193).
94 (p.161): Desmoscolecina: the 4 cephalic sensilla of the third circle stand on peduncles.
95 (p.163): Desmoscolecidae: the cephalic setae are arranged in the following pattern (individual pairs of setae may be absent):

1st	3rd	5th	7th	9th	11th	13th		16th	17th
2nd	4th	6th	8th	10th	12th	14th	15th		

96 (p.164): Meyliidae: there are more body setae on the ventral side of the body than on the dorsal side of the body.

292 FREELIVING NEMATODES

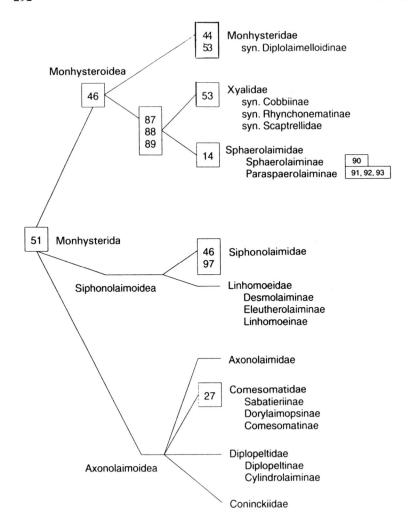

Fig. 40: The inter-relationships within the Monhysterida.

14 (p.46): Sphaerolaimidae: the 4 cephalic sensilla of the third sensilla circle are longer than the 6 of the second sensilla circle.
27 (p.67): Comesomatidae: amphids spiral with at least 2½ turns.
44 (p.95): among others Monhysteridae: only one anterior testis present.
46 (p.96): among others Monhysteroidea: only one anterior ovary present.
46 (p.96): among other Siphonolaimidae: only one anterior ovary present.
51 (p.98): among others Monhysterida: ovaries outstretched.
53 (p.101): among others Monhysteridae: anterior gonad always to the right of the intestine (posterior gonad absent).

53 (p.101): among others Xyalidae: almost always the anterior gonad is to the left and the posterior gonad to the right of the intestine.
87 (p.198): Xyalidae + Sphaerolaimidae: cuticle at least primarily striated.
88 (p.198): Xyalidae + Sphaerolaimidae: tail has, at least primarily, 2 or three terminal setae.
89 (p.198): Xyalidae + Sphaerolaimidae: there exists the genetic potential for the formation of 8 groups of subcephalic setae, the longest ones of which are longer than the longest cephalic setae.
90–93 (Sphaerolaiminae and Parasphaerolaiminae): see pp. 202–203.
98 (p.219): Siphonolaimidae: wall of the buccal cavity comes to a spear-like point anteriorly.

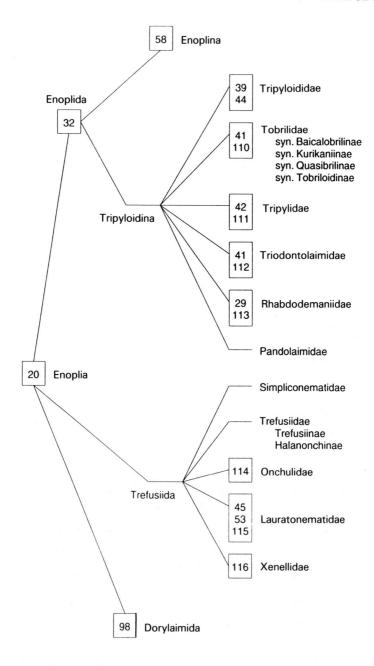

Fig. 41: The inter-relationships within the Enoplia.

20 (p.64): Enoplia: amphids at least primarily non-spiral.
29 (p.67): Rhabdodemaniidae: amphids have no aperture or fovea; instead there is a subcuticularly situated sinusoidal fibrous band (Fig. 10 on p.56).
32 (p.74): Enoplida: metanemes present.
39 (p.78): Tripyloididae: where metanemes occur, they are almost always ventrolateral loxometanemes of type II.
41 (p.79): among others Triodontolaimidae: only dorsolateral loxometanemes of type II present.
41 (p.79): among others Tobrilidae: where metanemes occur, they are always dorsolateral loxometanemes of type II.
42 (p.79): among others Tripylidae: where metanemes occur, they are alternating dorsolateral and ventrolateral orthometanemes, which are delicate in structure and have a very short caudal filament.
44 (p.95): among others Tripyloididae: only one anterior testis present.
45 (p.96): among others Lauratonematidae: only one posterior testis present (Fig. 16a on p.87).
53 (p.101): among others Lauratonematidae: the ovary (anterior gonad) constantly lies to the right and the testis (posterior gonad) to the left of the intestine.
58 (p.112): Enoplina: the caudal glands usually extend well into the precaudal region of the body.
98 (p.219): Dorylaimida: all pharyngeal glands open posterior to the nerve ring.
110 (p.262): Tobrilidae: adult males have bubble-shaped precloacal supplements.
111 (p.264): Tripylidae: at the end of the narrow buccal cavity there is a small dorsal tooth, through or near which the dorsal pharyngeal gland opens.
112 (p.266): Triodontolaimidae: buccal cavity has 3 characteristic teeth, which are equal in size.
113 (p.267): Rhabdodemaniidae: buccal cavity with 6 teeth on the anterior edge and 3 teeth in the middle region.
114 (p.272): Onchulidae: pharynx has an undulating outline.
115 (p.274): Lauratonematidae: the vulva lies very close to the anus or forms a cloaca with it.
116 (p.275): Xenellidae: the finely and clearly striated cuticle has several longitudinal rows of cuticular elevations.

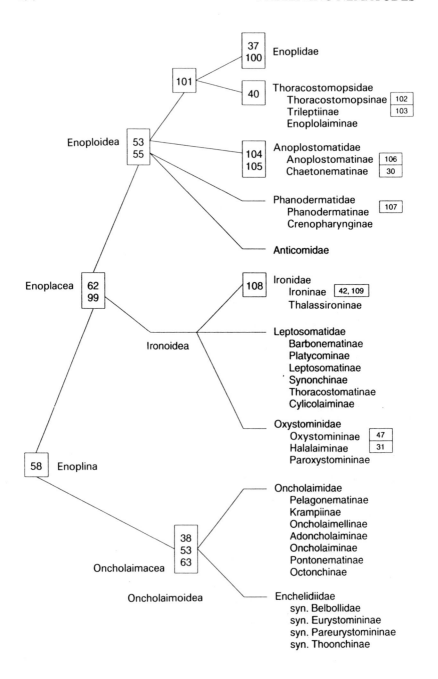

OUTLINE OF A PHYLOGENETIC SYSTEM

Fig. 42: The inter-relationships within the Enoplina.

30 (p.67): Chaetonematinae: amphids show extreme sexual dimorphism (Fig. 11 on p.58).
31 (p.67): Halalaiminae: the aperture of the amphids consists of a longitudinal duct.
37 (p.78): Enoplidae: adults have dorsolateral and ventrolateral loxometanemes of type II; juveniles only have dorsolateral loxometanemes of type I.
38 (p.78): Oncholaimoidea: delicately built dorsolateral and ventrolateral orthometanemes, which each have a caudal filament.
40 (p.78): Thoracostomopsidae: only dorsolateral orthometanemes, which have a well built scapulus but no caudal filament.
42 (p.79): among others Ironidae: delicately built dorsolateral and ventrolateral orthometanemes in strictly alternating sequence.
47 (p.96): among others Oxystomininae: only one posterior ovary present.
53 (p.101): among others Enoploidea: anterior and posterior gonad always to the left of the intestine.
53 (p.101): among others Oncholaimoidea: anterior and posterior gonad always to the right of the intestine.
55 (p.107): Enoploidea: adult males have, primarily at least, a precloacal tubule.
58 (p.112): Enoplina: the caudal glands usually extend well into the pre-caudal region of the body.
62 (p.115): Enoplacea: the cervical gland, where present, is usually situated completely in the pharyngeal body section.
63 (p.116): Oncholaimoidea: cervical gland is always situated to the right of the intestine.
99 (p.222): Enoplacea: the pharynx musculature inserts in the body wall in the region of the buccal cavity. Here the inner layer of the body wall is usually differentiated, forming a cephalic capsule.
100 (p.224): Enoplidae: buccal cavity only has mandibles and no onchia.
101 (p.225): Enoplidae + Thoracostomopsidae: buccal cavity, primarily at least, has mandibles and onchia.
102 (p.228): Thoracostomopsinae: buccal cavity has a characteristic spear (Fig. 30 on p.227).
103 (p.228): Trileptiinae: 3 onchia of equal size on the anterior edge of the buccal cavity (Fig. 31 on p.229).
104 (p.231): Anoplostomatidae: buccal cavity is spacious, does not contain teeth and is surrounded in pharyngeal tissue only at the posterior.
105 (p.231): Anoplostomatidae: a cephalic capsule is present but there is no pharyngeal musculature inserted into it.
106 (p.232): Anoplostomatinae: adult males have a copulatory bursa.
107 (p.233): Phanodermatinae: buccal cavity has 1 small dorsal tooth and 2 larger subventral teeth, which all face towards the anterior.
108 (p.236): Ironidae: the anterior edge of the buccal cavity has 3 or 4 movable teeth.
109 (p.238): Ironinae: freshwater habitat.

6. SUMMARY

1) It is the aim of this study to find a solution to the current dilemma of nematode systematics. A dilemma is seen to exist because it is not clear, from most systematic statements on nematodes, whether they are designed to provide nothing more than a more comprehensive survey of species diversity or whether they are designed to further develop a scientific theory on the phylogenetic relationships of species and species groups. The only solution to the dilemma of nematode systematics is seen to exist in the determined pursuit of the scientific method, and in the recognition and strict discrimination between scientific and purely pragmatic statements. Bearing these considerations in mind, an outline for a phylogenetic system of freeliving nematodes is presented. This system deals with evidence to establish the holophyly of many taxa and makes an explicit distinction between holophylectic and non-holophyletic taxa, i.e. between scientific and purely pragmatic statements.

2) The theory of phylogenetic systematics (cladistics) provides the basis on which scientific statements on the inter-relationships of species and of species groups can be studied. In its earlier form this theory contained logical deficiencies, which are revealed and rectified in chapter 2. The most important changes are as follows:

a) Ashlock's term "holophyly" is preferred to "monophyly" *sensu* Henning, because synapomorphies are not able to demonstrate more than monophyly in its old sense.

b) Previously it was erroneously assumed that a synapomorphy was a sufficient condition for the holophyly of a species group, while symplesiomorphies were meaningless where the establishment of holophylies are concerned. A new term "holapomorphy" is introduced; its use is sufficient for the establishment of holophylies. A holapomorphy is extracted by making decisions about synapomorphies and symplesiomorphies.

c) The following criterion is put forward: that a distinction should be made between primary and secondary absence of a feature.

3) The following features are analysed extensively and assessed phylogenetically:

a) New features:

- Presence, structure, arrangement and postembryonic development of metanemes (which are thought to be stretch receptors).
- Position of the gonads relative to the intestine (left, right or ventral to the intestine).
- Differentiation of the terminal duct of the epidermal glands in representatives of the Enoploidea.

b) Previously known features which had hardly any significance as systematic characters:

- Number of testes.
- Position of the caudal glands relative to the tail (caudal glands lying completely within the tail or extending far into the precaudal region of the body).
- Position of the cervical gland relative to the pharynx (cervical gland situated completely within the pharyngeal region of the body or extending well into the post-pharyngeal region of the body).

c) Previously known features which were recognized as being of considerable significance (although some new aspects are added):

- Number, position, structure and postembryonic development of the 6 labial and 6+4 cephalic sensilla (new aspect: jointing of the sensilla more widespread than previously assumed).
- Presence and position of the dereids (new aspect: dereids may also possibly occur in representatives of the Ironidae, Oxystominidae and Trefusiidae).
- structure, position and postembryonic development of the amphids (new aspect: a distinction is made between ventrally and dorsally-spiral amphids).
- Structure of the ovaries (new aspect: a distinction is made between antidromously and homodromously reflexed ovaries).
- Form and position of supplementary copulatory organs (new aspect: proximal end of the pre-cloacal tubules almost always situated to the right of the dorsoventral plane of the body).
- Occurrence of phasmids and phasmata.

4) A new classification of freeliving nematodes is developed with special emphasis on the Adenophorea (excluding the Dorylaimida). It is considered to be only an outline for a phylogenetic system mainly because the freeliving nematodes represent only a non-holophyletic part of all nematodes. The valid sub-families and genera are listed for all families (excluding the Dorylaimida). Compared with previous

SUMMARY 301

systems, some of the changes proposed are considerable. The following changes are particularly significant:

a) The 5 former orders "Araeolaimida", "Desmoscolecida", "Monhysterida", "Desmodorida" and "Chromadorida" are re-distributed between the two newly defined order Chromadorida and Monhysterida.

b) The former order "Enoplida" is re-divided to form the orders Enoplida (as defined in this work), Trefusiida, Dorylaimida and Chromadorida.

The new system is presented graphically in figures 37–42.

5) The following new taxa were erected by Lorenzen (1981):
 - Order: Trefusiida.
 - Suborder: Leptolaimina.
 - Taxa ranked between suborder and superfamily: Enoplacea, Oncholaimacea.
 - Families and subfamilies: Prodesmodorinae, Rhadinematidae, Tarvaiidae, Aegialoalaimidae, Tubolaimoididae, Paramicrolaimidae, Coninckiidae, Pandolaimidae.
 - Species: *Viscosia bandaensilis* Lorenzen 1981 pro *V. bandaensis* (Kreis 1932), a subjective homonym of *V. bandaensis* Kreis 1932; *Viscosia keiensilis* Lorenzen 1981 pro *V. keiensis* (Kreis 1932), a subjective homonym of *V. keiensis* Kreis 1932; *Viscosia papillatula* Lorenzen 1981 pro *V. papillata* Chitwood 1951, a subjective homonym of *V. papillata* (Kreis 1932); *Viscosia viscosula* Lorenzen 1981 pro *V. viscosa* (Kreis 1930), a subjective homonym of *V. viscosa* Bastian 1865.

6) Six subfamilies were raised to families in Lorenzen 1981. These are the Ethmolaimidae, Neotonchidae, Achromadoridae, Peresianidae, Ohridiidae, Odontolaimidae.

7) In Lorenzen (1981), the Harpagonchidae was reduced to a subfamily of the Chromadoridae, the Molgolaimidae to a subfamily of the Desmodoridae, the Prochaetosomatidae to a subfamily of the Draconematidae, and the Crenopharyngidae to a subfamily of the Phanodermatidae.

8) The following taxa were synonymized with other taxa in Lorenzen (1981):

Valid name:

New synonym:

a) Families and subfamilies:

Valid name	New synonym
Spilipherinae Filipjev 1918	Acantholaiminae Gerlach & Riemann 1975
Pomponematinae Gerlach and Riemann 1973	Longicyatholaiminae Andrassy 1976
Selachinematidae Cobb 1915	Choniolaimidae Stekhoven Adam 1931
	Richtersiidae Kreis 1929
Prochaetosomatinae Allen and Noffsinger 1978	Cygnonematinae Allen & Noffsinger 1978
	Dracognominae Allen & Noffsinger 1978
	Notochaetosomatinae Allen & Noffsinger 1978
Microlaimidae Micoletzky 1922	Aponematinae Jensen 1978
	Bolbolaiminae Jensen 1978
Leptolaiminae Örley 1880	Alaimellinae Andrassy 1976
Camacolaiminae Micoletzky 1924	Procamacolaiminae de Coninck 1965
Peresianidae Vitiello de Coninck 1973	Manunematinae Andrassy 1973
Ceramonematidae Cobb 1933	Dasynemellidae de Coninck 1965
	Leptodasynemellinae Haspeslagh 1973
	Metadasynemellinae de Coninck 1965
	Pselionematinae de Coninck 1965
Rhabdolaimidae Chitwood 1951	Monochromadorinae Andrassy 1958
	Rogerinae Andrassy 1976
	Tobriliinae Andrassy 1976
Greeffiellinae Filipjev 1929	Calligyridae Andrassy 1974
	Hapalominae Andrassy 1976
Desmolaiminae G. Schneider 1926	Terschellingiinae Andrassy 1976
Diplopeltinae Filipjev 1918	Campylaiminae Chitwood 1937
Enoplolaiminae de Conick 1965	Enoploidinae de Conick 1965
	Oxyonchinae de Conick 1965

SUMMARY

Crenopharynginae Platonova 1976
Thalassironinae Andrassy 1976
Platycominae Platonova 1975
Pelagonematinae de Conick 1965
Enchelidiidae Filipjev 1918

Tobrilidae de Conick 1965

Mononchidae Filipjev 1934

Dayellidae Platonova 1976
Coniliinae Coomans & van der Heiden 1979
Parabarbonematinae Platonova 1976
Curvolaiminae Andrassy 1976
Belbollinae Andrassy 1976
Eurystomininae Filipjev 1954
Pareurystomininae Andrassy 1976
Thoonchinae Gerlach and Riemann 1974
Baicalobrilinae Tsalolikhin 1976
Kurikaniinae Tsalolikhin 1976
Quasibrilinae Tsalolikhin 1976
Tobriloidinae Tsalolikhin 1976
Dioncholaiminae de Coninck 1965

b) Genera and subgenera:

Pomponema Cobb 1917

Leptolaimus de Man 1876
Euteratocephalus Andrassy 1958
Ceramonema Cobb 1920

Dasynemoides Chitwood 1936

Metadasynemella de Coninck 1942
Pselionema Cobb 1933
Domorganus T. Goodey 1947

Apodontium Cobb 1920
Araeolaimus de Man 1888

Parapomponema Ott 1972
Propomponema Ott 1972
Polylaimium Cobb 1914
Metateratocephalus Eroshenko 1973
Ceramonemoides Haspeslagh 1973
Cyattaronema Haspeslagh 1973
Dasynemelloides Haspeslagh 1973
Leptodasynemella Haspeslagh 1973
Dictyonemella Haspeslagh 1973

Pselionemoides Haspeslagh
Ohridius Gerlach and Riemann 1973
Leoberginema Tsalolikhin 1977
Mikinema Tschesunov 1978
Synodontoides Hopper 1963
Araeolaimoides de Man 1893

Southerniella Allgen 1932
Coninckia Gerlach 1956
Enoplus Dujardin 1845
Fenestrolaimus Filipjev 1927
Litinium Cobb 1920
Oxystomina Filipjev 1921
Halalaimus de Man 1888

Viscosia de Man 1890

Pontonema Leidy 1855
Quasibrilus Tsalolikhin 1976
Rhabdodemania Baylis and Daubney 1926

Parachromagasteriella Allgen 1933
Paratarvaia Wieser & Hopper 1967
Ruamowhitia Yeates 1967
Trichenoplus Mawson 1956
Alaimonemella Allgen 1935
Adorus Cobb in Thorne 1939
Nemacoma Hope and Murphy 1972
Nualaimus Juario 1974
Pachydora Wieser 1953
Meroviscosia Kreis 1932
Mononcholaimus Kreis 1924
Convexolaimus Kreis 1928
Lamuania Tsalolikhin 1976
Conistomella Stekhoven 1942

c) Species:

Microlaimus tenuispiculum de Man 1922
Enoplus schulzi Gerlach 1952

Halalaimus borealis (Steiner 1916)

Molgolaimus demani Jensen 1978
Ruamowhitia orae Yeatee 1967
Ruamowhitia halopila Guirado 1975
Halalaimus similis Allgen 1930

9) In Lorenzen (1981), most families of the free-living Adenophorea (except the Dorylaimida) were rearranged. The following subfamilies, genera and species were transferred from one to another family or subfamily (quotation marks indicate that the respective taxa had previously contained another set of taxa than in the present paper):

Acantholaiminae from "Comesomatidae" to Chromadoridae.
Nygmatonchus and *Trochamus* from Hypodontolaiminae to Euchromadorinae (Chromadoridae).
Kreisonematinae from "Leptolaimidae" to Achromadoridae.
Alaimonema from "Axonolaimidae" to Spiriniinae (Desmodoridae).
Prodesmodora from "Molgolaimidae" to Prodesmodorinae (Desmodoridae).
Aponema from "Molgolaimidae" to Microlaimidae.
Microlaimus tenuispiculum from "Molgolaimidae" to Microlaimidae.

SUMMARY

Polylaimium from "Axonolaimidae" to *Leptolaimus* (Leptolaimidae).
Assia from "Leptolaiminae" to Anonchinae (Leptolaimidae).
Mehdilaimus from "Axonolaimidae" to Aulolaimidae.
Thalassoalaimus aquaedulcis from "Oxystominidae" to *Aulolaimus* (Aulolaimidae).
Rhadinema from "Leptolaimidae" to Rhadinematidae.
Aegialoalaimus and *Diplopeltoides* from "Axonolaimidae" to Aegialoalaimidae.
Cyartonema and *Paraterschellingia* from "Siphonolaimidae" to Aegialoalaimidae.
Tubolaimoides from "Linhomoeidae" to Tubolaimoididae.
Chitwoodia from "Axonolaimidae" to Tubolaimoididae.
Paramicrolaimus from "Desmodoridae" to Paramicrolaimidae.
Domorganus and *Leoberginema* from "Axonolaimidae" to Ohridiidae.
Mikinema from "Linhomoeidae" to Ohridiidae.
Monochromadorinae and Tobriliinae from "Tripylidae" to Rhabdolaimidae.
Tricominae from "Desmoscolecidae" to Meyliidae.
Antarcticonema from "Desmoscolecidae" to Meyliidae.
Parachromagasteriella annelifer from "Cylindrolaiminae" to *Axonolaimus* (Axonolaimidae).
Coninckia from "Linhomoeidae" to Coninckiidae.
Paratarvaia from "Axonolaimidae" to Coninckiidae.
Thoracostomopsinae, Trileptiinae, Enoplolaiminae (syn. Enoploidinae, Oxyonchinae) from "Enoplidae" to Thoracostomopsidae.
Chaetonematinae from "Enoplidae" to Anoplostomatidae.
Dayellus, Klugea, Micoletzkyia, Paraphanoderma, Phanodermella, Phanodermopsis from Phanodermatinae to Crenopharynginae (Phanodermatidae).
Syringolaimus from "Rhabdolaimidae" to Ironidae.
All genera except *Ironus* from "Ironinae" to Thalassironinae (Ironidae).
Alaimonemella from Trefusiidae to Oxystomininae (Oxystominidae).
Nemacoma from Halalaiminae to Oxystomininae (Oxystominidae).
Cacolaimus from Oncholaiminae to Oncholaimellinae (Oncholaimidae).
Gairleanema from "Enoplidae" to Tripyloididae.
Pandolaimus from "Anoplostomatidae" to Pandolaimidae.
Porocoma from "Oxystominidae" to Xenellidae.
Thalassogenus from Oncholaimidae to Mononchidae.
Diplopeltula incisa, D. indica, D. intermedia and *D. onusta* are transferred to the genus *Diplopeltis* (Diplopeltidae).
Phanodermopsis necta is transferred to *Phanoderma* (Phanodermatidae).

10) In Lorenzen (1981) the following 9 genera were thought to be doubtful:

Pristionema Cobb, 1933 (Ceramonematidae), *Nijhoffia* Allgen, 1935 (Linhomoeidae), *Paraegialoalaimus* Allgen, 1934 (Linhomoeidae), *Acmaeolaimus* Filipjev, 1918 (Adenophorea incertae sedis), *Angustinema* Cobb, 1933 (Adenophorea incertae sedis), *Frostina* Gerlach & Riemann, 1974 (Adenophorea insertae sedis), *Pseudodilaimus* Kreis, 1928 (Adenophorea incertae sedis), *Pseudorhabdolaimus* Soos, 1937 (Adenpohorea incertae sedis), *Kreisia* Allgen, 1929 (Adenophorea incertae sedis).

11) In Lorenzen (1981) the following 4 species were thought to be doubtful:

Litinium simplex (Allgen 1935) (Oxystominidae), *Pontonema filicaudatum* (Kreis 1928) and *P. teissieri* (Vitiello 1967) (both Oncholaimidae), *Rhabdodemania brevicaudata* (Stekhoven 1942) (Rhabdodemaniidae).

7. BIBLIOGRAPHY

Preliminary remark: many of the studies quoted, which were published before (and including) 1972, are not listed in the following bibliography. These studies are cited in the detailed bibliography of the Bremerhaven Checklist by Gerlach & Riemann (1973/1974). The following list deals mainly with the works which have been published since 1973, but also contains a few older studies, which either do not appear in the Bremerhaven Checklist or are quoted frequently in the present study.

Alekseyev, V. M. (1979): Evolution of supplementary organs in Leptolaimus and establishment of the genus *Boveelaimus* (Nematoda, Leptolaimidae). Zool. Zh. *58*: 1296–1301 (Russian with English summary).
Alekseyev, V. M. and I. V. Rassadnikova (1977): A new species of the genus *Leptolaimus* (Nematoda, Araeolaimida). Zool. Zh. *56*: 1766–1774 (Russian with English summary).
Ali, S. M., M. N. Farooqui and S. Tejpal (1969): *Neotylocephalus annonae* n. gen., n. sp. (Nematoda: Wilsonematinae) from Marathawada, India. Riv. Parassit. *30*: 287–290.
Allen, M. W. and E. M. Noffsinger (1978): A revision of the marine nematodes of the superfamily Draconematoidea Filipjev, 1918 (Nematoda: Draconematina). Univ. Calif. Press, Berkeley, Los Angeles, London, 133 pp.
Andrassy, I. (1973a): Nematioden aus Strand- und Höhlenbiotopen von Kuba. Acta zool. hung. *19*: 233–270.
— (1973b): Ein Meeresrelikt und einige andere bemerkenswerte Nematodenarten aus Neuguinea. Opusc. zool. (Bpest) *12*: 3–19.
— (1974): A nematodák evaolúciója és rendeszerezése. Magy. tudom. Akad. biol. Csoporty. Közl. *17*: 13–58.
— (1976): Evolution as a basis for the systematization of Nematodes. Pitman – London, San Francisco, Melbourne, 288 pp.
— (1977): Die Gattungen *Amphidelus* Thorne, 1939, *Paramphidelus* n. gen. und *Etamphidelus* n. gen. (Nematoda: Alaimidae). Opusc. zool. (Bpest) *14*: 3–43.
Ashlock, P. D. (1971): Monophyly and associated terms. Syst. Zool. *20*: 63–69.
Belogurov, O. I. (1977): Morphology of *Pseudonchus furugelmus* sp. n. and *Bradybucca pontica* (Sergeeva, 1974) comb. n. and the establishment of Bradybuccini trib. n. (Nematoda, Oncholaimidae). Zool. Zh. *56*: 1597–1605 (Russian with English summary).
Belogurov, O. I. and L. S. Belogurova (1978): Systematics and evolution of Oncholaiminae (Nematoda: Oncholaiminae). III. The system of Oncholaiminae. Biologia Moria *2*: 22–31 (Russian with English summary).

— and — (1978a): Die Morphologie von drei Arten der Meeresnematoden der Gattung *Adoncholaimus* und Errichtung von *Admirandus multicavus* gen. et sp. b. (Nematoda: Oncholaimidae). In: Freilebende und parasitische Würmer. Fernöstliche staatliche Universität (Russian, not published).
Belogurova, L. S. and O. I. Belogurov (1974): *Fotolaimus marinus* gen. et sp. n. (Nematoda, Oncholaimidae) from the Schikotan Island (Kuril Islands). Zool. Zh. *53*: 1566–1568 (Russian with English summary).
Blome, D. (1974): Zur Systematik von Nematoden aus dem Sandstrand der Nordseeinsel Sylt. Mikrofauna Meeresboden *33*: 1–25.
Blome, D. and M. Schrage (1985): Freilebende Nematoden aus der Antarktis. Mit einer Beschreibung der Simpliconematidae nov. fam. (Trefusiida) und einer Revision von *Filipjeva* Ditlevsen 1928 (Monhysterida, Xyalidae). Veröff. Inst. Meeresforsch. Bremerh. *21*: 71–96.
Bock, W. (1974): Philosophical foundations of classical evolutionary classification. Syst. Zool. *22*: 375–492.
Boucher, G. (1974): Nématodes des sables fins infralittoraux de la Pierre Noire (Manche occidentale). I. Desmodorida. Bull. Mus. natn. Hist. nat., Paris, 3. Sér. *285* (Zool. *195*): 101–128.
— (1976): Nématodes des sables fins infralittoraux de la Pierre Noire (Manche occidentale). II. Chromadorida. Bull. Mus. natn. Hist. nat., Paris, 3. Sér. *352* (Zool. *245*): 25–61.
Boucher, G. and F. de Bovée (1972): *Halaphanolaimus harpaga* n. sp., espèce nouvelle de Leptolaiminae (Nematoda). Vie Milieu *23*: 127–132.
Boucher, G. and M.-N. Helléouet (1977): Nématodes des sables fins infralittoraux de la pierre Noire (Manche occidentale). III. Araeolaimida et Monhysterida. Bull. Mus. natn. Hist. nat., Paris, 3 Sér. *427* (Zool. *297*): 85–122.
Burr, A. H. and C. Burr (1975): The amphid of the nematode *Oncholaimus vesicarius*: Ultrastructural evidence for a dual function as chemoreceptor and photoreceptor. J. ultrastruct. Res. *51*: 1–15.
Chawla, M. L., E. Khan and M. Saha (1975): *Wilsereptus andersoni* gen. n., sp. n. (Wilsonematinae: Plectidae, Nematoda) from soil around roots of grass at Dalhousi, India with a key to the genera of the subfamily. Indian J. Nematol. *5*: 176–179.
Chitwood, B. G. and M. B. Chitwood (1950): An introduction to nematology. 2. Auflage, Monumental Printing, Baltimore, 213 pp.
Chitwood, B. G. and E. E. Wehr (1934): The value of cephalic structures as characters in nematode classification with special reference to the superfamily Spiruroidea. Z. ParasitKde *7*: 273–335.
Clark, W. C. (1961): A revised classification of the order Enoplida (Nematoda). N. Z. Jl. Sci. *4*: 123–150.
— (1962): The systematic position of the Alaimidae and the Diphtherophoroidea (Enoplida, Nematoda). Nematologica *7*: 119–121.
Clasing, E. (1980): Postembryonic development in species of Desmodoridae, Epsilonematidae and Draconematidae. Zool. Anz. *204*: 337–344.
Cobb, N. A. (1920): One hundred new nemas (type species of 100 new genera). Contributions to a Science of Nematology (Baltimore) *9*: 217–343.
Coles, J. W. (1977): Freeliving marine nematodes from Southern Africa. Bull. Br. Mus. nat. Hist. (Zool.) *32*, 1: 1–49.
de Coninck, L. A. (1942a): De symmetrie-verhoudingen aan het vooreinde der (vrijlevende) Nematoden. Natuurw. Tijdschr. *24*: 29–68.

— (1942b): Sur quelques espèces nouvelles de Nématodes libres (Ceramonematinae COBB, 1933), avec quelques remarques de systématique. Bull. Mus. r. Hist. nat. Belg. *18* (22): 1–37.
— (1965) Classe des Némotades – Systématique des Nématodes et sousclasse des Adenophorea. Traité de Zoologie (éd. Grassé) *4* (2): 586–681.
— (1965a): Classe des Nématodes – Géneralités. Tratité de Zoologie (éd. Grassé) *4* (2): 1–217.
Coomans, A. (1977): Evolution as a basis for the systematization of Nematodes – a critical review and exposé. Nematologica *23*: 129–136.
— (1979): The anterior sensilla of nematodes. Revue Nématol. *2*: 259–283.
Coomans, A. and D. de Waele (1979): Species of *Aphanolaimus* (Nematoda, Araeolaimida) from Africa. Zool. Scr. *8*: 171–180.
Coomans, A. and A. van der Heiden (1979): The systematic position of the family Ironidae and its relation to the Dorylaimida. Annls Soc. r. zool. Belg. *108*: 5–11.
Coomans, A. and P. A. Loof (1970): Morphology and taxonomy of Bathyodontina (Dorylaimida). Nematologica *16*: 180–196.
Croll, N. A. and J. M. Smith (1974): Nematode setae as mechanoreceptors. Nematologica *20*: 291–296.
Decraemer, W. (1974a): Scientific report on the Belgian expedition to The Great Barrier Reef in 1967. Nematodes II. *Desmoscolex* species (Nematodes, Desmoscolecida) from Yonge Reef, Lizard Island and Nymphe Island. Zool. Scr. *3*: 167–176.
— (1974b): Scientific report on the Belgian expedition to The Great Barrier Reef in 1967. Nematodes V. Observations on *Desmoscolex* (Nematoda, Desmoscolecida) with description of three new species. Zool. Scr. *3*: 243–255.
— (1975): Scientific report on the Belgian expedition to The Great Barrier Reef in 1967. Nematodes I: *Desmoscolex*-species (Nematoda – Desmoscolecida) from Yonge Reef, Lizard Island and Nymphe Island with general characteristics of the genus *Desmoscolex*. Annls Soc. r. zool. Belg. *104*: 105–130.
— (1976): Scientific report on the Belgian expedition to The Great Barrier Reef in 1967. Nematodes IV. Morphological observations on a new genus Quadricomoides of marine Desmoscolecida. Aust. J. mar. Freshwat. Res. *27*: 89–115.
— (1977): Origin and evolution of the Desmoscolecida, an aberrant group of nematodes. Z. zool. Syst. Evolut.-forsch. *15*: 232–236.
— (1978): Morphological and taxonomic study of the genus *Tricoma* Cobb (Nematoda: Desmoscolecida), with the description of new species from The Great Barrier Reef of Australia. Aust. J. Zool., Suppl. Ser. *55*: 1–121.
Decraemer, W. and Jensen, P. (1982): Revision of the subfamily Meyliinae De Coninck, 1965 (Nematoda: Desmoscolecoidea) with a discussion of its systematic position. Zool. J. Linn. Soc. *75*: 317–325.
Eroshenko, A. S. (1973): New data on taxonomy of the family Teratocephalidae Andrassy (Nematoda). Zool Zh. *52*: 1768–1776.
Fielding, M. J. (1950): Three new predacious nemas. Gt. Basin Nat. *10*: 45–50.
Filipjev, I. (1918/1921): Free-living marine nematodes of the Sevastopol area. Part I (1918) and part II (1921). English translation by M. Raveh, Israel

Program for Scientific Translations, Jerusalem, 1968 (part 1, pp. 1–255), 1970 (part 2, pp. 1–203).
— (1934): The classification of the free-living nematodes and their relation to the parasitic nematodes, Smithson. misc. Collns *89* (6): 1–63.
Freudenhammer, I. (1975): Desmoscolecida aus der Iberischen Tiefsee, zugleich eine Revision dieser Nematoden-Ordnung. "Meteor" Forsch.-Ergebnisse, Reihe D, No. *20*: 1–65.
Gadea, E. (1973): Sobre la filogenia interna de los Nematodos. Publns Inst. Biol. apl. (Barcelona) *54*: 87–92.
Gagarin, V. G. (1975): A contribution to the taxonomy and phylogeny of the superfamily Plectoidea (Nematoda). Zool. Zh. *54*: 503–509 (Russian with English summary).
— (1978): The description of a male of *Prodesmodora circulata* (Nematoda, Spiriniidae). Zool. Zh. *57*: 1261–1262 (Russian with English summary).
Gagarin, V. G. and L. L. Kuzmin (1972): New data on taxonomy of the genus *Prismatolaimus* (Nematoda, Onchulidae). Zool. Zh. *51*: 1879–1881 (Russian with English summary).
Gerlach, S. A. and F. Riemann (1973/1974): The Bremerhaven Checklist of aquatic Nematodes. A catalogue of Nematoda Adenophorea excluding the Dorylaimida. Veröff. Inst. Meeresforsch. Bremerh. Suppl. *4*: 1–404 (1973) and 405–734 (1974).
Goodey, T. (1963): Soil and freshwater nematodes, 2. Ed., rewritten by J. B. Goodey. Methuen – London, John Willey – New York, 544 pp.
Gourbault, N. and W. Decraemer (1986): Nématodes marins de Guadeloupe III. Epsilonematidae des generes nouveaux *Metaglochinema* (Glochinematinae) et *Keratonema* n.g. (Keratonematinae n. subfam.). Bull. Mus. natn. Hist. nat., 4ᵉ sér., *8*, sect. A, No. 1: 171–183.
de Grisse, A. T., P. L. Lippens and A. Coomans (1974): The cephalic sensory system of *Rotylenchus robustus* and a comparison with some other tylenchids. Nematologica *20*: 88–95.
Guirado, D. Jeminez (1975): *Ruamwhitia halophila* sp. n. de la costa de Almeria con observaciones sobre el genero *Ruamowhitia* Yeates, 1967 (Nematoda: Adenophorea). Cuad. Cienc. biol. Univ. Granada *4*: 159–165.
Haspeslagh, G. (1973): Superfamille des Ceramonematoidea (Cobb, 1933) (Nematoda), évolution et systématique. Annls Soc. r. zool. Belg. *102*: 235–251.
— (1979): Superfamily Ceramonematoidea (Cobb, 1933). General Morphology. Annls Soc. r. zool. Belg. *108*: 65–74.
Hendelberg, M. (1979): Occurrence and taxonomical significance of additional cephalic setae in some Linhomoeidae (Nematoda). Annls Soc. r. zool. Belg. *108*: 57–64.
Henning, W. (1965) Phylogenetic systematics. A. Rev. Ent., Palo Alto, *10*: 97–116.
— (1966): Phylogenetic systematics. Univeristy of Illinois Press – Urbana, Chicago, London, 263 pp.
Hirschmann, H. (1971): Comparative morphology and anatomy. In: Plant Parasitic Nematodes (eds. Zuckerman, Mai, Rhode), Vol. *1*: 11–63.
Hope, W. D. (1977): Gutless nematodes of the deep-sea. Mikrofauna Meeresboden *61*: 307–308.

Hope, W. D. and D. G. Murphy (1972): A taxonomic hierarchy and checklist of the genera and higher taxa of marine Nematodes. Smithson. Contr. Zool. *137*: 1–101.
Hopper, B. E. (1977): *Marylynnia*, a new name for *Marilynia* of Hopper, 1972. Zool. Anz. *198*: 139–140.
Hopper, B. E. and E. J. Cairns (1967): Taxonomic keys to plant, soil and aquatic nematodes. Alabama Polytechnic Inst., 176 pp.
Inglis, W. G. (1964): The marine Enoplida (Nematoda): A comparative study of the head. Bull. Br. Mus. nat. Hist. (Zool.) *11*: 265–376.
Inglish, G. W. (1983): An outline classification of the phylum Nematoda. Aust. J. Zool. *31*: 243–255.
Jensen, P. (1976): Redescription of the marine nematode *Pandolaimus latilaimus* (Allgen, 1929), its synonyms and relationships to the Oncholaimidae. Zool. Scr. *5*: 257–263.
— (1978a): Revision of Microlaimidae, erection of Molgolaimidae fam. n., and remarks on the systematic position of *Paramicrolaimus* (Nematoda, Desmodorida). Zool. Scr. *7*: 159–173.
— (1978b): Four Nematoda Araeolaimida from the Öresund, Denmark, with remarks on the oesophageal structures in *Aegialoalaimus*. Cah. Biol. mar. *19*: 221–231.
— (1979a): Revision of Comesomatidae (Nematoda). Zool. Scr. *8*: 81–105.
—(1979b): Nematodes from the brackish waters of the southern archipelago of Finland. Benthic species. Ann. Zool. Fenn. *16*: 151–168.
— (1979c): Description of the aquatic nematode *Adorus astridae* sp. n., with notes in the systematic position and geographical distribution of *Adorus* species. Ann. Zool. Fenn. *16*: 84–88.
Jensen, P. and S. A. Gerlach (1977): Three new Nematoda-Comesomatidae from Bermuda. Ophelia *16*: 59–76.
Juario, J. V. (1973): *Cyartonema germanicum* sp. n. (Nematoda: Siphonolaimidae) aus dem Sublitoral der Deutschen Bucht und Bemerkungen zur Gattung *Cyartonema* Cobb 1920. Veröff. Inst. Meeresforsch. Bremerh. *4*: 81–86.
— (1974): Neue freilebende Nematoden aus dem Sublitoral der Deutschen Bucht. Veröff. Inst. Meeresforsch. Bremerh. *14*: 275–303.
van der Heiden, A. (1975): The structure of the anterior feeding apparatus in members of the Ironidae (Nematoda: Enoplida). Nematologica *20*: 419–436.
Kaiser, H. (1977): Untersuchungen zur Morphologie, Biometrie, Biologie und Systematik von Mermithiden. Ein Beitrag zum Problem der Trennung morphologisch schwer unterscheidbarer Arten. Zool. Jb. Syst. *104*: 20–71.
Lambshead, P. J. D. and H. M. Platt (1979): *Bathyeurystomina*, a new genus of freeliving marine Nematodes (Enchelidiidae) from the Rockall Trough. Cah. Biol. mar. *20*: 371–380.
Lorenzen, S. (1972a): *Desmoscolez*-Arten (freilebende Nematoden) von der Nord- und Ostsee. Veröff. Inst. Meeresforsch. Bremerh. *13*:307–316.
— (1972b): Die Nematodenfauna im Verklappungsgebiet für Industrieabwässer nordwestlich von Helgoland. I. Araeolaimida und Monhysterida. Zool. Anz. *187*: 223–248.
— (1972c): Die Nematodenfauna im Verklappungsgebiet für Industrieabwässer nordwestlich von Helgoland. III. Cyatholaimidae, mit einer Revision

von *Pomponema* Cobb 1917, Veröff. Inst. Meeresforsch. Bremerh. *13*: 285-306.

— (1972d): *Leptolaimus*-Arten (freilebende Nematoden) aus der Nord- und Ostsee, Kieler Meeresforsch. *28*: 92-97.

— (1973a): Freilebende Meeresnematoden aus dem Sublitoral der Nordsee und der Kieler Bucht. Veröff. Inst. Meeresforsch. Bremerh. *14*: 103-130.

— (1973b): Die Familie Epsilonematidae (Nematodes). Mikrofauna Meeresboden *25*: 1-86.

— (1974a): Die Nematodenfauna der sublitoralen Region der Deutschen Bucht, insbesondere im Titan-Abwassergebiet bei Helgoland. Veröff. Inst. Meeresforsch. Bremerh. *14*: 305-327.

— (1974b): *Glochinema* nov. gen. (Nematodes, Epsilonematidae) aus Südchile. Mikrofauna Meeresboden *47*: 1-22.

— (1975) *Rhynchonema*-Arten (Nematodes, Monhysteridae) aus Südamerika und Europa. Mikrofauna Meeresboden *55*: 1-29.

– (1976a): Zur Theorie der phylogenetischen Systematik. Verh. dt. zool. Ges. *1976*: 229.

— (1976b): Desmodoridae (Nematoden) mit extrem langen Spicula aus Südamerika. Mitt. Inst. Colombo-Alemán Invest. cient. *8*: 63-78.

— (1977a): Revision der Xyalidae (freilebende Nematoden) auf der Grundlage einer kritischen Analyse von 56 Arten aus Nord- und Ostsee. Veröff. Inst. Meeresforsch. Bremerh. *16*: 197-261.

— (1977b): Haftborsten bei dem Nematoden *Haptotricoma arenaria* gen. n., sp. n. (Desmoscolecidae) aus sublitoralem Sand bei Helgoland, Veröff. Inst. Meeresforsch. Bremerh. *16*: 117-124.

— (1978a): New and known gonadal characters in free-living nematodes and the phylogenetic implications. Z. zool. Syst. Evolut.-forsch. 16: 108-115.

Lorenzen, S. (1978b): Triodontolaimidae (freilebende Nematoden): Wiederfund der einzigen Art nach 85 Jahren. Veröff. Inst. Meeresforsch. Bremerh. *17*: 87-93.

— (1978c): The system of the Monhysteroidea (Nematodes) – a new approach. Zool. Jb. Syst. *105*: 515-536.

— (1978d): Discovery of stretch receptor organs in nematodes – structure, arrangement and functional analysis. Zool. Scr. 7: 175-178.

— (1978e): Postembryonalentwicklung von *Steineria*- und Spaerolaimidenarten (Nematoden) und ihre Konsequenzen für die Systematik. Zool. Anz. *200*: 53-78.

— (1979a): Marine Monhysteridae (sensu stricto, Nematodes) von der südchilenischen Küste und aus dem küstenfernen Sublitoral der Nordsee. Stud. neotrop. Fauna Envoron. *14*: 203-214.

— (1979b): Entwurf eines phylogenetischen Systems der freilebenden Nematoden. Habilitationsschrift, Univ. Kiel, 400 pp.

Lorenzen, S. (1981): Entwurf eines phylogenetischen Systems der freilebenden Nematoden. Veröff. Inst. Meeresforsch. Bremerh. Suppl. *7*:1-472.

— (1981a): Bau, Anordnung und postembryonale Entwicklung von Metanemen bei Nematoden der Ordnung Enoplia. Veröff. Inst. Meeresforsch. Bremerh., im Druck.

— (1981b): Revised theory of phylogenetic systematics. Manuskript in Bearbeitung.

Lorenzen, S. (1982): Phylogenetic systematics: problems, achievements and its application to the Nematoda. In "Concepts in nematode systematics" (eds. A. R. Stone, H. M. Platt and L. F. Khalil), Academic Press, London and New York, pp. 11–23.

McLaren, D. J. (1976): Sense organs and their secretions. In: The organization of nematodes (ed. N. A. Croll). Academic Press – London, New York, San Francisco, pp. 139–161.

Maggenti, A. R. (1963): Comparative morphology in nemic phylogeny. In: The lower metazoa (eds. Dougherty, Brown, Hanson and Hartmann). Univ. of California Press – Berkeley, pp. 273–282.

— (1970): System analysis and nematode phylogeny. J. Nemat. 2: 7–15.

Malakhov, V. (1986): Nematodi. Moscow "Nauka".

Malakhov, V., K. Ryzhikov and M. Sonin (1982): The system of large taxa of nematodes: subclasses, orders, suborders. Zool. Zh. 61: 1125–1134 (in Russian with English summary).

Muchina, T. I. (1978): *Multidens montanus* gen. et sp. n. (Nematoda, Tripylidae) from Primorsky district. Zool. Zh. 57: 1087–1090 (Russian with English summary).

Nair, P. and A. Coomans (1973): The genus *Axonchium* (Nematoda: Belondiridae). I. Morphology, juvenile stages, diagnosis and list of species. Nematologica 18: 494–513.

Osche, G. (1952): Systematik und Phylogenie der Gattung *Rhabditis* (Nematoda). Zool. Jb. Syst. 81: 190–280.

— (1955): Der dreihöckerige Schwanz, ein ursprüngliches Merkmal im Bauplan der Nematoden. Zool. Anz. 154: 135–148.

Osche, G. (1958): Die Bursa- und Schwanzstrukturen und ihre Aberrationen bei den Strongylina (Nematoda). Morphologische Studien zum Problem der Pluri- und Paripotenzerscheinungen. Z. Morph. Ökol. Tiere 46: 571–635.

Platonova, T. A. (1974): Volume and systematic status of the genus *Rhabdodemania* (Nematoda, Enoplida). Zool. Zh. 53: 1295–1303 (Russian with English summary).

— (1976): Neue Enoplida (freilebende Meeresnematoden) aus den Meeren der USSR. In: Akad. Nauk CCP, Zool. Inst., Issledovanija Fauny morjei 15 (23): 3–164. Leningrad (Russian).

— (1979): Structure of the head end and system of the subfamily Synonchinae (Nematoda, Enoplida, Leptosomatidae). Zool. Zh. 58: 1117–1129 (Russian with English summary).

Platonova, T. A. and V. V. Potin (1972): On new genera *Harpagonchus* and *Harpagonchoides* (Nematoda, Chromadorida, Harpagonchidae fam. n.) living on the parapodia and gills of the antarctic polychaetes *Aglaophamus* Kinberg and *Hemipodus* Quatrefages, Issledovanija Fauny morjei 11 (18), Resultatibiologitscheskich Issledovanii sovjetskich antarktitscheskich Ekspedizii 5: 81–85 (Russian with English summary).

Platt, H. M. (1982): Revision of the Ethmolaimidae (Nematoda: Chromadorida). Bull. Br. Mus. nat. Hist. 43: 185–252.

Platt, H. M. (1985): The freeliving marine nematode genus *Sabatieria* (Nematoda: Comesomatidae). Taxonomic revision and pictorial keys. Zool. J. Linn. Soc. 83: 27–78.

Poinar, G. O. and R. Leutenegger (1968): Anatomy of the infective and normal third-stage juveniles of *Neoaplectana carpocapsae* Weiser (Steinernematidae: Nematoda). J. Parasit. 54: 340–350.

Prabha, M. J. (1974): *Mehdilaimus aurangabadensis* n. gen., n. sp. (Axonolaimidae: Cylindrolaiminae) from Marathwada, India. Nematologica *19*: 477–480.
Raski, D. J., N. O. Jones and D. R. Roggen (1969): On the morphology and ultrastructure of the oesophageal region of *Trichodorus allius* Jensen. Proc. helminth. Soc. Wash. *36*: 106–118.
Remane, A. (1936): Gastrotricha und Kinorhyncha. In: Dr. H. G. Bronn's Klassen und Ordnungen des Oierreichs, 4. Bd (Vermes), II. Abt. (Askhelminthes, Trochhelminthes), 1. Buch, 2. Teil, 385 pp.
Riemann, F. (1970a): Freilebende Nematoden aus dem Grenzbereich Meer-Süß-Wasser in Kolumbien, Südamerika. Veröff. Inst. Meeresforsch. Bremerh. *12*: 365–412.
— (1970b): Das Kiemenlückensystem von Krebsen als Lebensraum der Meiofauna, mit Beschreibung freilebender Nematoden aus karibischen Decapoden. Veröff. Inst. Meeresforsch. Bremerh. *12*: 413–428.
— (1972): Corpus gelatum und ciliäre Strukturen als lichtmikroskopisch sichtbare Bauelemente des Seitenorgans freilebender Nematoden. Z. Morph. Tiere *72*: 46–76.
Riemann, F. (1974): *Trefusialaimus* nov. gen. (Nematoda) aus der Iberischen Tiefsee mit Diskussion des männlichen Genitalapparates von Enoplida Tripyloidea. "Meteor" Forsch.-Ergebnisse, Reihe D, No. *18*: 39–43.
Riemann, F. (1976): Meereshematoden (Chromadorida) mit lateralen Flossensäumen (Alae) und dorsoventraler Abplattung. Zool. Jb. Syst. *103*: 290–308.
— (1977a): Oesophagusdrüsen bei Linhomoeidae (Monhysterida, Siphonolaimoidea). Beitrag zur Systematik freilebender Nematoden. Veröff. Inst. Meeresforsch. Bremerh. *16*: 105–116.
— (1977b): On the excretory organ of *Sabatieria* (Nematoda, Chromadorida). Veröff. Inst. Meeresforsch. Bremerh. *16*: 263–267.
— (1977c): Causal aspects of nematode evolution: Relations between structure, function, habitat and evolution. Mikrofauna Meeresboden *61*: 217–230.
Riemann, F. (1986): *Nicascolaimus punctatus* gen. et sp. n. (Nematoda, Axonolaimoidea) with notes on sperm dimorphism in free-living marine nematodes. Zool. Scr. *15*: 119–124.
Ritter, M. and J. Théodoridès (1965): Famille des Diplogasteridae Steiner, 1929. In: Taité de Zoologie (éd. Grassé) 4 (3): 765–774. Famille des Cephalobidae Chitwood et Chitwood, 1934. Ibid. pp. 776–793.
Schiemer, F. (1978): Verteilung und Systematik der freilebenden Nematoden des Neusiedlersees. Hydrobiologia *58*: 167–194.
Schrage, M. and S. A. Gerlach (1975): Über Greeffiellinae (Nematoda, Desmoscolecida). Veröff. Inst. Meeresforsch. Bremerh. *15*: 37–64.
Sergeeva, N. G. (1973): New species of free-living nematodes from the order Chromadorida in the Black Sea. Zool. Zh. *52*: 1238–1241 (Russian with English summary).
— (1974): New free living nematodes (Enoplida) from the Black Sea. 2. Zool. Zh. *53*: 120–125.
Siddiqi, M. R. (1973): Systematics of the genus *Trichodorus* Cobb, 1913 (Nematoda: Dorylaimida), with descriptions of three new species. Nematologica *19*: 259–278.
Stekhoven, J. H. Schuurmans and L. A. de Coninck (1933): Morphologische

Fragen zur Systematik der freilebenden Nematoden. Verh. dt. zool. Ges. *1933*: 138–143.
Storch, V. and F. Riemann (1973): Zur Ultrastruktur der Seitenorgane (Amphiden) des limnischen Nematoden *Tobrilus aberrans*. Z. Morh. Tiere *74*: 163–170.
Sudhaus, W. (1974): Nematoden (insbesondere Rhabditiden) des Strandanwurfs und ihre Beziehungen zu Krebsen. Faun.-ökol. Mitt. *4*: 365–400.
— (1976): Vergleichende Untersuchungen zur Phylogenie, Systematik, Ökologie, Biologie und Ethologie der Rhabditidae (Nematoda). Zoologica *43*, 2. Lieferung, Heft 125: 1–229.
Teuchert, G. (1968): Zur Fortpflanzung und Entwicklung der Macrodasyoidea (Gastrotricha). Z. Morph. Tiere *63*: 343–418.
— (1977): The ultrastructure of the marine gastrotrich *Turbanella cornuta* (Macrodasyoidea) and its functional and phylogenetic importance. Zoomorphologie *88*: 189–246.
Thorne, G. (1935): Notes on free-living and plant-parasitic nematodes, 2. Proc. helminth. Soc. Wash. *2*: 96–98.
— (1939): A monograph of the nematodes of the superfamily Dorylaimoidea. Capita zool. *8* (5): 1–261.
Timm, R. (1978): Marine nematodes of the order Desmoscolecida from McMurdo Sound, Antarctica. Antarctic Research Series *26*: 225–236.
Tsalolikhin, S. J. (1976): New species of nematodes of the order Enoplida in the Baikal Lake. Zool. Sh. *55*: 346–353 (Russian with English summary).
— (1977): New species of nematodes from the Baikal Lake. Zool. Zh. *56*: 989–995 (Russian with English summary).
Tschesunov, A. V. (1978): Free-living nematodes of the family Linhomoeidae from the Caspian Sea. Zool. Zh. *57*: 1623–1631 (Russian with English summary).
Vitiello, P. (1973): Nouvelles espèces de Desmodorida (Nematoda) des côtes de Provence. Téthys *5*: 137–146.
— (1974): Considérations sur la systématique des nématodes Araeolaimida et description d'espéces nouvelles ou peu connues. Archs Zool. exp. gén. *115*: 651–669.
Ward, S., N. Thomson, J. G. White and S. Brenner (1975): Electron microscopical reconstruction of the anterior sensory anatomy of the nematode *Caenorhabditis elegans*. J. comp. Neurol. *160*: 313–338.
Warwick, R. M. and H. M. Platt (1973): New and little known marine nematodes from a Scottish sandy beach. Cah. Biol. mar. *14*: 135–158.
Weingärtner, I. (1955): Versuch einer Neuordnung der Gattung *Diplogaster* Schulze 1857 (Nematoda). Zool. Jb. Syst. *83*: 248–317.
Wright, K. A. (1980): Nematode sense organs. In "Nematodes as biological models. 2. Aging and other model systems" (ed. B. M. Zuckerman), Academic Press, London and New York, pp. 237–295.
Yeates, G. W. (1967a): Studies on nematodes from dune sands. 2. Araeolaimida. N. Z. Jl Sci. *10*: 287–293.
— (1967b): Studies on nematodes from dune sands. 6. Dorylaimoidea. N. Z. Jl Sci. *10*: 752–784.

8. APPENDIX

8.1 Summary in tabulated form of the characteristics of the labial and cephalic sensilla in freeliving nematodes (Table 10).

The following summary in tabulated form has been compiled using data in the literature and the author's own data. The taxa of the freeliving Adenophorea (excluding Dorylaimida and Monhysteroidea) are listed in the same extent and in the same sequence as in Gerlach & Riemann (1973/1974); exceptions are marked. The Monhysteroidea are listed as in Lorenzen (1978c). Abbreviations. se, pa, *se, pa* : type of labial and cephalic sensilla : setiform (se) or papilliform (pa). The circle in which the longest sensilla are situated, is shown by underlining the symbols (*se, pa*). Cases where sensilla are hardly detectable or not detectable at all are shown by the symbol –. //, / : these symbols indicate whether the second and third sensilla circles are separated (//) or at more or less the same level (/).

Jointing in the sensilla : where jointed sensilla occur in at least one species of a taxon listed, this is shown by entering 1, 2 or 3 (first, second or third circle) in the third column of the table, to indicate in which sensilla circle the jointed sensilla are situated.

C: sensilla circle.

Table 10: See text 8.1.

"ARAEOLAIMIDA"

Taxon	Sensilla type in 1stC	2ndC		3rdC	Jointed Sensilla
Teratocephalidae, usually	–	–		–	–
seldom	–	–		se	–
"Plectidae", usually	–	–		se	–
seldom	–	pa	//	se	–
"Rhabdolaimidae"					
Rogerus orientalis	–	se	/	se	–
Syringolaimus, either	pa	pa	/	pa	–
or	–	pa	//	pa	–
Isolaimiidae	pa	pa	/	pa	–
Aulolaimidae	–	–		–	–
Haliplectidae, usually	–	–		–	–
Setoplectus	–	–		se	–
"Leptolaimidae"					
Peresianinae	–	–		se	–
"Leptolaiminae" usually	–	–	//	se	–
very seldom	–	pa	//	se	–
"Leptolaiminae", seldom	pa	pa	//	se	–
Stephanolaimus, partim	–	se	//	se	–
"Procamacolaiminae"	–	pa	//	se	–
"Camacolaiminae", partim	–	–		se	–
partim	–	pa	//	se	–
Kreisonematinae	–	–		–	–
Bastianiidae, partim	pa	se	//	se	2
partim	pa	se	/	se	?
"Axonolaimidae", often	pa	pa	//	se	–
often	–	pa	//	se	–
seldom (eg. *Pararaeolaimus nudus*)	–	pa	//	pa	–
Chitwoodia ("Diplopeltinae")	pa	se	//	se	–
Paratarvaia ("Campylaiminae")	pa	se	//	se	–

"DESMOSCOLECIDA"

Taxon	Sensilla type in 1stC	2ndC		3rdC	Jointed Sensilla
"Desmoscolecida", usually	–	–		se	3
very seldom	–	pa	//	se	3
Calligyrus	–	–		–	–
Desmoscolex rostratus	–	–		–	–

APPENDIX

Table 10 *continued*

"DESMODORIDA"

Taxon	Sensilla type in 1stC	2ndC		3rdC	Jointed Sensilla
Aponchiidae	–	pa	//	se	–
"Desmodoridae"					–
"Microlaiminae", partim	pa	se	//	se	–
partim	–	pa	//	se	–
Paramicrolaimus, partim	–	se	//	se	–
"Spiriniinae", partim	pa	pa	//	se	–
partim	pa	se	//	se	–
Pseudonchinae	pa	pa	/	se	–
Stilbonematinae, usually	–	pa	//	se	–
Eubostrichus cobbi	pa	–	//	se	–
Desmodorinae, partim	pa	se	//	se	–
partim	pa	pa	//	se	–
"Xenellidae"	pa	se	//	se	–
Ceramonematidae[4]					
"Dasynemellinae"	–	se	//	se	–
"Ceramonematinae", partim	–	se	//	se	–
partim	–	se	//	se	–
"Leptodasynemellinae", partim	–	se	//	se	–
partim	–	se	/	se	–
"Metadasynemellinae", partim	–	se	/	se	–
partim	–	se	/	se	–
"Pselionematinae",	–	–		se	–
Monoposthiidae, usually	–	pa	//	se	–
Nudora omercooperi	–	se	//	se	–
Nudora besnardi	pa	se	//	se	–
"Richtersiidae",	se	se	/	se	1,2,3
Draconematidae[5]	–	se	//	se	–
Epsilonematidae	–	–		se	–

"CHROMADORIDA"

Taxon	Sensilla type in 1stC	2ndC		3rdC	Jointed Sensilla
"Comesomatidae"					
"Acantholaiminae", partim	pa	pa	//	se	–
partim	pa	pa	/	se	–
partim	pa	se	//	se	–
"Dorylaimopsinae"	pa	pa	//	se	–
"Comesomatinae, usually	pa	pa	//	se	–
Comesoma heterosetosum	pa	se	//	se	–
Mesonchium	pa	se	/	se	–
Paracomesoma	pa	se	//	se	–
"Sabatieriinae" usually	pa	pa	//	se	–
Cervonema	pa	se	//	se	2
Paramesonchium	pa	se	//	se	–

Table 10 *continued*

Taxon	Sensilla type in				Jointed Sensilla
	1stC	2ndC		3rdC	
Laimella	pa	se	/	se	–
Sabatieria supplicans	pa	se	//	se	–
Pierrickia	pa	se	//	se	–
"Chromadoridae"					
"Spilipherinae",	pa	pa	//	se	–
"Chromadorinae"	pa	pa	//	se	–
Euchromadorinae					
Actinonema	pa	se	/	se	–
Adeuchromadora	pa	se	/	se	–
Austranema	pa	se	/	se	–
Rhips	pa	se	/	se	–
Euchromadora, partim	pa	se	/	se	–
partim	pa	pa	//	se	–
Graphonema	pa	pa	//	se	–
Steineridora	pa	pa	//	se	–
Hypodontolaiminae	pa	pa	//	se	–
Spilophorella papillata	pa	pa	/	pa	–
Nygmatonchus, partim	pa	se	/		–
partim	pa	se	//		–
"Cyatholaimidae"					
"Neotonchinae", partim	pa	pa	//	se	–
partim	pa	pa	//	pa	–
partim	pa	se	//	se	–
Gomphionema, partim	pa	pa	/	pa	–
"Achromadorinae",	–	se	/	se	–
	–	se	//	se	–
Achromadora buikensis	–	pa	//	se	–
Pomponematinae	se	se	//	se	2
	pa	se	/	se	2
Nannolaimus volutus	pa	se	//	se	–
Paracanthonchinae, partim	se	se	/	se	2
partim	pa	se	/	se	2
"Xenocyatholaiminae"	se	se	/	se	2
"Cyatholaiminae", partim	se	se	/	se	2
partim	pa	se	/	se	2
"Choniolaimidae"	pa	se	/	se	2
Gammanema cancellatum	pa	se	/	se	–
partim	se	se	/	se	1,2
Latronema, partim	se	se	/	se	1,2,3
partim	se	se	/	se	1,2,3
Choanolaimus	pa	pa	/	pa	–
Halichoanolaimus, partim	pa	pa	/	*pa*	–
"Selachinematidae", partim	pa	pa	/	se	–
partim	pa	se	/	se	–
partim	se	se	/	se	2
partim	pa	pa	/	pa	–

APPENDIX

Table 10 *continued*

Taxon	Sensilla type in 1stC	2ndC		3rdC	Jointed Sensilla
"Tripylidae"					
"Tripylinae", partim	pa	*se*	//	se	2
partim	pa	pa	//	pa	–
Trischistoma, partim	pa	*se*	/	*se*	–
"Tobrilinae", usually	pa	*se*	/	se	2
Tobrilia	pa	pa	//	pa	–
Tobriloides	pa	*se*	//	se	–
Monochromadorinae, partim	pa	pa	//	pa	–
Sinanema	pa	*se*	/	*se*	–
"Prismatolaimidae"					
Onchulinae, partim	se	*se*	//	se	2
partim	pa	*se*	//	se	2
Prismatolaiminae	pa	*se*	/	*se*	2
Odontolaiminae	pa	*se*	/	se	–
"Ironidae"					
Ironella	se	se	//	*se*	–
Thalassironus	pa	*se*	/	se	–
Trissonchulus	pa	pa	/	pa	–
Pheronus	pa	pa	/	pa	–
Dolicholaimus	pa	pa	/	*pa*	–
Conilia	pa	se	/	*se*	–
Parironus	pa	se	/	*se*	–
Ironus	pa	pa	/	*se*	–
"Cryptonchidae"	pa	pa	/	pa	
Tripyloididae, partim	pa	*se*	/	se	2
partim	se	*se*	/	se	1,2,3
Trefusiidae, usually	pa	*se*	//	se	2
Rhabdocoma sp.	se	*se*	//	se	1,2,3
Trefusialaimus	pa	*se*	/	se	2,3
"Alaimidae", usually	–	pa	/	pa	–
Adorus	pa	pa	//	*pa*	–
"Oxytominidae"					
"Oxystomininae"					
Litinium	se	*se*	//	*se*	–
Thalassoalaimus	se	*se*	//	*se*	–
Wieseria, partim	se	*se*	//	*se*	–
partim	se	se	//	*se*	–
Nemanema, usually	–	*se*	//	*se*	–
Oxystomina, usually	–	*se*	//	*se*	–
Porocoma, usually	–	*se*	//	*se*	–
seldom	–	pa	//	pa	–
Halalaiminae, usually	pa	*se*	//	*se*	–
seldom	pa	*se*	//	se	–
seldom	pa	se	//	*se*	–
seldom	se	*se*	//	*se*	–
Paroxystomininae, partim	–	se	//	*se*	–
partim	–	pa	//	pa	–

Table 10 *continued*

Taxon	Sensilla type in 1stC 2ndC 3rdC				Jointed Sensilla
Lauratonematidae	pa	*se*	/	*se*	–
Leptosomatidae, partim	pa	*se*	/	*se*	–
partim	pa	pa	/	pa	–
Barbonema, partim	se	*se*	/	*se*	–
Triodontolaimidae	pa	*se*	/	*se*	–
Anticomidae	pa	*se*	/	*se*	–
Phanodermatidae	pa	*se*	/	*se*	–
"Enoplidae"					
Thoracostomopsinae	se	*se*	/	*se*	–
Chaetonematinae[6]	pa	*se*	/	*se*	–
Trileptiinae	se	*se*	/	*se*	–
"Oxyonchinae", usually	se	*se*	/	*se*	–
Fenestrolaimus, partim	pa	*se*	/	*se*	–
Trichenoplus	pa	*se*	/	*se*	–
"Enoplolaiminae"	se	*se*	/	*se*	–
Enoplinae	pa	*se*	/	*se*	–
Rhabdodemaniidae[6]	pa	*se*	/	*se*	–
"Anoplostomatidae"					
Anoplostoma, partim	pa	*se*	/	*se*	–
partim	pa	*se*	//	*se*	–
Pandolaimus[7]	pa	*se*	//	*se*	2,3
Oncholaimidae, usually	pa	*se*	/	*se*	–
Pontonema ardens	se	*se*	/	*se*	–
Enchelidiidae, usually	pa	*se*	/	*se*	–
Ditlevsenella, partim	se	*se*	/	*se*	–
Rhaptothyreidae	–	pa	/	pa	–

DORYLAIMIDA

Dorylaimida, usually	pa	pa	/	pa	–
Encholaimidae	pa	*se*	/	*se*	–
Cephalodorylaiminae	*se*	pa	/	pa	–
Helmabia	pa	*se*	/	pa	–

RHABDITIDA

Diplogasteroidea, partim: ♂♂	pa	*se*	//	*se*	–
♀♀	pa	*se*		–	–
partim: ♂♂	pa	pa	//	pa	–
♀♀	pa	pa		–	–
Rhabditoidea, partim	pa	*se*	//	*se*	–
partim	pa	pa	//	pa	–

APPENDIX

Table 10   *continued*

TYLENCHIDA

Taxon	Sensilla type in			Jointed Sensilla
	1stC	2ndC	3rdC	
Tylenchida, usually	–	–	–	–
Atylenchus	–	–	*se*	–
Eutylenchus	–	–	*se*	–

Footnotes to Table 10

1) At the anterior end, besides the 2 circles, closely situated one behind the other, each with 4 setae, no additional labial or cephalic sensilla can be detected in *Sarsonia* ("Desmolaininae"), *Eleutherolaimus* (Eleutherolaiminae) and *Linhomoeus filaris* Lorenzen 1973 ("Linhomoeinae"); the anterior 4 setae are interpreted as cephalic sensilla of the third sensilla circle and the posterior 4 as subcephalic setae (Gerlach, 1963: 603, Lorenzen, 1973a: 119).
2) The cephalic sensilla in the Monhysteridae are almost always short.
3) Often only 4, and not 6+4 cephalic sensilla are recorded; the 6 cephalic sensilla are usually very small and are therefore probably often overlooked.
4) Division of Ceramonematidae into sub-families according to Haspeslagh (1973).
5) It is slightly uncertain whether the 6 sensilla belong to the first or second circle.
6) In the Chaetonematinae and Rhabdodemaniidae the 4 cephalic setae of the third sensilla circle are often situated anterior, and not posterior to the 6 cephalic setae of the second sensilla circle.
7) *Pandolaimus* was removed from the Linhomeidae and put in the Anoplostomatidae by Jensen (1976).

8.2 Summary in tabulated form of the gonad characteristics in freeliving nematodes (Table 11).

8.2.1. Personal data
The following table is a summary of the results of the author's research on the gonads of 400 freeliving species of the Adenophorea and 20 freeliving species of the Secernentea. The table is constructed in the following way:

Column 1: Except where otherwise stated, the species studied are listed in the same sequence is in Gerlach & Riemann (1973/1974 : freeliving Adenophorea excluding Dorylaimida and Monhysteroidea), Lorenzen (1978c : Monhysteroidea), de Coninck (1965 : Dorylaimida) and Goddey (1963 : Rhabditida and Tylenchida). The authority for individual taxa are only cited if the taxa concerned are not contained in the studies listed.

Column 2: Testes
2 opp : 2 testes are present, one of which faces towards the anterior and one of which faces towards the posterior.
2 ant. : 2 testes are present, both of which face towards the anterior; the posterior testis may be reflexted, but the reflexed part extends no further than the point where the anterior and posterior testis join.
1 ant. : there is only one testis present and it faces towards the anterior.
1 post. : there is only one testis present and it faces towards the posterior.
? : ♂♂ were not available, but are recorded in the literature.
– : ♂♂ have never been recorded.

Column 3: Ovaries
2 : 2 ovaries are present.
1 ant. : Only one anterior ovary is present.
1 post : Only one posterior ovary is present.
antidr. : Ovaries are antidromously reflexed.
homodr. : Ovaries are homodromously reflexed.
out. : Ovaries are outstretched.

Column 4 : Gonad position relative to the intestine
l : Gonad lies to the left or subventrally to the left of the intestine.
r : Gonad lies to the right or subventrally to the right of the intestine.
    A letter above and a letter below together form a symbol; the symbols can be strung together one after another. For example:
l : Within species, the anterior gonads lies constantly to the left
r   and the posterior gonad to the right of the intestine (laterally or sublaterally).

APPENDIX 325

lr : Within species, either the anterior gonad lies to the left and the
rl   posterior gonad to the right of the intestine, or the anterior gonad
      lies to the right and the posterior gonad to the left of the intestine
      (laterally or sublaterally).

lr lr : Within species there are 4 possible gonad positions, anterior left
rl lr   and posterior to the right, anterior right and posterior to the
         left, anterior and posterior to the left, and anterior and posterior
         to the right of the intestine (laterally and sublaterally).

r : The anterior gonad lies to the right of the intestine, and the
−   posterior gonad is absent in both sexes.

ventr. : Both gonads lie ventral to the intestine.
ventr.
± ventr. :
± ventr.   Both gonad lie more or less ventral to the intestine.

Column 5 : Biotope.
The biotope is given as "woods", "flats" etc. The use of these headings to refer to the different habitats is identical to that in Chapter 3 (material pp 25–27).

Column 6 : The figures in this column refer to the numbers of males ♂♂ and females ♀♀ studied.

Table 11. See text 82.

"ARAEOLAIMIDA"

Species	Testes	Ovaries	Gonad position	Biotope	♂♂	♀♀
Teratocephalidae						
Euteratocephalus palustris	?	2 antidr	lr lr	Moorland	–	9
Teratocephalus sp.	1 ant.[1]	1 ant. antidr	r –	Woods	1	5
"Plectidae"						
"Plectinae"						
Anaplectus granulosus	2 opp	2 antidr	r l	Woods	5	4
A. submersus	2 opp	2 antidr	r l	Salt marshes	6	9
Plectus cirratus	?	2 antidr	r l	Moors and Woods	–	22
Plectus spp. (3 spp.)	?	2 antidr	r l	Moors and Woods	–	16
Wilsonematinae						
Wilsonema sp.	?	2 antidr	r l	Woods	–	3
"Rhabdolaimidae"						
Rhabdolaimus minor	–	2 antidr	lr lr	[2]	–	5
R. nannus	–	2 out	lr lr	[2]	–	8
R. terrestris	–	2 out	lr lr	Woods	–	10
Syringolaimus striatocaudatus	1 ant.	2 antidr	lrlr rllr	Flats	8	5
Isolaimiidae						
No species studied						
Aulolamidae						
Aulolaimus aff. costatus (cuticle longitudinally striped)	?	2 antidr	lr rl	Woods	–	2
Haliplectidae						
Haliplectus dorsalis	2 opp	2 antidr	lr rl	Salt marshes	13	13
H. schulzi	2 opp	2 antidr	lr rl	Salt marshes	2	2

APPENDIX

Table 11 *continued*

Species	Testes	Ovaries	Gonad position	Biotope	♂♂	♀♀
"Leptolaimidae" Peresianinae						
Manunema annulata and *proboscidis*	1 post	2 out	ventr ventr	Helgoland	11	7
"Leptolaiminae"						
Antomicron elegans	1 post	2 antidr	r l	Flats	1	1
A. sp.	?	2 antidr	r l	Sublittoral, Chile	–	1
Aphanolaimus aquaticus	1 post	2 antidr	r r	Stream in a wood	1	2
Chronogaster typica	–	1 ant antidr	r –	Stream, Malaysia	–	11
Dagda bipapillata	2 ant[3)]		l r	Beach, Chile	1	–
Deontolaimus papillatus	2 ant	2 antidr	r l	Salt marshes	9	4
Laptolaimoides thermastris	1 post	2 antidr	r l	Salt marshes	5	1
Leptolaimus acicula	2 ant	2 antidr	r l	Salt marshes	4	4
L. elegans	2 opp[4)]	2 antidr	r l ($\frac{l}{r}$)	[5)]	7	3
L. limicola	1 post	2 antidr	r l ($\frac{r}{r}$)	Flats	5	4
L. mixtus	2 opp		r l	Baltic	1	–
L. papilliger	1 post	2 antidr	r l ($\frac{l}{l}$)	Salt marshes	8	8
L. puccinelliae	2 opp[4)]	2 antidr	r l	Salt marshes	5	4
L. venustus	2 opp	2 antidr	r l	Helgoland	3	1
L. sp.	2 opp	2 antidr	r l	Kattegatt	1	3
Rhadinema flexile	2 ant	2 antidr	lrlr rllr	Helgoland	6	2
Stephanolaimus elegans	2 [6)]	2 antidr	r l	Baltic + Helgoland	6	5

Table 11 *continued*

"DESMOSCOLECIDA"

Species	Testes	Ovaries	Gonad position	Biotope	♂♂	♀♀
"Desmoscolecidae"						
Desmoscolecinae						
Desmoscolex fennicus	1 ant	2 out	$\frac{r}{l}$ ($\frac{r}{r}$) 11)	Salt marshes, Finland	1	13
Pareudesmoscolex laciniosus	1 ant	2 out	$\frac{r}{l}$ 11)	Salt marshes	1	2
Greeffiellinae						
Calligyrus gerlachi	1 ant	2 out	$\frac{l}{r}$ 11)	Salt marshes	3	3
Greeffiella sp.	1 ant	2 out	ventr. ventr.	Chile	6	2
Hapalomus terrestris	?	2 out	$\frac{r}{l}$ 11)	Salt marshes	–	1
Tricominae						
Tricoma sp. (37 rings)	2 opp		$\frac{r}{l}$ 11)	Helgoland	2	–
T. sp. (c. 80 rings)	2 opp	2 out	$\frac{r}{l}$ 11)	Helgoland	7	4
"Meyliidae"						
Meylia spinosa	?	2 out	$\frac{r}{l}$	Skagerrak	–	1

"MONHYSTERIDA"

Species	Testes	Ovaries	Gonad position	Biotope	♂♂	♀♀
"Siphonolaimidae"						
Cyartonema germanicum	?	2 antidr	$\frac{r}{l}$	Helgoland	–	1
C. zosterae	2 opp	2 antidr	$\frac{lr}{rl}$ ($\frac{l}{l}$)	Helgoland	5	7
C. sp.	2 opp	2 antidr	$\frac{lr}{rl}$ ($\frac{l}{l}$)	Helgoland	3	5
Siphonolaimus cobbi	1 vo	1 ant out	$\frac{l}{-}$	Helgoland	5	5
"Linhomoeidae"						
"Desmolaiminae"						
Coninckia circularis	2 opp	2 out	$\frac{l}{r}$	Helgoland	1	3
Desmolaimus aff. *bulbulus*	2 opp	2 out	$\frac{lr}{rl}$	Helgoland	3	7
D. zeelandicus	2 opp	2 out	$\frac{lr}{rl}$	Flats & Baltic	11	6
Metalinhomoeus aff. *typicus*	2 opp	2 out	$\frac{lr}{rl}$	Helgoland	3	3
Terschellingia distalamphida Juario 1974	2 opp	2 out	$\frac{l}{r}$ ($\frac{l}{l}$)	Helgoland	4	2

APPENDIX

Table 11 *continued*

Species	Testes	Ovaries	Gonad position	Biotope	♂♂	♀♀
"Axonolaiminae"						
Apodontium pacificum	2 ant			Beach, Chile	1	–
Ascolaimus elongatus	2 ant[9]	2 out	lrlr rllr	Flats	5	9
Axonolaimus helgolandicus	2 ant[9]	2 out	lr rl (^r_r)	Helgoland	4	14
A. paraspinosus	2 ant[9]	2 out	lr rl	Flats	9	5
Odontophora armata	2 ant[9]	2 out	lr rl	Flats	4	3
O. ornata	2 ant[9]	2 out	lr rl	Helgoland	3	2
O. peritricha	2 ant[9]	2 out	l r (^l_l)	Sublittoral, Chile	4	4
O. phalarata	2 ant[9]	2 out	l r	Helgoland	2	2
Parodontophora quadristicha	2 ant[9]	2 out	lr rl	Sublittoral, Chile	10	6
Odontophoroides monhystera (=*Synodontium m.*)	2 ant[9]	1 post out	lr rl	Helgoland	3	5
Synodontoides procerus	2 ant	2 out	l r	Beach, Chile	2	2
"Campylaiminae"						
Campylaimus gerlachi	2 opp	2 out	lr rl	Helgoland & Kattegatt	4	8
C. spp.	2 opp	2 out	lr rl	Kattegatt	7	8
Diplopeltoides ornatus	2 opp	2 antidr	lr rl	Baltic	2	1
Diplopeltula incisa	2 opp	2 out	lr rl	Baltic & Kattegatt	4	7
D. aff. *nuda*	2 ant[10]	2 antidr	lr rl	Sublittoral, Chile	1	1
Tarvaia sp. 1	2 ant	2 antidr	r l	Baltic	1	1

FREELIVING NEMATODES

Table 11 *continued*

Species	Testes	Ovaries	Gonad position	Biotope	♂♂	♀♀
"Campylaiminae" (continued)						
Stephanolaimus flevensis	2 ant	1 post antidr	r l	Baltic	6	2
S. paraflevensis	1 ant	1 post antidr	r l	Baltic	1	2
S. spartinae	2 ant	1 post antidr	r l	Baltic	1	2
"Procamacolaiminae"						
Procamacolaimus sp.	2 ant		r l	Sublittoral, Chile	4	–
"Camacolaiminae"						
Camacolaimus tardus	2 ant	2 antidr	♂:$\frac{l}{l}$ ♀:$\frac{r}{l}$	Salt marshes	6	1
C. barbatus	2 ant	2 antidr	r l	Flats	2	3
C. sp. 1	2 ant		r l	Sublittoral, Chile	2	–
C. sp. 2		1 post antidr	– l	Beach, Colombia	–	1
Bastianiidae						
Bastiania gracilis	2 opp	2 antidr	lrlr rllr	Woods	3	4
"Axonolaimidae"						
"Cylindrolaiminae"						
Aegialoalaimus elegans	2 ant	2 antidr	lrlr rllr	Flats	4	4
A. sp. 1	2 ant	2 antidr	lrlr rllr	Baltic & Kattegatt	5	2
A. sp. 2	2 ant	2 antidr	lrlr rllr	Sublittoral, Chile	3	4
Cylindrolaimus sp. 1	2 ant[7]	2 out	lr rl	Salt marshes	7	5
Domorganus oligochaetophilus	2 opp		r l	Salt marshes	1	–
Parachromagasteriella sp.	2 ant	2 out	r l	Beach, Chile	2	1
"Diplopeltinae"						
Araeolaimus elegans	2 opp[8]	2 out	lr rl	Baltic & Helgoland	4	4
A. sp. 1	2 opp[8]	2 out	lr rl ($\frac{r}{r}$)	Chile	2	5
Pararaeolaimus nudus	2 opp	2 out	lr rl	Helgoland & Kattegatt	6	7
Terschellingia longicaudata	2 opp	2 out	l r	Helgoland	3	2

APPENDIX

Table 11 *continued*

Species	Testes	Ovaries	Gonad position	Biotope	♂♂	♀♀
Eleutherolaiminae						
Eleutherolaimus sp. 1	2 opp	2 out	lr / rl	Helgoland	3	2
"Linhomoeinae"						
Linhomoeus filaris	2 opp	2 out	± ventr. / ± ventr.	Helgoland	3	3
Tubolaimoides 2 spp.	2 opp	2 antidr	lr / lr (^1_r)	Helgoland	1	8
Monhysteridae sensu Lorenzen 1978c						
Diplolaimella ocellata	1 ant	1 ant out	r / –	Brackish water	5	6
D. stagnosa	1 ant	1 ant out	r / –	Salt marshes	4	3
Diplolaimelloides altherri	1 ant	1 ant out	r / –	Brackish water	2	2
D. islandica	1 ant	1 ant out	r / –	Salt marshes	4	4
D. oschei	1 ant	1 ant out	r / –	Salt marshes	4	2
Gammarinema cardisoma	1 ant	1 ant out	r / –	12)	1	1
G. gammari	1 ant	1 ant out	r / –	Baltic, on Gammarus	10	10
G. ligiae	1 ant	1 ant out	r / –	Helgoland, on Ligia	8	10
Monhystera anophthalma	1 ant	1 ant out	r / –	Salt marshes	6	6
M. cuspidospiculum	1 ant	1 ant out	r / –	Chile	4	1
M. disjuncta	1 ant	1 ant out	r / –	Salt marshes	7	7
M. islandica	1 ant	1 ant out	r / –	Salt marshes	5	4
M. multisetosa	1 ant	1 ant out	r / –	Salt marshes	5	1
M. paludicola	1 ant	1 ant out	r / –	Fresh water	5	4
M. paramacramphis	1 ant	1 ant out	r / –	Salt marshes	2	2
M. parasimplex	1 ant	1 ant out	r / –	Salt marshes	4	4
M. parva	1 ant	1 ant out	r / –	Salt marshes	7	4
M. pusilla Boucher and Helléouet 1977	1 ant	1 ant out	r / –	Helgoland	17	19

Table 11 *continued*

Species	Testes	Ovaries	Gonad position	Biotope	♂♂	♀♀
Monhysteridae (continued)						
Monhystera stagnalis	1 ant	1 ant out	r –	Fresh water	12	10
M. venusta Lorenzen 1979	1 ant	1 ant out	r –	Helgoland	12	23
M. sp. 1	–	1 ant out	r –	Woods	–	13
M. sp. 2	–	1 ant out	r –	Fresh water	–	8
Monhystrella inaequispiculum Lorenzen 1979	1 ant	1 ant out	r –	Chile	12	15
M. paramacrura	1 ant	1 ant out	r –	13)	–	7
Tripylium carcinicola	1 ant	1 ant out	r –	14)	5	2
Sphaerolaimidae						
Doliolaimus agilis	?	1 ant out	lr ??	Salt marshes	1	3
Parasphaerolaimus paradoxus	1 ant	1 ant out	lr –	Kattegatt	21	15
Sphaerolaimus gracilis	2 opp	1 ant out	lr rl	Salt marshes	13	7
Subsphaerolaimus litoralis Lorenzen 1978e	2 opp	1 ant out	lr rl	Flats, Chile	16	10
Xyalidae sensu Lorenzen 1978c						
Hofmaenneria niddensis	1 ant	1 ant out	r –	15)	3	1
Rhynchonema 7 (Lorenzen 1975)	2 opp	1 ant out	l r	Europe and South America	22	12
Rhynchonema 2 species (Lorenzen 1975)	?	1 ant out	l ?	Europe and South America	5	4
Steineria ericia	2 opp	1 ant out	l r	Beach, Colombia	7	2
S. pilosa	2 opp	1 ant out	r l	Sublittoral, Chile	11	15
Theristus wegelinae	1 ant		l –	Rhine banks	1	–

APPENDIX

Table 11 *continued*

Species	Testes	Ovaries	Gonad position	Biotope	♂♂	♀♀
Xyalidae (continued)						
40 species (Lorenzen 1977a)	2 opp	1 ant out	l r	North Sea & Baltic	170	99
11 species (Lorenzen 1977a)	1 ant	1 ant out	l r	North Sea & Baltic	59	29
15 species (Publication in preparation)	2 opp	1 ant out	l r	South America	123	87
10 species (Publication in preparation)	1 ant	1 ant out	l –	South America	77	69

"DESMODORIDA"

Species	Testes	Ovaries	Gonad position	Biotope	♂♂	♀♀
Aponchiidae						
No species studied						
"Desmodoridae"						
"Microlaiminae"						
Calomicrolaimus rugatus Lorenzen 1976	2 opp	2 out	± ventr. ± ventr.	Beach Colombia	4	7
Ixonema sp.	?	2 out	lr rl	Sublittoral, Chile	–	3
Microlaimus citrus	1 ant	2 antidr	lrlr rllr	Salt marshes	10	7
M. conspicuus Lorenzen 1973a	2 opp		l r	Helgoland	3	–
M. dentatus	2 opp	2 out	lr rl	Helgoland	4	6
M. globiceps	2 opp	2 out	lr rl	Salt marshes	4	6
M. aff. *honestus*	2 opp	2 out	lr rl ($^{lr}_{lr}$)	Salt marshes	7	15
M. tenuispiculum	2 opp	2 out	l r ($^{r}_{l}$)	Baltic	7	14
M. torosus Lorenzen 1973a	1 ant	2 out	lr rl	Helgoland	9	7
M. turgofrons	1 ant	2 antidr	l r ($^{l}_{l}$)	Helgoland	9	6

Table 11 continued

Species	Testes	Ovaries	Gonad position	Biotope	♂♂	♀♀
"Microlaiminae" (continued)						
Molgolaimus tenuispiculum	1 ant	2 antidr	lrlr rllr	Sublittoral, Chile	15	7
Ohridius bathybius	2 opp	2 antidr	$\frac{r}{l}$ (♂)	Type material	1	1
Paramicrolaimus primus	2 opp	2 antidr	lr rl	Sublittoral, Chile	5	5
Prodesmodora circulata	?	2 antidr	lrlr rllr	Fresh water	–	9
"Spiriniinae"						
Chromaspirina thieryi	1 ant	2 antidr	lr rl	Salt marshes	5	5
C. sp. 1	1 ant	2 antidr	lrlr rllr	Beach, Colombia	6	3
Metachromadora remanei	1 ant	2 antidr	lrlr rllr	Flats	12	11
Onyx septempapillatus	1 ant	2 antidr	lrlr rllr	Beach, Chile	9	7
Sigmophoranema rufum	1 ant	2 antidr	l r	Kattegatt	2	3
Spirinia laevis	1 ant	2 antidr	lrlr rllr	Helgoland	5	3
Pseudonchinae						
Pseudonchus sp	1 ant	2 antidr	♂ $\frac{ventr}{-}$ ♀ $\frac{r}{l}$	Beach, Colombia	1	1
Stilbonematinae						
Leptonemella aphanothecae	1 ant	2 antidr	♂ $\frac{ventr}{-}$	Helgoland	5	1
L. gorgo	1 ant	2 antidr	♀ lrlr rllr	Baltic	7	5
Desmodorinae						
Desmodora filispiculum Lorenzen 1976	1 ant	2 antidr	♀ lrlr rllr	Sublittoral, Chile	6	5
D. aff. *perforata*	1 ant	2 antidr	♂ subv. 1 lrlr rllr	Sublittoral, Chile	4	6
D. sinuata Lorenzen 1976	1 ant	2 antidr	♀ $\frac{l}{r}$ ♂ ventr	Beach, Colombia	6	2
D. aff *tenuispiculum*	1 ant	2 antidr	lrlr rllr	Sublit. Chile, Baltic	4	4
D. sp. 1	1 ant	2 antidr	lrlr rllr	Sublittoral, Chile	5	6
Paradesmodora immersa	1 ant	2 antidr	lrlr rllr	Beach, Chile	4	3

APPENDIX

Table 11 *continued*

Species	Testes	Ovaries	Gonad position	Biotope	♂♂	♀♀
"Xenellidae"						
Xenella suecica	?	1 post antidr	– / l	Beach, Chile	–	2
Ceramonematidae						
4 species	?	2 antidr		South America	3	4
Monoposthiidae						
Monoposthia costata	2 opp	1 ant antidr	l / r	Baltic	2	5
M. mirabilis	2 opp	1 ant antidr	l / r	Baltic	2	3
Nudora crepidata	2 opp	1 ant antidr	l ($_{rl}^{lr}$) / l	Sublittoral, Chile	9	7
"Richtersiidae"						
Richtersia inaequalis	1 ant	2 antidr	♂ ventr. ♀ lrlr rllr	Helgoland	8	8
Draconematidae						
Draconema sp.	1 ant	2 antidr	♂ ventr. ♀ lrlr rllr	Sublittoral, Chile	6	5
Epsilonematidae						
Epsilonema byssicola Lorenzen 1973b	1 ant	2 antidr	♂ ventr. ♀ lr lr	Chile	4	3
E. dentatum Lorenzen 1973b	1 ant	2 antidr	lrlr rllr	Beach, Chile	11	6
Metepsilonema hagmeieri	1 ant	2 antidr	ventr. ventr.	Helgoland	2	1
Perepsilonema crassum Lorenzen 1973b	1 ant	2 antidr	ventr. ventr.	Helgoland	8	12
"Comesomatidae"						
"Acantholaiminae"						
Acantholaimus sp. 1	1 ant	2 antidr	r / l	Beach, Chile	8	7
A. sp. 2	1 ant	2 antidr	r / l	Sublittoral, Chile	1	4
"Dorylaimopsinae"						
Dorylaimopsis punctatus	2 opp	2 out	l / r ($_{l}^{r}$)	Kattegatt	4	5
"Comesomatinae"						
Comesoma minima	2 opp	2 out	l / r	Sublittoral, Chile	2	1

Table 11 *continued*

Species	Testes	Ovaries	Gonad position	Biotope	♂♂	♀♀
"Sabatieriinae"						
Cervonema allometrica	2 opp	2 out	$\begin{smallmatrix}l\\r\end{smallmatrix}$ $(\begin{smallmatrix}r\\l\end{smallmatrix})$	Sublittoral, Chile	7	3
Laimella longicauda	2 opp	2 out	lr rl	Sublittoral, Chile	1	1
Sabatieria celtica	2 opp	2 out	$\begin{smallmatrix}l\\r\end{smallmatrix}$ $(\begin{smallmatrix}rlr\\llr\end{smallmatrix})$	Helgoland	19	10
S. hilarula	2 opp	2 out	$\begin{smallmatrix}l\\r\end{smallmatrix}$ $(\begin{smallmatrix}r\\l\end{smallmatrix})$	Kattegart	12	11
"Chromadoridae"						
"Spilipherinae"						
Ethmolaimus pratensis	2 opp	2 antidr	l l r l	Lake Plöner	1	4
Chromadorinae						
Atrochromadora microlaima	1 ant	2 antidr	r l	Flats	4	9
Chromadora kreisi	1 ant	2 antidr	r l	Helgoland	2	1
C. nudicapitata	1 ant	2 antidr	r l	Beach, Chile	6	9
Chromadorina bioculata	1 ant	2 antidr	r l	Lake Plöner	2	1
C. supralitoralis	1 ant	2 antidr	r l	Salt marshes	5	1
Prochromadorella attenuatea	1 ant	2 antidr	r l	Helgoland	5	5
P. paramucrodonta	1 ant	2 antidr	r l	Sublittoral, Chile	10	10
Punctodora ratzeburgensis	1 ant	2 antidr	r l	Lake Plöner	6	2
Euchromadorinae						
Euchromadora sp.	1 ant	2 antidr	r l	Sublittoral, Chile	3	2

APPENDIX

Table 11 *continued*

Species	Testes	Ovaries	Gonad position	Biotope	♂♂	♀♀
Hypodontolaiminae						
Chromadorita abnormis	1 ant	2 antidr	r l	Salt marshes	3	3
C. leuckarti	1 ant	2 antidr	r l	Lake Plöner	8	5
C. nana	1 ant	2 antidr	r l	Helgoland	6	3
Dichromadora cucullata Lorenzen 1973a	1 ant	2 antidr	r l	Helgoland	5	4
D. scandula	1 ant	2 antidr	r l	Salt marshes	7	1
Innocuonema tentabunda	1 ant	2 antidr	r l	Flats	1	5
Neochromadora poecilosoma	1 ant	2 antidr	r l	Helgoland	2	1
Spilophorella papillata	1 ant	2 antidr	r l	Salt marshes	10	5
S. paradoxa	1 ant	2 antidr	r l	Salt marshes	5	3
"Cyatholaimidae"						
Neotonchinae						
Neotonchus aff. *corcundus*	1 ant		lr ??	Helgoland	3	–
N. meeki	2 opp		l r	Helgoland	1	–
N. sp. 1	2 opp	2 antidr	lr rl	Sublittoral, Chile	1	1
N. sp. 2	2 opp	2 antidr	lr rl	Sublittoral, Chile	1	2
"Achromadorinae"						
Achromadora sp. 1	?	2 antidr	lr lr	River bank, Africa	–	2
A. sp. 2	?	2 antidr	lr lr	Woods	–	3
Pomponematinae						
Pomponema ammophilum	2 opp	2 antidr	lr rl	Helgoland	2	2
P. astrodes	2 opp	2 antidr	lr rl	Kattegatt	6	16
P. compactum	1 ant	2 antidr	lr rl	Helgoland	4	5
P. debile	2 opp	2 antidr	r l	Helgoland	4	1
P. elegans	2 opp	2 antidr	r l	Helgoland	8	1
P. multipapillatum	1 post	1 ant antidr	l l	Kattegatt	10	9

Table 11 continued

Species	Testes	Ovaries	Gonad position	Biotope	♂♂	♀♀
Paracanthonchinae						
Paracanthonchus caecus	1 ant	2 antidr	$\frac{r}{l}$	Helgoland	10	7
P. macrodon	2 opp	2 antidr	$\frac{r}{l}$ ($\frac{l}{r}$)	Helgoland	9	8
P. sp. 1	1 ant	2 antidr	$\frac{r}{l}$	Beach, Chile	5	8
Paracyatholaimus intermedius	2 opp	2 antidr	$\frac{r}{l}$	Salt marshes	1	4
P. occultus	2 opp	2 antidr	$\frac{r}{l}$	Helgoland	6	5
P. pentodon	2 opp	2 antidr	$\frac{r}{l}$	Helgoland	8	9
P. sp. 1	1 ant	2 antidr	$\frac{r}{l}$ ($\frac{l}{r}$)	River bank, Africa	7	6
Cyatholaiminae						
Longicyatholaimus complexus	2 opp	2 antidr	$\frac{r}{l}$ ($\frac{l}{l}$)	Kattegatt	9	9
Paralongicyatholaimus minutus	1 ant	2 antidr	$\frac{lr}{rl}$	Kattegatt	7	6
Praeacanthonchus aff. kreisi	2 opp	2 antidr	$\frac{r}{l}$	Kattegatt	8	10
"Choniolaimidae"						
Choniolaimus panicus	2 opp	2 antidr	$\frac{lrlr}{rllr}$	Kattegatt	8	9
C. papillatus	2 opp	2 antidr	$\frac{lr}{rl}$	Kattegatt	7	1
Halichoanolaimus robustus	2 opp	2 antidr	$\frac{lr}{rl}$	Flats	7	4
H. longicaudatus	2 opp	2 antidr	$\frac{lr}{rl}$	Kattegatt	2	3
"Selachinematidae"						
Synonchiella riemanni	2 opp	2 antidr	$\frac{lrl}{rll}$	Kattegatt	1	4

APPENDIX

Table 11 *continued*

"ENOPOLIDA"

Species	Testes	Ovaries	Gonad position	Biotope	♂♂	♀♀
"Tripylidae"						
"Tripylinae"						
Tripyla glomerans	2 opp	2 antidr	± ventr ± ventr	Lake Plöner	4	11
"Tobrilinae"						
Tobrilus gracilis	?	2 antidr	± ventr ± ventr	Lake Plöner	–	7
T. grandipapillatus	2 opp	2 antidr	± ventr ± ventr	Lake Selenter	8	11
"Prismatolaimidae"						
Onchulinae						
Onchulus nolli	?	2 antidr	lr ??	Rhine banks	–	2
Stenonchulus troglodytes	2 opp	2 antidr	lrlr rllr	terr. south Germany and the Azores	2	9
Prismatolaiminae						
Prismatolaimus dolichurus	?	2 antidr	l l	Moorland	–	6
P. aff. *verrucosus*	–	2 antidr	r r	Salt marshes	–	2
P. sp. 1	–	2 antidr	l l	River bank, Colombia	–	2
Odontolaiminae						
Odontolaimus aquaticus	–	1 post antidr	– ventr	River bank, USA	–	1
"Ironidae"						
Ironus ignavus	2 opp	2 antidr	llr rlr	Lake Plöner	12	12
I. sp.	?	2 antidr	lrlr rllr	River bank, Africa	–	18
Parironus bicuspis	1 ant	2 antidr	r r	Kattegatt	5	10
Trissonchulus acutus	2 opp		r r	Beach, Chile	1	–
"Cryptonchidae"						
Cryptonchus tristis	?	1 ant antidr	l –	River bank, USA	–	1
Tripyloididae						
Bathylaimus australis	1 ant	2 antidr	ventr ventr	Beach, Chile	4	1
B. inermis	1 ant	2 antidr	ventr ventr	Helgoland	4	3
B. parafilicaudatus	1 ant	2 antidr	ventr ventr	Helgoland	10	5
Tripyloides marinus	1 ant	2 antidr	ventr ventr	Flats	2	3

Table 11 continued

Species	Testes	Ovaries	Gonad position	Biotope	♂♂	♀♀
Trefusiidae						
Trefusiinae						
Cytolaimium exile	2 opp	2 out	? ?	Sublittoral, Chile	1	3
Rhabdocoma americana	2 opp		ventr l	Sublittoral, Chile	1	–
R. sp. 1	2 opp	1 post antidr	ant: ventr post: l or r	Sublittoral, Chile	4	2
R. sp. 2	?	2 antidr	rl ll	Helgoland	–	2
Trefusia longicauda	2 opp	2 antidr	lr lr	Kattegatt	2	3
T. sp. 1 (*c.* 7mm)	?	2 antidr	ventr ventr	Kattegatt	–	2
Trefusiinae, sp.	1 ant	1 ant antidr	lr –	Beach, Chile	1	2
Halanonchinae						
Halanonchus sp. 1	2 opp	1 post antidr	ant: ventr post: l or r	Sublittoral, Chile	4	2
"Alaimidae"						
Adorus tenuis	1 ant	1 post antidr	♂: l or r ♀:	Salt marshes	2	6
Alaimus primitivus	1 ant	1 post antidr	r r	Woods	1	8
Amphidelus puccinelliae	1 ant	1 post antidr	♂: r	Salt marshes	6	1
A. sp.	1 ant	1 post antidr	± ventr ± ventr	Salt marshes	2	3
"Oxystominidae"						
"Oxystomininae"						
Litinium bananum	2 opp	1 post antidr	♂: l l ♀: – r	Helgoland	1	1
Nemanema rotundicaudatum	1 ant		♂ ventr	Kattegatt	1	–
Oxystomina elongata	1 ant	1 post antidr	♂: l or r ♀: l or r	Flats and Kattegatt	5	8
Thalassoalaimus egregius	2 opp	1 post antidr	♂: ll ♀ – rl rl	Kattegatt	2	4
T. pirum	2 opp	1 post antidr	lrlr rllr	Salt marshes	7	5
T. tardus	2 opp	1 post antidr	l lr r lr	Flats and Kattegatt	4	6
T. septentrionalis	2 opp	1 post antidr	♂: l ♀: – l lr	Helgoland	1	3
Halalaiminae						
Halalaimus terrestris	2 opp	2 antidr	lrlr rllr	Salt marshes	3	2
H. sp. 1	2 opp	2 antidr	lrlr rllr	Kattegatt	9	6

APPENDIX

Table 11 *continued*

Species	Testes	Ovaries	Gonad position	Biotope	♂♂	♀♀
Lauratonematidae						
Lauratonema aff. *spiculifer*	1 post	1 post antidr	♀: l/– ♂: –/r	Beach, Colombia	8	9
L. hospitum	1 post	1 post antidr	♀: l/– ♂: –/r	Beach, Chile	2	2
Leptosomatidae						
Leptosomatinae						
Platycoma sudafricana	2 opp	2 antidr	lrlr rllr	Beach, Colombia	1	4
Thoracostomatinae						
Deontostoma arcticum	2 opp	2 antidr	l (lr) l (rl)	Chile	8	7
Triodontolaimidae						
Triodontolaimus acutus	?	2 antidr	ventr ventr	Beach, Colombia	1	2
Anticomidae						
Anticoma acuminata	1 ant	2 antidr	l l	Baltic	4	7
A. trichura	2 ant	2 antidr	l l	Red Sea	3	8
Phanodermatidae						
Crenopharynx marioni	2 opp		l l	Kattegatt	2	–
Phanoderma campbelli	2 opp		l l	Chile	2	–
Phanodermopsis necta	2 opp	2 antidr	l l	Beach, Colombia	3	6
"Enoplidae"						
"Thoracostomopsinae"						
Thoracostomopsis barbata	?	2 antidr	l l	Helgoland	–	4
Chaetonematinae						
Chaetonema riemanni Platt 1973	2 ant[16]	2 antidr	l l	Helgoland	2	7
Trileptiinae						
Trileptium sp.	?	2 antidr	l l	Beach, Chile	–	2
"Oxyonchinae"						
Fenestrolaimus sp.	–	2 antidr	l l	Kattegatt	–	5
Oxyonchus dentatus	2 opp	2 antidr	l l	Helgoland and Kattegatt	6	8
Saveljevia sp.	?	2 antidr	l l	Kattegatt	–	3

Table 11 *continued*

Species	Testes	Ovaries	Gonad position	Biotope	♂♂	♀♀
"Enoplolaiminae"						
Enoplolaimus connexus	2 opp	2 antidr	l l	Beach, Chile	2	2
E. propinquus	2 opp	2 antidr	l l	Beach, Baltic	5	7
Mesacanthion diplechma	2 opp	2 antidr	l l	Helgoland	5	5
"Enoploidinae"						
Enoploides labrostriatus	2 opp	2 antidr	l l	Helgoland	6	3
Epacanthion buetschlii	2 opp	2 antidr	l l	Helgoland	4	2
Metenoploides aff. *alatus*	2 opp	1 post antidr	l l	Beach, Chile	2	3
Enoplinae						
Enoplus brevis	2 opp	2 antidr	l l	Salt marshes	4	10
E. michaelseni	2 opp		l l	Chile	2	–
Rhabdodemaniidae						
Rhabdodemania minor	1 ant	2 antidr	lrlr rllr	Helgoland, Kattegatt	5	10
"Anoplostomatidae"						
Anoplostoma vivipara	2 opp[17]	2 antidr	l l	Salt marshes and brackish water	13	19
Pandolaimus latilaimus[18]	1 ant	2 antidr	± ventr ± ventr	Kattegatt	7	7
Oncholaimidae						
Pelagonematinae						
Pelagonema sp.	?	2 antidr	r r	Kattegatt	–	1
Oncholaimellinae						
Oncholaimelloides vonhaffneri	?	1 ant antidr	r –	Sublittoral Chile	1	4
Viscosia cobbi	2 opp	2 antidr	r r	Helgoland and Kattegatt	4	9
V. rustica	2 opp	2 antidr	r r	Helgoland and Kattegatt	6	7
Adoncholaiminae						
Adoncholaimus thalassophygas	1 post	2 antidr	r r	Brackish water Finland	3	3
Oncholaiminae						
Oncholaimus brachycercus	2 opp	1 ant antidr	r r	Beach Baltic	5	7
Oncholaimus skawensis	2 opp	1 ant antidr	r r	Kattegatt	1	2

APPENDIX

Table 11 *continued*

Species	Testes	Ovaries	Gonad position	Biotope	♂♂	♀♀
Pontonematinae						
Pontonema ditlevseni	2 opp	2 antidr	r r	Helgoland	3	1
Enchelidiidae "Thoonchinae"						
Ditlevsenella danica	2 opp	2 antidr	r r	Helgoland and Kattegatt	1	1
"Eurystomininae"						
Belbolla asupplementata (Juario 1974)	2 opp	2 antidr	lr lr	Kattegatt	2	–
Eurystomina assimilis	2 opp	2 antidr	r r (^l_l)	Beach, Colombia	3	1
Ledovitia sp.	2 opp	2 antidr	r r	Beach, Chile	2	2
DORYLAIMIDA						
Dorylaimina						
Aporcelaimus sp.	?	2 antidr	lr rl	River bank, Africa	–	7
Dorylaimus stagnalis	2 opp	2 antidr	lr lr	Lake Plöner	2	2
Doryllium uniforme	2 opp	1 post antidr	lrlr rllr	Salt marshes	10	12
Eudorylaimus carteri	2 opp	2 antidr	lrlr rllr	Salt marshes	4	8
E. aff. *carteri*	2 opp	2 antidr	lrlr rllr	Salt marshes	7	4
E. doryuris	2 opp	2 antidr	lr rl (^r_r)	Salt marshes	6	7
Longidorus sp.	?	2 antidr	lr lr	River bank Brasil	–	4
Paractinolaimus sp.	?	2 antidr	lr lr	River bank, Africa	–	28
Dorylaimina, sp. 1	2 opp	2 antidr	lr lr	Woods	9	10
Dorylaimina, sp. 2	2 opp	2 antidr	lrlr rllr	River bank, Brasil	6	5
Dorylaimina, sp. 3	?	1 post antidr	– lr	River bank, Africa	–	4
Mononchina						
Mononchus sp.	?	2 antidr	lr rl	Rhine banks	–	3
Mylonchulus sp. 1	?	2 antidr	lr rl	River bank, USA	–	9
M. sp. 2	?	2 antidr	lr rl	Woods	–	2
Prionchulus sp.	?	2 antidr	lr rl	River bank, Brasil	–	2

Table 11 *continued*

RHABDITIDA

Species	Testes	Ovaries	Gonad position	Biotope	♂♂	♀♀
Diplogasteridae						
Diplogaster rivalis	1 ant[19]	2 antidr	♂: ventr ♀: r/l	Fresh water	9	8
2 additional species	1 ant[19]	2 antidr	♂: ventr ♀: r/l	Salt marshes	4	6
Rhabditidae						
Protorhabditis sp.	1 ant[20]	2 homodr	r/l	Woods	2	1
Rhabditis (P.) marina	1 ant[20]	2 homodr	r/l	Waste waters	5	18
Bunonematidae						
Bunonema (B.) sp.		2 homodr	rlr / llr	Woods	–	4
B. sp.	1 ant[21]		ventr	Woods	1	–
Panagrolaimidae						
Panagrolaimus sp.	1 ant[22]	1 ant homodr	r / –	Salt marshes	14	13
Cephalobidae	[23]					
Acrobeles ciliatus	1 ant	1 ant homodr	r / –	Salt marshes	1	1
Cephalobus sp. 1	1 ant	1 ant homodr	r / –	Woods	6	8
C. sp. 2	1 ant	1 ant homodr	r / –	Woods	3	3
Eucephalobus paracornutus	1 ant	1 ant homodr	r / – (^{1})	Salt marshes	12	18

TYLENCHIDA

Species	Testes	Ovaries	Gonad position	Biotope	♂♂	♀♀
Tylenchoidea	[24]					
Helicotylenchus varicaudatus	1 ant	2 out	lr / lr	Salt marshes	5	16
Pratylenchus aff. *macrophallus*	?	1 ant out	lr / –	Salt marshes	–	9
Pratylenchus pratensis	1 ant	1 ant out	lr / –	Salt marshes	8	9
Tylenchorhynchus sp. 1	1 ant	2 out	lr / lr (^1_r)	Salt marshes	11	8
T. sp. 2	1 ant	2 out	lr / lr	Salt marshes	3	3
Tylenchoidea sp. (small)	1 ant	1 ant out	l / –	Salt marshes	1	6

APPENDIX

TYLENCHIDA *continued*

Species	Testes	Ovaries	Gonad position	Biotope	♂♂	♀♀
Aphelenchoidea	[24)]					
Aphelenchoides sp. 1	1 ant	1 ant out	r –	Salt marshes	8	10
A. sp. 2	1 ant	1 ant out	lr –	Woods	5	13

Footnotes to Table 11

1) The testis lies almost ventrally and has a short fold that faces to the right.
2) Banks of the River Magdalena, Colombia; material from Riemann (1970a).
3) The posterior testis has a small fold.
4) The posterior testis is smaller than the anterior testis.
5) Soft, sublittoral ground in Northumberland; material from Warwick & Buchanan (1970) and Lorenzen (1972).
6) The posterior testis is considerably smaller than the anterior testis and was not recognisable in some ♂♂; it continues anteriorly a short way and then folds over.
7) The posterior testis has a small fold.
8) The posterior testis continues anteriorly a short way and then folds over for a longer distance.
9) The posterior testis is folded over in the short anterior third.
10) The germinal zone is very short; neither testis is folded over.
11) The gonads are situated ventrally in each case.
12) Gill cavity of the Land Crab *Cardisoma guanhumi*; material from Riemann (1968).
13) Fresh water on the island of Ischia; material from Gerlach & Riemann (1971).
14) Gill chamber of the tropical Land Crab *Gecarcinus lateralis*; material from Riemann (1970).
15) Sandy beach on the Elbe at Hamburg; material from Gerlach & Reimann (1971).
16) The posterior testis has a small fold.
17) The posterior testis continues anteriorly and then has a long fold over.
18) Systematic position according to Jensen (1976).
19) The testis has a short fold to the right or left.
20) The testis has a fold to the ventral side.
21) The testis has a fold to the right.
22) In 12 ♂♂ the testis has a fold to the ventral side and in 2 ♂♂ it has a fold to the dorsal side.
23) In the animals studied from this family the testis has a fold to the ventral or dorsal side.
24) The testis is extended.

8.2.2. Concluding summary of data (Table 12).

Personal observations and data from the literature on gonad characteristics in freeliving nematodes are brought together in the following table. The table is constructed in the same way as the first four columns of the table in section 8.2.1; the symbols and abbreviations are also the same. There is one additional symbol:

* : A * in front of the name of a taxon indicates that the gonad characteristics of at least one species of this taxon were studied for the purposes of the current work, and that the species is therefore contained in the table in section 9.2.1.

APPENDIX

Table 12. See text 8.2.2.

"ARAEOLAIMIDA"

Taxon	Testes	Ovaries		Gonad position
* **Teratocephalidae**				
* *Euteratocephalus*	1 ant	2	antidr	lr lr
* *Teratocephalus*	1 ant	1 ant	"	r –
*"**Plectidae**"				
*"Plectinae ", in most cases	2 opp	2	antidr	r l
Plectus acuminatus	1 ant	2	"	
* Wilsonematinae, in most cases	–	2	"	"
Wilsonema otophorum	1 ant	2	"	
*"**Rhabdolaimidae**"				
* *Rhabdolaimus minor*	–	2	antidr	lr lr
* *R. nannus*	–	2	out	"
* *R. terrestris*	–	2	"	"
Rogerus	1 ant	2	"	
* *Syringolaimus* [1]	1 ant	2	antidr	lrlr lrlr
Isolaimiidae	2 opp	2	antidr	
* **Aulolaimidae**	2 opp	2	antidr	lr rl
* **Haliplectidae**	2 opp	2	antidr	lr rl
*"**Leptolaimidae**" [2]				
* Peresianinae	1 post	2	out	ventr ventr
*"Leptolaiminae "				
Alaimella	2 ant	1 post	antidr	
Anomonema	1 ant	2	antidr	
Anonchus, partim	1 ant	1 ant	"	
partim	1 ant	2	"	
* *Antomicron*, partim	2 opp	2	"	
partim	1 ant	2	"	r l
* *Aphanolaimus*, partim	2 opp	2	antidr	
partim	1 ant	2	"	
* *A. aquaticus*	1 post	2	"	r r
Aplectus	2 opp	2	"	
Assia	?	2	"	
* *Chronogaster*	2 opp	1 ant	"	r ?

Table 12 continued

Taxon	Testes	Ovaries		Gonad position
Cricolaimus	?	2	"	
Cynura	2 ant	2	"	
* Dagda	2 ant	2	"	l r
* Deontolaimus	2 ant	2	"	r l
Diodontolaimus	?	2	"	
Halaphanolaimus	?	2	"	
* Leptolaimoides	1 post	2	"	r l
* Leptolaimus, partim	2 opp	2	"	"
partim	2 ant	2	"	"
partim	1 post	2	"	"
Paraphanolaimus	2	2	"	
Paraplectonema	1 ant	2	"	
Plectolaimus	?	2	"	
* Rhadinema	2 ant	2	"	lrlr rllr
* Stephanolaimus, partim	2 ant	2	"	r l
partim	2 ant	1 post	"	"
partim	1 ant	1 post	"	"
*"Procamacolaiminae"	2 ant	2	"	"
*"Camacolaiminae", in most cases	2 ant	2	"	"
Camacolaimus monohystera	?	1 post	"	
* C. sp. 2	?	1 post	"	
Kreisonematinae	–	2	"	
* **Bastianiidae** 3)	2 opp	2	antidr	lrlr rllr
*"**Axonolaimidae**" 4)				
*"Cylindrolaiminae"				
* Aegialoalaimus, in most cases	2 ant	2	antidr	lrlr rllr
A. tenuis	?	2	out	
A. tenuicaudatus	1 ant	2	antidr	
Cylindrolaimus, in most cases	1 ant	2	out	
* C. sp. 1	2 ant	2	"	lr rl

APPENDIX

Table 12 *continued*

Taxon	Testes	Ovaries	Gonad position
C. monhystera	?	1 ant "	
* *Domorganus*	2 opp	2 antidr	
Linolaimus	?	2 "	
* *Parachromagasteriella*	2 ant	2 out	r l
Polylaimium	?	2 antidr	
*"Diplopeltinae ", in most cases	2 opp	2 out	lr rl
Alaimonema	1 ant	?	
Chitwoodia	?	?	
Pseudaraeolaimus	?	2	
*"Axonolaiminae ", in most cases	2 ant	2 out	
* *Apodontium*	2 ant	?	
Margonema	1 ant	2 "	lrlr or rllr
Odontophora villoti	1 ant	2 "	lr or l
O. wieseri	1 ant	2 "	rl r
* *Odontophoroides*	2 ant	1 post "	
Synodontium	2 ant	1 post "	
*"Campylaiminae "			
* *Campylaimus*	2 opp	2 "	lr rl
* *Diplopeltoides*	2 opp	2 antidr	"
* *Diplopeltula*, partim	2 opp	2 out	"
* *D.* aff. *nuda*	2 ant	2 antidr	
Paratarvaia	?	2 ?	
Striatodora	?	?	
* *Tarvaia*	?	2 antidr	r l

"DESMOSCOLECIDA"

Taxon	Testes	Ovaries	Gonad position
*"**Desmoscolecidae**" (sensu Lorenzen 1969) [5)]			
* Desmoscolecinae	1 ant	2 out	r less l' frequently
* Greeffiellinae	1 ant	2 "	l or ventr
* Tricominae	2 opp	2 "	r ventr
*"**Meyliidae**"	?	2 out	r l

Table 12 continued

"MONHYSTERIDA"

Taxon	Testes	Ovaries	Gonad position
*"Siphonolaimidae"			
Cyartonema flexile [6]	1 ant	2 antidr	
* C. germanicum [6]	?	2 "	
* C. zosterae [6]	2 opp	2 "	lr rl
Paraterschellingia [6]	?	2 "	
* Siphonolaimus	1 ant	1 ant out	l –
*"Linhomoeidae" [7]			
*"Desmolaiminae ", in most cases	2 opp	2 out	lr or l rl r
Metalinhomoeus variabilis	?	1 ant "	
Sarsonia	?	1 ant "	
Terschellingia monhystera	?	1 ant "	
Terschellingioides filiformis	1 ant	2 "	
* Eleutherolaiminae, partim	2 opp	2 "	lr
partim	1 ant	2 "	rl
*"Linhomoeinae ", in most cases	2 opp	2 out	
Anticyclus exilis	2 opp	1 ant "	
Paralinhomoeus fuscacephalum	1 ant	1 ant "	
P. uniovarium	?	1 ant "	
Sphaerocephalum crassicauda	?	1 ant "	
* Tubolaimoides	2 opp	2 antidr	lr lr
* **Monhysteridae** (sensu Lorenzen 1978c)	1 ant	1 ant out	r –
* **Sphaerolaimidae**, in most cases	2 opp	1 ant out	lr rl
* Parasphaerolaimus	1 ant	1 ant "	lr - -
Sphaerolaimus hadalis Freudenhammer, 1975	1 ant	1 ant "	
S. peruanus Freudenhammer, 1975	1 ant	1 ant "	
* **Xyalidae** (sensu Lorenzen 1978c), in most cases	2 opp	1 ant out	l r
22 species from different genera (see Table 11)	1 ant	1 ant "	l –

APPENDIX

Table 12 *continued*

"DESMODORIDA"

Taxon	Testes	Ovaries		Gonad position
Aponchiidae, partim	2?	1 ant	out	
partim	1 ant	1 ant	"	
***"Desmodoridae"** [8)]				
*"Microlaiminae", in most cases	2 opp	2	out	$\genfrac{}{}{0pt}{}{lr}{rl}$ or $\genfrac{}{}{0pt}{}{l}{r}$
* *Microlaimus torosus* Lorenzen, 1973	1 ant	2	"	
* *Molgolaimus*	1 ant	2	antidr	lrlr
* *Prodesmodora*	1 ant	2	"	rllr
* *Ohridius*	2 opp	2	"	
* *Paramicrolaimus primus*	2 opp	2	"	$\genfrac{}{}{0pt}{}{lr}{rl}$
*"Spiriniinae", in most cases	1 ant	2	antidr	$\genfrac{}{}{0pt}{}{lrlr}{rllr}$ or
				$\genfrac{}{}{0pt}{}{lr}{rl}$ or $\genfrac{}{}{0pt}{}{l}{r}$
Bolbolaimus	1 ant	2	out	
* Pseudonchinae	1 ant	2	antidr	
* Stilbonematinae	1 ant	2	"	
* Desmodorinae	1 ant	2	"	mostly $\genfrac{}{}{0pt}{}{lrlr}{rllr}$
Desmodora (X.) torquens		?1 ant	"	
***"Xenellidae"**	1 ant	1 post	antidr	
* **Ceramonematidae**, partim	1 ant	2	antidr	
partim	2 opp	2	"	
* **Monoposthiidae** [9)], in most cases	2 opp	1 ant	antidr	$\genfrac{}{}{0pt}{}{l}{r}$ or $\genfrac{}{}{0pt}{}{l(rl)}{l(lr)}$
Rhinema	2 opp	2	antidr	
***"Richtersiidae"**	1 ant	2	antidr	lrlr (in each
* **Draconematidae** [10)]	1 ant	2	antidr	rllr case subventral)
* **Epsilonematidae** [10)]	1 ant	2	antidr	or ventr ventr

Table 12 *continued*

"CHROMADORIDA"

Taxon	Testes	Ovaries	Gonad position
*"**Comesomatidae**" [11], in most cases	2 opp	2 out	$\frac{l}{r}$, less often $\frac{lr}{rl}$
*"Acantholaiminae"	1 ant	2 antidr	$\frac{r}{l}$
*"**Chromadoridae**" [12], in most cases	1 ant	2 antidr	$\frac{r}{l}$
* *Ethmolaimus*	2 opp	2 "	$\frac{rl}{ll}$
*"**Cyatholaimidae**" [13]			
* Neotonchinae, partim	2 opp	2 antidr	$\left. \begin{array}{l} rl \\ lr \end{array} \right\}$
partim	1 ant	2 "	
*"Achromadorinae "	2 opp	2 "	$\begin{array}{l} lr \\ lr \end{array}$
* Pomponematinae, in most cases	2 opp	2 "	
* *Pomponema compactum*	1 ant	2 "	
P. litorium	1 ant	2 "	$\frac{lr}{rl}$ or $\frac{l}{r}$
* *P. multipapillatum*	1 post	1 vo "	
P. syltense	?	1 vo "	
* Paracanthonchinae, in most cases	2 opp	2 "	$\left. \begin{array}{l} r \\ l \end{array} \right\}$
* 3 species (see Table 11)	1 ant	2 "	
Acanthonchus gracilis	1 post	2 "	
Dentatonema	2	1 vo "	
Xenocyatholaiminae	?	2 "	
* Cyatholaiminae, partim	2 opp	2 "	$\left. \begin{array}{l} lr \\ rl \end{array} \right\}$ or $\frac{r}{l}$
partim	1 ant	2 "	
*"**Choniolaimidae**"	2 opp	2 antidr	$\frac{lrlr}{rllr}$ or $\frac{lr}{rl}$
*"**Selachinematidae** " [14]	2 opp	2 antidr	$\frac{lrl}{rll}$

APPENDIX

Table 12 *continued*

"ENOPLIDA"

Taxon	Testes	Ovaries	Gonad position
*"**Tripylidae**"			
*"Tripylinae "			
Abunema	?	1 ant antidr	
Paratripyla	?	2 "	
* *Tripyla*	2 opp	2 "	± ventr ± ventr
Trischistoma[15)]	1 ant	1 ant "	
*"Tobrilinae "	2 opp	2 antidr	± ventr ± ventr
Tobrilia longicaudata	?	?1 ant "	
Monochromadorinae			
Monochromadora[16)]	?	1 ant "	
Sinanema	?	1 ant "	
Udonchus	?	1 ant "	
*"**Prismatolaimidae**"			
* Onchulinae	2 opp	2 antidr	lrlr rllr
* Prismatolaiminae, partim	2 opp	2 "	l or r l r
partim	2 opp	1 ant "	
* Odontolaiminae			
Odontolaimus aquaticus	?	1 post "	
O. chlorurus	?	2 "	
*"**Ironidae**", in most cases	2 opp	2 antidr	*Ironus:* lrlr rllr otherwise r/r
* *Parironus bicuspis*	1 ant	2 "	
Trissonchulus oceanus	2 opp	1 post "	
T. raskii	?	1 post "	
*"**Cryptonchidae**"	2 opp	1 ant antidr	
* **Tripyloididae**	1 ant	2 antidr	ventr ventr
* **Trefusiidae**			
Trefusialaimus Riemann 1974	1 ant	?	
* Trefusiinae, in most cases	2 opp	2 antidr	± ventr ± ventr
* *Rhabdocoma americana*	2 opp	1 post "	
* *R.* sp. 1	2 opp	1 post "	
* *Cytolaimium*	2 opp	2 out	
* Trefusiinae, sp.	1 ant	1 ant antidr	
* Halanonchinae	2 opp	1 post "	

Table 12 *continued*

Taxon	Testes	Ovaries	Gonad position
*"Alaimidae"			
* *Adorus*	1 ant	1 post antidr	♂ : lr ♀ : --
* *Alaimus*, partim	1 ant	1 post "	-- lr
partim	1 ant	2 "	
* *Amphidelus*, partim	1 ant	1 post "	
partim	1 ant	2 "	
A. effilatus	?	1 ant "	
A. papuanus Andrassy 1973	?	1 ant "	
*"Oxystominidae" [17]			
*"Oxystomininae", in most cases	2 opp	1 post antidr	
Nemanema filiforme	1 ant	1 post "	lrlr
* *N. rotundicaudatum*	1 ant	1 post "	rllr
* *Oxystomina elongata*	1 ant	1 post "	
Porocoma striata[18]	?	2 post	
* Halalaiminae	2 opp	2 "	lrlr
			rllr
Paroxystomininae	?	2 "	
* **Lauratonematidae** [19]	1 post	1 ant antidr	l
			r
* **Leptosomatidae** [20]	2 opp	2 antidr	lrlr
			rllr
* **Triodontolaimidae**	2 opp	2 antidr	ventr
			ventr
* **Anticomidae**			l
			l
Anticoma, partim	2	2 antidr	
* *A. acuminata*	1 ant	2 "	l
			l
A. lata	2 ant	2 "	
* *A. trichura*	2 ant	2 "	l
			l
A. typica	2 ant	2 "	
Paranticoma	1 ant	2 "	
* **Phanodermatidae** [21]	2 opp	2 antidr	l
			l
*"**Enoplidae**", in most cases	2 opp	2 antidr	l
			l
Mesacanthion monhystera	?	1 post "	
* *Metenoploides* aff. *alatus*	2 opp	1 post "	
Ruamowhitia orae	1 ant	2 "	
* **Rhabdodemaniidae** [22]	1 ant	2 antidr	lrlr
			rllr

APPENDIX

Table 12 *continued*

Taxon	Testes	Ovaries	Gonad position
*"**Anoplostomatidae**" [23], mostly	2 opp	2 antidr	l l
* *Pandolaimus*	1 ant	2 "	± ventr ± ventr
* **Oncholaimidae** [24]			
* Pelagonematinae, in most cases	?	2 antidr	r r
Curvolaimus	?	1 ant "	
Krampiinae	?	1 ant "	
* Oncholaimellinae, in most cases	2 opp	2 "	} r
* *Oncholaimelloides*	?	1 ant "	} r
* Adoncholaiminae, in most cases	2 opp	2 "	} r
* *Adoncholaimus thalassophygas*	1 post	2 "	} r
* Oncholaiminae	2 opp	1 ant "	r r
* Pontonematinae	2 opp	2 "	r r
Octonchinae	?	?	
"Dioncholaiminae"	?	?	
* **Enchelidiidae** [25], mostly	2 opp	2 antidr	r r
Calyptronema (Catalaimus) sabulicola	?	1 post "	
Symplocostoma tenuicolle	1	2 "	
Rhaptothyreidae	?	?	

"DORYLAIMIDA"

Taxon	Testes	Ovaries	Gonad position
* **Mononchidae**, partim	2 opp	2 antidr	lr rl
partim	2 opp	1 ant "	
Cobbonchulus, partim	2 opp	1 post "	
Mylonchulus, partim	2 opp	1 post "	
Bathyodontidae			
Bathyodontinae		2 antidr	
Mononchulinae		1 ant "	
Nygolaimidae, in most cases	2 opp	2 antidr	
Nygellus, partim		1 post "	
Campydoridae			
Campydora	?	1 ant antidr	
Nygolaimellus	?	2 "	

Table 12 *continued*

Taxon	Testes	Ovaries	Gonad position
* **Dorylaimidae**, in most cases	2 opp	2 antidr	see 8.2.1.
Discolaimus, partim		1 post "	
Eudorylaimus, partim		1 post "	
Meylonema		1 ant "	
Pungentus, partim		1 post "	
Thornenema		1 post "	
Actinolaiminae,	2 opp	2 antidr	
Actinolaimus, partim		1 post "	
* **Longidoridae**, in most cases	2 opp	2 antidr	
Discomyctus		1 ant "	
Mumtazium		1 post "	
Tylencholaimus, partim		1 ant "	
T. zeelandicus		1 post "	
Xiphinema, partim		1 post "	
Belondiridae, partim	2 opp	2 antidr	
partim[26)]	2 opp	1 post "	
Dorylaimellus, partim		1 ant "	
* **Leptonchidae**	2 opp		
Dorylaimoides, Leptonchus		2 antidr	
Proleptonchus		1 ant "	
Aulolaimoides, Doryllium, Tylencholaimellus, Tyleptus		1 post "	
Encholaimidae Golden and Murphy 1967	2 opp	1 post antidr	
Opailaimidae	?	2 antidr	
Diphtherophoridae, mostly	1 ant	2 antidr	
Trichodorus monohystera		1 ant "	

"RHABDITIDA"

Taxon	Testes	Ovaries	Gonad position
* **Diplogasteridae**			
* **Diplogasterinae**, in most cases	1 ant	2 antidr	$\genfrac{}{}{0pt}{}{r}{l}$ in ♀♀
partim	1 ant	1 ant "	
Odontopharynginae	1 ant	1 ant "	
Tylopharynginae	1 ant	2 "	

APPENDIX

Table 12 *continued*

Taxon	Testes	Ovaries	Gonad position
Cylindrocorporidae			
Cylindrocorpus	1 ant	2 ?	
Goodeyus	1 ant	1 ant ?	
Myctolaimus	1 ant	2 ?	
Pseudodiplogasteroididae	1 ant	2 ?	
* **Rhabditidae**, in most cases	1 ant	2 homodr	r l
partim	1 ant	1 ant "	
* **Bunonematidae**	1 ant	2 homodr	rlr llr
Brevibuccidae	1 ant	2 ?	
* **Panagrolaimidae**	1 ant	1 ant homodr	r −
Myolaimidae	1 ant	1 ant homodr	
Chambersiellidae, partim	1 ant	2 ?	
partim	1 ant	1 ant ?	
* **Cephalobidae**	1 ant	1 ant homodr	r −

"TYLENCHIDA"

Taxon	Testes	Ovaries	Gonad position
* **Tylenchoidea**			mostly lr / lr
Tylenchidae, partim	1 ant	2 out	
partim	1 ant	1 ant "	
Heteroderidae	1 ant	2 out	
Hoplolaimidae, partim	1 ant	2 out	
partim	1 ant	1 ant "	
Tylenchulidae	1 ant	1 ant out	
Criconematidae	1 ant	1 ant out	
Neotylenchidae	1 ant	1 ant out	
Allantonematidae	1 ant	1 ant out	
* **Aphelenchoidea**			lr / -- or r / −
Aphelenchidae	1 ant	1 ant out	
Aphelenchoididae	1 ant	1 ant out	
Paraphelenchidae	1 ant	1 ant out	
Anomyctidae	?	1 ant out	

Footnotes to Table 12

1) Dubious datum: ovaries described as outstretched in *Syringolaimus* aff. *breviacaudatus* by Timm (1963).
2) Dubious data: ovaries described as outstretched in *Anonchus mangrovi* by Gerlach (1957), in *Antomicron donsi* by Allgen (1947), in *Leptolaimus ampullacea* by Warwick (1970). Testes described as facing in opposite directions by *Camacolaimus barbatus* by Warwick (1969).
3) Dubious datum: ovaries described as outstretched in *Bastiania parexilis* by de Coninck (1935).
4) Dubious data: ovaries described as reflexed in *Axonolaimus orcombensis* by Warwick (1970), in *A. ponticus* by Gerlach (1951), in *Odontophora fatisca* by Vitiello (1970), in *Diplopeltula incisa* by Ditlevsen (1928). Cobb (1920) records only one anterior testis in *Apodontium*, whereas there are two anterior testes according to personal observations.
5) Dubious data: ovaries described as reflexed in *Greeffiella dasyura* by Cobb (1920), in *Tricoma spuria* by Inglis (1967) and generally throughout the Desmoscolecidae by Chitwood & Chitwood (1950:22) and Chitwood (1951:642).
6) Dubious data: ovaries described as outstretched in *Cyartonema flexile* by Cobb (1920), in *C. germanicum* by Juario (1974), in *Paraterschellingia brevicaudatus* by Kreis (1924) and in *P. fusiforme* by Gerlach (1951). According to personal observations, the fold of the ovaries is sometimes very difficult to detect in *Cyartonema*. According to Juario (1974), *Southernia* is a synonym for *Cyartonema*.
7) Dubious data: ovaries described as reflexed in *Coninckia mediterranea* by Vitiello (1973), in *Disconema minutum* by Vitiello (1969), in *Metalinhomoeus gracilior* by Allgen (1959), in *M. variabilis* by Murphy (1965), in *Paralinhomoeus tenuicaudatus* by Allgen (1935, as *P. mirabilis*), in *Terschellingia longicaudata* by Vitiello (1969), in *T. parva* by Vitiello (1969). *Pandolaimus*: see "Anoplostomatidae".
8) Dubious data: ovaries described as reflexed in *Microlaimus cyatholaimoides* by Hopper & Meyers (1967), in *M. pecticauda* by Murphy (1966), in *M. spirifer* by Warwick (1970). Additional "*Microlaimus*" species with reflexed ovaries belong to the genera *Molgolaimus* and *Prodesmodora*.
Additional dubious data: ovaries described as outstretched in *Acathopharynx japonicus* by Steiner & Hoeppli (1926), in *Chromaspirina denticulata* by Gerlach (1953), in *Desmodora (D.) hirsuta* by Luc & de Coninck (1959), in *D. (Croconema) sphaerica* by Kreis (1928).
The presence of two testes is assumed for some species, e.g. for *Desmodora (D). communis* by Schulz (1932, as *leucocephala*), *D. (D.) pilosa* by Filipjev (1946, as *gorbunovi*), for *Onyx perfectus* by Filipjev (1918). More definite evidence of two testes is not given.
9) Chitwood & Chitwood (1950: 156) erroneously record two testes for *Monoposthia*, although it was already known that *M. costata* and *M. mielcki* only have one testis.
10) Incorrect datum: ovaries recorded as outstretched in the Draconematoidea (Draconematidae plus Epsilonematidae) by Lorenzen (1974b: 407).
11) Dubious data: the ovaries are described as reflexed in *Dorylaimopsis hawaiiensis* by Allgen (1951), in *Hopperia massiliensis* by Vitiello (1969), in *Mesonchium* by Chitwood & Chitwood (1950: 143), in *Comesoma minimum* by Chitwood & Chitwood (1950: 143). Presumably folds in the germinal zone have been interpreted as the reflexed condition.
12) Dubious data: occasionally two testes are recorded, e.g. in *Chromadorita mucrodonta* by Steiner (1916), in *Dichromadora cephalata* by Filipjev (1922, as *cricophana*), in *D. microdonta* by Kreis (1929), in *D. scandula* by Hopper (1969). In this last species, personal investigations clearly revealed the presence of only one anterior testis.
13) Dubious data: ovaries described as outstretched in *Cyatholaimus canariensis* by Steiner (1921), in *Microlaimus cervoides* by Vitiello (1970), in *Paracanthonchus falklandiae* by Allgen (1959), in *P. maior* by Kreis (1928).
14) Dubious data: ovaries described as outstretched in *Synonchiella major* by Murphy (1965) and in *S. minor* by Murphy (1965).
15) The observation of only one testis originates from Riemann (1974: 42).
16) The type species of *Monochromadora*, *M. monohystera*, has a reflexed, and not an outstretched ovary, as shown by a re-investigation of the types by Schiemer (1978: 189).
17) Dubious data: ovaries described as outstretched in *Halalaimus gracilis* by Allgen (1935) in *Nemanena filiforme* by Luc & de Coninck (1959), in *N. rotundicaudatum* by Ditlevsen (1926, the fig. shows quite clearly, however, an antidromous reflex), in *Oxystomina brevicaudata* by Kreis (1929), in *O. novozemeliae* by Filipjev (1927), in *O. oxycaudata* by Allgen (1929, the fig. shows, however, an antidromous reflex in the ovary). One anterior ovary is said to occur in *Thalassoalaimus oxycauda*. *Wieseria* is said to have 2 ovaries, although V = 30, 57% is recorded.
18) The original anterior ovary continues anteriorly for a short way at first and then folds over

towards the posterior, with the result that the entire germinal and growth zone lies behind the vulva (Cobb 1920, Gerlach 1962).
19) Dubious data: according to Gerlach (1953), *Lauratonema reductum* has one testis that is extended towards the anterior.
20) Dubious data: according to Bresslau & Stekhoven (1940), *Synonchus longisetosus* has only one ovary, although V = 62, 5%.
21) Dubious data: ovaries described as outstretched in *Phanoderma (P.) etha* by Gaden (1960) and in *Phanodermella longicaudata* by Kreis (1928; however the fig. clearly shows an antidromous reflex in the ovaries).
22) Dubious datum: according to Filipjev (1927), *Rhabdodemania minor* has 2 testes. Personal investigations revealed the presence of only one anterior testis.
23) Dubious datum: outstretched ovaries described in *Anoplostoma elegans* by Kreis (1928).
24) Dubious data: ovaries described as outstretched in *Kreisoncholaimus, Mononcholaimus, Pelagonema* and *Pseudopelagonema* by Kreis (1932 and 1934), in *Krampia opaca* by Allgen (1930), in *Metaparoncholaimus campylocercus* by de Coninck & Stekhoven (1933), in *Metoncholaimus pristiurus* by Stekhoven & Adam (1931, as *denticulatus*), in *Mononcholaimus rusticus* by Kreis (1929), in *Oncholaimellus labiatus* by Kreis (1932), in *Pelagonema omalum* by Vitiello (1970, the smaller female), in *Pontonema balticum* by Schulz (1932). A posterior ovary was recorded in *Oncholaimus keiensis* by Kreis (1934). Two ovaries, instead of only one anterior ovary are recorded for the following, inadequately described *Oncholaimus* species: *Oncholaimus microdon, O. notoviridis* and *O. tobagoensis*; their membership of the genus *Oncholaimus* is dubious.

In those instances, where Kreis (1932 and 1934) deals with the number of testes, (in his numerous species descriptions) he always records the presence of only one anterior testis. Since two testes have almost always been found by other authors and during personal research, Kreis' data should be regarded as dubious.
25) Dubious datum: ovaries described as outstretched in *Symplocostoma tenuicolle* by Luc & de Coninck (1959).
26) In *Axonchium* and *Belondira*, there is only a rudiment of the anterior ovary; this does not produce eggs (Nair & Coomans 1973, Goodey 1963).

Index

[Figures marked in **bold type**]
[<u>Underlined</u> numbers indicate taxon classification]

Abunema 266, 353
Acantholaiminae 67, 81, 86, 92–3, 100, 137–8, 211, 302, 304, 319, 335, 352
Acantholaimus 95, 101, 133, 137–8, 335
Acanthonchus 145
A. gracilis 352
Acanthopharynginae 150
Acanthopharyngoides 150
Acanthopharynx 150
A. japonicus 358
Achromadodoridae 168
Achromadorinae 337, 352
Achromadora 34, 143, 337
A. buikensis 320
Achromadoridae 136, <u>142</u>–3, 178, **288**, 301, 304
Achromadorinae 46, 47, 92, 101, 143–4, **288**, 320
Acmaeolaimus 280, 306
Acontiolaimus 170
Acrobeles ciliatus 344
Actarjana 212
Actinolaiminae 356
Actinolaimus 74, 356
A. microdentatum 84
Actinonema 139, 320
A. pachydermatum 134
Aculeonchus 150
Adenophorea 16, 29, 31, 33, 38, 40, 42–3, 47–8, 62, 64–5, 80, 81, 83, 85–6, 88, 91, 95, 98, 103–07, 111, 113–16, 120–1, 123, **125**–6, 171–2, 174, 222, 226, 245–6, 273, 280–81, **286**, 300, 304, 306, 317
Adeuchromadora 139, 320
adhesive glands 163–4
Adoncholaiminae 73, 247–8, <u>252</u>–3, **296**, 342, 355
Adoncholaimus 247, 253
A. thalassophygas 37, 73, 88, 246, 274, 342, 355
Adorus 76, 108–9, 114–15, 243, 279, 304, 321, 354
A. tenuis 340
Adulolaimidae 65

Aegialoalaimidae 167, <u>179</u>, **180**, 204, 213, **290**, 301, 305
Aegialoalaimus 53, 92, 180–81, 213, 305, 330, 348
A. elegans 330
A. tenuis 348
A. tenuicaudatus 348
Aegialospirina 177
Africanthion 230
Alaimella 167–8, 347
A. truncata 88
Alaimellinae 168, 302
Alaimidae 56, 71, 76, 86, 93, 108–09, 114–15, 217–19, 243, 276–7, <u>279</u>, 321, 340, 354
Alaimina 217–18, 277, 280
Alaiminae 217
Alaimonema 150–51, 213, 304, 349
Alaimonemella [see also *Litinium*] 243, 272, 304–5
A. simplex [see also *Litinium bananum*] 243
Alaimus 279, 354
A. primitivus 72, 340
Allantonematidae 357
Allgenia 269
Allgeniella 139
Allomonhystera 201
Alyncoides 234
Alveolaimus 169
Amblyura 280
Ammotheristus 200
Amphidelus 72, 279, 340, 354
A. effilatus 354
A. papuanus 354
A. puccinelliae 340
amphids 14, 30, 31, 36, 38, 40–2, 47, 55–62, 129, 137, 158, 214, 232, 244, 269, 301
  bubble-shaped/blister-like 67, 124, 130
  circular 176, 271
  fovea 47, 49, 53, 55, 57, 64, 66, 216, 231, 261, 263, 267
  homology with gastrotrich lateral sensory organs 60, 62

INDEX 361

inverted U-shaped 66, 180, 182
loop-shaped 196–7, 208, **212**–13
multi-spiral 141–2, 144, 211, **212**
non-spiral 32, **49**, **50**, 52–3, 55, 57,
  59–60, 62, 64–5, 124, 127, 132, 159,
  161, 166–7, 172, 177, 192–3, 195,
  216, 246, 254, **255**, 263, 270–71,
  276
O-shaped 215
papilliform 62
plectoid 66
pocket-shaped (non-spiral) 49, 51,
  55, 60, 171, 182, 192, 194, 239, 244,
  263, 265–6, 273–6
pore-shaped (non-spiral) 49, 51, 173,
  208
position of 63, 137–8, 173, 224, 226,
  244, 300
round 49, 53, 66, 152, 157, 173–175,
  178, 180, 187, 196–8, 204, 205
sensory filaments 48
sexual dimorphism 53, 57–60, 67
spiral 14, 47, 49, 60, 64, 129–30, 141,
  143, 146, 207, 209, 213, 215–16,
  261, 270
  dorsally **49**, **50**, 51, 53–5, 57,
    63–6, 127, 162, 189, 191–2,
    246–**257**, 260, 271, 273
  laterally 55
  ventrally **49**–**51**, 52–5, 59, 63–7,
    126–7, 132, 135–6, 140, 149, 162,
    166–7, 169, 172, 179, 182, 185, 187,
    189, 192, 196, 260, 271
structure of 47, 99, 100, 107, 110–11,
  130–31, 173, 188, 193, 203, 232,
  242, 257, 259, **260**, 300
Amphimonhystrella 200
Amphimonhystera 200
A. anechma 37
Amphispira 150
Amphistenus [see *Symplocostoma*]
  258
Anaplectus 51, 70, 127, 166, 172–3
  A. granulosus 326
  A. submersus 326
Anaxonchium 144
  A. litorium see *Pomponema litorium*
Andrassya 266
Anguillulata 120
Anguillulidae 281
Anguinoides 170
Angustinema 243, 280, 306
Anivanema 241
Anomonema 51, 167–8, 176, 347
Anomyctidae 357

Anonchinae 169–70, **290**, 305
Anoncholaimus 250
Anonchus 167, 169, 347
  A. maculatus 104, 167
  A. mangrovi 104, 358
  A. mirabilis 104
Anoplostoma 105, 107, 109, 232, 246,
  322
  A. camus 231
  A. elegans 359
  A. vivipara 38, 73, 75, 231, 342
Anoplostomatidae 34, 38, 56, 73, 75–6,
  78, 93, 102, 105, 107–09, 115, 205,
  223, 231–2, 244, 246, 270, **296**, 305,
  322, 342, 355
Anoplostomatinae 231, 232, **296**
Antarcticonema 164–5, 305
Anthonema 168, 172–3
Anthraconema 204
Anticoma 89, 235–6, 354
  A. acuminata 72, 89, 235, 341, 354
  A. allgeni 235
  A. lata 88, 354
  A. strandi 235
  A. trichura 72, 77, 89, 235, 341, 354
  A. typica 88, 354
Anticomidae 56, 72, 75–8, 88, 97,
  102–3, 107, 109, 115, 221, 223, 231,
  235–6, 240, 276, **296**, 322, 341, 354
Anticomopsis 236
  A. typicus 235
Anticyathus 206
Anticyclus 74, 205–6
  A. exilis 350
  A. junctus 197, 205
Antomicron 168, 327, 347
  A. donsi 358
  A. elegans 88, 327
  A. pratense 104
  A. profundum 104
Antopus 236
  A. serialis 235
anus 79, 97, 274
Apenodraconema 156
Aphanolaiminae 168
Aphanolaimus 70, 114, 167–8, 347
  A. aquaticus 88, 104, 327, 347
Aphasmidia 113, 120
Aphelenchidae 357
Aphelenchoidea 93, 96, 218, 285, 345,
  357
Aphelenchoididae 357
Aphelenchoides 74, 345
Aplectus 169, 347
Apodontium 208, 209, 303, 349, 358

A. pacificum 329
A. procerum 209
Aponchiidae 81, 93, 96, 98–9, 126, 133–5, 149, 156–158, 159, 197, **288**, 319, 333, 351
Aponchium 159
A. cylindricolle 159
Aponcholaimus 207
Aponema 152, 157–8, 304
Aponematinae 152, 156–7, **288**, 302
Aporcelaimus 74, 343
Araeolaimica 130–31
Araeolaimida 33, 35, 37, 42, 45, 51, 53, 61, 70, 76, 81–2, 84–6, 89, 92, 96–9, 101, 103–05, 108, 123, 127, 133, 166, 176, 179, 186, 188, 194, 197, 237, 259, 261, 279, 301, 318, 324, 347
Araeolaimidae 14, 33, 54, 83
Araeolaimoidea 168
Araeolaimoides 214, 303
A. microphthalamus 214
A. ovalis 214
A. paucisetosus 214
Araeolaimus 214, 303, 330
A. bioculatus 214
A. elegans 330
A. gajevskii 88
Archepsilonema 154
Archionchus 279
Arenasoma 261
Ascolaimus 53, 70, 91–2, 98, 130, 208–9, 211
A. elongatus 70, 329
Assia 51, 167–9, 305, 347
Asymmetrella 259
Asymmetrica 243
atavism 144
Atlantadorus 279
Atractidae 274
Atrochromadora 138
A. microlaima 336
Atylenchus 322
Aulolaimidae 51, 81, 92, 108, 126–7, 133, 167, 177–8, 181, 243, **290**, 305, 318, 326, 347
Aulolaimoides 356
Aulolaimus 166, 177–8
A. aquaedulcis 178, 243, 305
A. costatus 326
Aulacomya ater 26
Austranema 139, 320
Austronema 202
auxillary organs, male see copulatory organs

Axonchium 359
Axonolaimidae 43, 52–3, 60–1, 66, 81–2, 84, 86, 88–9, 91–2, 97, 108, 126, 133, 168, 177, 179, 181, 186, 188, 196–7, 207, 208–211, 213, 215, **292**, 304–5, 318, 330, 348
Axonolaiminae 81, 329, 349
Axonolaimoidea 128, 179, 197, 205, 207, 213, **292**
Axonolaimus 60, 70, 131, 208–9, 305
A. annelifera 209
A. helgolandicus 37, 240
A. orcombensis 358
A. paraspinosus 329
A. ponticus 358

Baicalobrilinae 261, **294**, 303
Baicalobrilus 264
Barbonema 30, 235, 240, 322
B. flagrum 239
B. horridum 44, 239
Barbonematinae 235, 240, **296**
Bastiania sp. 65–6, 104, 191
B. gracilis 33, 37, 53, **54, 330**
B. parexilis 358
Bastianiidae 33, 37, 45, 53, 63–6, 81, 104, 127, 133, 167, 189, 191–3, **290**, 318, 330, 348
Bathyepsilonema 153–4
Bathyeurystomina 258
Bathylaimella 207
Bathylaimoides 261
Bathylaimus 34, 44, 169, 209, 260–61
B. australis 71, 75, 339
B. inermis 339
B. parafilicaudatus 339
Bathyodontida 276–8, 355
Bathyodontina 217, 277, 278
Bathyodontinae 355
Bathyodontus 278
Bathyonchus 167–8
Belbolla 55, 247, 254, 256, 258
B. asupplementata 73, 343
Belbollidae 248, **296**
Belbollinae 254, 303
Belondira 359
Belondiridae 356
Biarmifer 145
biotope
  freshwater 26, 127
  marine 25, 127
  terrestrial 27, 127
Bitholinema 173
Bla 150

body rings, ornamentation of 179, 181, 209
Bognenia 272
Bolbella [see *Belbolla*] 258
Bolbolaiminae 156–8, **288**, 302
Bolbolaimus 95, 151, 158, 351
Bolbollinae 256
Bolbonema 150
Boucherius 165
Boreomicrolaimoides 146
Boveelaimus 169
Bradybucca 250
B. rhopolurus 250
Bradylaimoides 150
Bradystoma [see *Calyptronema*] 258
Branchinema 199
Bremerhaven Checklist, the [see also reference Gerlach and Riemann 1973/1974] 16, 213, 249, 307
Brevibuccidae 282, 284, 357
buccal cavity 14, 47, 66, 99, 127, 133–4, 139, 142–4, 146–8, 153, 155–6, 158–9, 161, 166–7, 169, 171–3, 175–6, 178–80, 182, 185, 187–9, 191–6, 198–9, 202–05, 207–08, 210–11, 213, 215–16, 218–19, 221–2, 224, 226, 228, 231–40, 242–3, 246, 249–251, 254–7, 259–60, 262–4, 266–7, 269–79, 282–4
Buccolaimus 202
Bulbopharyngiella 148
Bunonema 98, 344
Bunonematidae 91, 93, 282, 284, 344, 357
bursae, copulatory see copulatory organs

Cacolaimus 252, 305
C. papillatus 252
Caenorhabditis see *Rhabditis*
Calligyridae 164, 302
Calligyrus 100, 164, 318
C. gerlachi 328
Calomicrolaimus 158
C. rugatus 333
Calyptronema 247, 254, 258
C. maxweberi 73, 248, 254, **255**
C. sabulicola 246, 255, 355
Camacolaiminae 93, 126, 166–169, 170, **290**, 302, 318, 330, 348
Camacolaimoides 170
Camacolaimus 170, 330, 348
C. barbatus 330, 358

C. monohystera 167, 348
C. tardus 91, 330
Campydora 355
Campydoridae 355
Campylaiminae 81, 179, 181, 213, 214, 302, 318, 329–30, 349
Campylaimus 53, 214, 329, 349
C. gerlachi 329
canalis, of amphids 48–9, 51–53, 56–7, 60, 63–67, 179, 191
cardia, structure of 129–131, 136, 189, 191–2, 194–5, 204, 263, 265
cardial glands 194, 263
Caribplectus 169
Catalaimus [see *Calyptronema*] 258
C. acuminatus [see *Calyptronema paradoxum*] 258
Catanema 151
catecholamine 40
caudal glands 14, 40, 107–111, 113, 124, 126–7, 177, 192, 198–9, 202, 208, 213, 216, 221–2, 225, 228, 231, 233, 237, 239, 243, 245, 259, 263, 265, 273, 276, 281
occurence in tail 108, 110, 112, 127, 185, 196, 221, 224, 233, 235, 237, 242, 259, 261, 263, 266–70, 272, 275–6, 300
caudal papillae 123, 174
Cavilaimus 250
cell
epidermal 35, 49, 114, 116, 268
glandular [see also epidermal glands] 113–14, 116
intestinal 147, 203
pocket 35, 49
sensory 35, 68–9
sheath 49
socket 34, 48–9
Cenolaimus 201
Cephalanticoma 236
cephalic capsule 221–4, 226, 231–2, 234–5, 239–40, 242, 259, 266–267, 269, 275
cephalic organs (=cephalic slits) 224, 226
Cephalobidae 93, 174, 284, 345, 358
Cephalobus 74, 344
Cephalodorylaiminae 30, 322
Cephalonema 238
Ceramonema 127, 183, 303
C. chitwoodi 182
Ceramonematidae 93, 127, 149, 166–7, 179, 181–6, **290**, 302, 306, 319, 335, 351

Ceramonematinae 319
Ceramonematoidea 179, 182
Ceramonemoides 183, 303
cervical gland 114–16, 127, 159, 167, 171–2, 191–2, 199, 200, 202, 210, 222–4, 226, 231–2, 235, 239, 242–3, 245–6, 259–60, 263, 265–6, 268–9, 271, 273, 275–6, 300
cervical pore 36, 115, 158, 187–8, 226, 231, 235, 237, 239, 272–3, 275–6
Cervonema 210–212, 319
C. allometrica **30**, 33, 336
Chabertia sp. 74
Chaetonema 49, 55, 57, 60, 63, 107, 225, 232, 244
C. amphora 231
C. captator 231
C. riemanni 38, 57, **58**, 73, 75, 109, 231, 341
Chaetonematidae 225
Chaetonematinae 57, 67–8, 73, 78–9, 108–9, 225, 231, 232, **296**, 305, 322, 341
Chaetosoma 154–5
Chaetosomatidae 154
Chambersiellidae 57, 282, 284, 357
Chaolaimus 279
Cheilopseudonchus 151
Cheilorhabdions 134
Cheironchus 148
Chitwoodia 185–6, 213, 305, 318, 349
Chloronemella 207
Choanolaimidae 146
Choanolaiminae 135
Choanolaimus 148, 320
Choniolaimidae 30, 34, 47, 92, 103, 146–47, **288**, 302, 320, 338, 352
Choniolaimus 146, 148
C. panicus 338
C. papillatus 338
Chromadora 74, 138
C. kreisi 336
C. nudicapitata 336
Chromadorata 120, 127, 129, 131, 133
Chromadorea 123
Chromadorella 138
C. salicaniensis 134
Chromadoria 42, 45, 47, 49, 51–3, 55, 62–4, 66–7, 91, 101, 108, 121, 123, **125**, 126, 127–32, 161, 177, 184–5, 188–9, 192–5, 210, 216, 218–19, 223, 243, 275, **286**, 319
Chromadorica 130–1
Chromadorida pharetra 104
Chromadorida 14, 30, 33–5, 37, 42–3, 45, 74, 81, 83, 85–6, 88, 92, 95, 98, 101–03, 105–6, 124, **125**–129, 132–3, 136, 139, 149, 157–160, 166, 170, 196–7, 210–11, 219, 274, 276, 279, **286**, 301
Chromadoridae 53, 86, 93, 95, 101, 103–4, 108, 132–7, 139–43, 170, 211, 276, **288**, 301, 304, 320, 352
Chromadorimorpha 121
Chromadorina 47, 99, 124, **125**–130, 133–5, 142–3, 146, 148, 156, 159, 166, 196–7, 210, 261, **286**, **288**, **290**
Chromadorina 35
C. bioculata 336
C. supralitoralis 336
Chromadorinae 135–6, 138, 213, **288**, 320, 336
Chromadorissa 139
Chromadorita 139
C. abnormis 337
C. leuckarti 337
C. mucrodonta 358
C. nana 337
C. tenuis 108
Chromadoroidae 136
Chromadoroidea 135–7, 143, 147, 149, 156, 211, **288**, 336
Chromadoropsis 138, 151
Chromagaster 204
Chromaspirina 150–51, 334
C. denticulata 358
C. thieryi 334
Chronogaster 51, 168, 347
C. alata 171
C. boettgeri 171
C. magnifica 171
C. typica **122**, 171, 327
Chronogasteridae 166, 170, **290**
cilia, sensory 35, 41, 111
Cinctonema 158
cladistics see phylogenetic systematics
Claparediella 155
Claparediellidae 154
cloaca 79, 97, 103, 274
Cobbia 201
Cobbiacathonchus 145
Cobbiinae 200, **292**
Cobbionema 148
Cobbonchulus 355
Cophonchus 233
Coinonema 214
Colpurella 280
Comesa 142
Comesoma 74, 100, 211
C. heterosetum 43, 319

INDEX

C. minimum **212**, 335, 358
Comesomatidae 30, 31, 33–4, 37, 43, 53, 60, 67, 81, 91–3, 100, 127, 131, 133–4, 136–7, 207–<u>209</u>–211, 213, **292**, 304, 319, 335, 352
Comesomatinae <u>211</u>, **292**, 319, 335
Comesomoides 211
Conilia 239, 321
Coniliinae <u>238</u>, 303
Coninckia 205, 215–16, 304–5
 C. circularis 215, 328
 C. mediterranea 358
Coninckiidae 126, 205, 207, 210, 213, <u>215</u>, **292**, 301, 305
Conistomella [see also *Rhabdodemania*] 268–9, 304
Conolaimella 251
Conolaimus 209
Convexolaimus [see also *Pontonema*] 253–4, 304
copulatory apparatus, male 139, 158, 279
copulatory organs, supplementary 14, 102–07, 232, 246, 272, 300
corpus gelatum, of amphids 47–8, 51, 62, 67, 161–2, 165, 188
Corythostoma 241
Cothonolaimus 261
Craspoderma 144
Crassolaibium australe 84
Crassolaimus 158
Crenopharyngidae <u>234</u>, 301
Crenopharynginae <u>234</u>, **296**, 303, 305
Crenopharynx 232–4
 C. marioni 72, **117**, 233, 341
Cricolaimus 52, 169, 348
Croconema 150
Criconematidae 161, 357
Criconemoides 161
Cryptenoplus 230
Cryptolaimus 207
Cryptonchidae 56, 71, 76, 105, 109, 114–15, 177–8, 219, 276–7, <u>278</u>, 321, 339, 353
Cryptonchulus 278
Cryptonchus 278
 C. tristis 71, 105, 339
Crystallonema 206
Curvolaiminae <u>250</u>, 303
Curvolaimus 250, 355
cuticle, structure of 14, 126, 129, 134–7, 140–2, 144, 146–50, 153, 155–6, 158–9, 161, 166–7, 169, 171, 173, 175, 177–80, 182, 185–7, 189, 191–2, 194–6, 199–200, 202–05,

207–09, 213–16, 221, 224, 226, 262–3, 265–6, 268, 271, 274–5
colouring by cotton blue 135–6, 140–41, 146–8, 152, 157, 159, 176, 186, 207
Cyartonema 81, 96, 133, 180–1, 204, 305, 328
 C. flexile 350, 358
 C. germanicum 328, 350, 358
 C. zosterae 328, 350
Cyatholaimidae 30, 33–4, 37, 44–6, 53, 60, 88, 92–3, 103, 105, 132, 134–6, 142, <u>143</u>–7, 170, **288**, 320, 337, 352
Cyatholaiminae 135, 141, <u>145</u>, **288**, 320, 338, 352
Cyatholaimus 145
 C. canariensis 358
Cyathonchus 274
Cygnonema 156
Cygnonematinae 156, 302
Cylicolaiminae 72, <u>241</u>, **296**
Cylicolaimus 241
Cylindrocorporidae 282, 284, 357
Cylindrocorpus 357
Cylindrolaimidae 168
Cylindrolaiminae 81, 177, 181, 188, 209, 213, <u>215</u>, **292**, 305, 330, 348
Cylindrolaimus sp. **82**, 215, 330, 348
 C. monhystera 213, 349
Cylindrotheristus 201
Cynura 167–9, 348
 C. klunderi 88
Cytolaimium 81, 98–9, 105, 272, 353
 C. exile 55, 57, 216, 271–2, 340
Cyttaronema 183, 303
 C. attenuatum 183
 C. carinatum 183
 C. chitwoodi 183
 C. filipjev 183
 C. pisanum 183
 C. racovitzai 183
 C. recticulum 183
 C. rectum 183
 C. salsicum 183
 C. sculpturatum 183
 C. undulatum 183

Dactylaimus 201
Dactylonema 240
Dadaya 169
Dagda 101, 166–7, 169–70, 348
 D. bipapillata 327
Daptonema 74, 201
Dasylaimus 140

Dasynema 183
Dasynemella 181, 183, 185
  D. phalangida 183
  D. sexalineata 182–3
Dasynemellidae 181, **290**, 302, 319
Dasynemelloides 183, 303
Dasynemoides 183, 303
  D. albaensis 183
  D. cinctus 183
  D. conicus 183
  D. falciphallus 183
  D. filum 183
  D. pselionemoides 183
  D. rhombus 183
  D. reimanni 183
  D. setosus 183
  D. spinosus 183
Dayellidae <u>234</u>, 303
Dayellus 234, 305
Deltanema 139
Demanema 165
Demania 269
demanian organ 248, 250–53, 255
Demonema 148
Denophorea 173
Dentatonema 144, 146, 352
Denticulella 139
Deontolaimus 70, 170, 348
  D. papillatus 104, 327
Deontostoma 68, 74, 239, 241
  D. arcticum 72, 75, 77, 239, 341
  D. washingtonense 267
dereids 14, 29, 31–2, 38–40, 42, 44, 65, 114, 125, 166, 172, 271, 281, 300
Dermatolaimus 169
desmen 165
Desmodora 74, 150, 334
  D. (D.) communis 358
  D. filispiculum 334
  D. (D.) hirsuta 358
  D. aff. *perforata* 334
  D. (D.) pilosa 358
  D. sinuata 334
  D. (Croconema) sphaerica 358
  D. tenuispiculum 334
  D. (X.) torquens 351
Desmodorella 150
Desmodorida 30, 33, 35, 37, 42, 45, 51, 74, 81–6, 92, 95, 98, 100, 102–04, 127, 133, 147, 149, 166, 179, 188, 197, 276, 301, 319, 333
Desmodoridae 53, 84, 86, 90, 92, 95–6, 103–04, 134, 148, <u>149</u>, 152–3, 155, 157–8, 180, 186, 213, **288**, 301, 304–5, 319, 333, 351

Desmodorinae <u>150</u>, **288**, 319, 334, 351
Desmodoroidea 95, 133, 135, 147, <u>148</u>–9, 153, 155–6, 179, 182, 211, **288**
Desmogerlachia 165
Desmolaiminae <u>206</u>, **292**, 302, 328, 350
Desmolaimus 74, 204, 206
  D. aff. *bulbulus* 328
  D. zeelandicus 328
Desmolorenzenia 163
Desmoscolecata 120, 127, 129, 133
Desmoscolecida 33, 42, 45, 51, 74, 81, 83–86, 89, 92, 95, 99, 100, 127, 130, 133, 162, 176, 301, 318, 349
Desmoscolecidae 33, 51, 53, 93, 95, 133, 161, **162**, 164–5, 198, **290**, 305, 328, 349, 358
Desmoscolecina 124, **125**–6, 129, 132–3, <u>160</u>, 166, **286**, **290**
Desmoscolecinae 33, 42, 86, 95, 111, 161–<u>163</u>, 165, **290**, 328, 349
Desmoscolecoidea 62, 67, 127–31, 160, <u>161</u>–2, 164, 176, 185, 197, 219
Desmoscolex 74, 163
  D. fennicus 328
  D. rostratus 318
Desmotimma 165
Dichromadora 139
  D. cephalata 140, 358
  D. cucullata 337
  D. microdonta 358
  D. scandula 140, 337, 358
Dicriconema 139
Dictyocaulus 74
Dictyonemella 183, 303
Didelta 197, 206
Didetta 205
Digitonchus 170
Dignathonema 148
Dilaimus pauli [see *Calyptronema paradoxum*] 258
Dintheria 191, 193
Dioctophymatina 108
Dioctophymatoidea 95, 113–14, 216
Dioctophymida 123
Diodontolaimus 166–7, 169–70, 348
Dioncholaiminae [see also Mononchidae] 73, 246, 280, 303, 355
Dioncholaimus see *Mononchus*
Dipeltis see *Diplopeltis*
Diphtherophora 279
Diphtherophoridae 86, 95, 274, 277, <u>279</u>, 356

Diphtherophorina 217, 277
Diphtherophoroidea 217
Diplobathylaimus 280
Diplogaster 74, 283-4
  D. rivalis **59**, 344
Diplogasteridae 57, 59, 81, 83, 90, 93, 105, 273, 282-3, 344, 356
Diplogasterididae 284
Diplogasterinae 356
Diplogasteroidea 35, 46, 174, 281-4, **286**, 322
Diplohystera 280
Diplolaimella 199
  D. ocellata 331
  D. stagnosa 331
Diplolaimelloides 105, 199
  D. altherri 331
  D. islandica 331
  D. oschei 331
Diplolaimelloidinae 198, **290**
Diplolaimita 199
Diplopeltidae 126, 179, 196, 207, 213, **292**, 305
Diplopeltinae 81, 151, 213-14, **292**, 302, 318, 330, 349
Diplopeltis **52**, 214, 305
  D. incisus 214
  D. indicus 214
  D. intermedius 214
  D. onustus 214
  D. typicus 214
Diplopeltoides 180-81, 213, 305, 349
  D. ornatus **180**, 329
Diplopeltula 214-15, 349
  D. breviceps 108
  D. incisa 108, 112, 305, 329, 358
  D. indica 305
  D. intermedia 305
  D. aff. nuda 329, 349
  D. onusta 305
Discolaimus 356
Discomyctus 356
Disconema 197, 205-06
  D. minutum 358
Discophora 215
Dispira 146
Dispirella 146
Ditlevsenella 55, 246-7, 254-5, 257-8, 322
  D. danica 73, 248, 343
  D. aff. murmanica **257**
Ditlevsenia 278
Dolicholaimus 109, 236-7, 239, 321
  D. marioni 275
Dolichosomatum 213

Doliolaimus 202
  D. agilis 332
Domorganus 188, 213, 303, 305, 349
  D. acutus 188-9
  D. bathybius 188-**190**
  D. oligochaetophilus 127, 187-8, 330
  D. macronephriticus 188
  D. supplementatus 188
Donsinema 232
Donsinemella 170
Dorylaimellus 356
Dorylaimida 16, 30, 31, 33, 35, 45-6, 57, 60, 80-1, 83-6, 92, 95, 105, 123, 177, 216-**220**, 247, 263, 270, 274-80, **294**, 300-01, 304, 317, 322, 343, 355
Dorylaimidae 217, 356
Dorylaimina 74, 92, 107, 217-18, 276-7, 280, 343
Dorylaiminae 217
Dorylaimoidea 217
Dorylaimoides 356
Dorylaimopsinae 210, 212, **292**, 319, 335
Dorylaimopsis 74, 212
  D. hawaiiensis 358
  D. punctatus 335
Dorylaimus 74
  D. stagnalis 343
Doryllium 356
  D. uniforme 343
Doryonchus 259
Dracogalerus 156
Dracognominae 156, 302
Dracognomus 155-6
Dracograllus 155
Draconactus 156
Draconema 155, 335
Draconematidae 86, 92, 95, 134, 149-50, 154-5, **288**, 301, 319, 335, 351
Draconematina 155
Draconematinae 155, **288**
Draconematoidea 358
Dracotoranema 156
Drepanonema 155
Drepanomatidae 154

Echinotheristus 201, 262
Eleutherolaiminae 206, **292**, 331, 350
Eleutherolaimus 74, 204, 206, 323, 331
Elzalia 201
Enchelidiella 259

Enchelidiidae   30, 55, 57, 73, 75–6, 78, 93, 102, 108–09, 115, 216, 222, 245–8, 254–255–257–258, 261, 268, **296**, 303, 322, 343, 355
Enchelidiinae   57, 73, 247–8, 254, 256
Enchelidioidea   248
Enchelidium   247, 258
Encholaimidae   323, 357
Encholaimus   275
Endeolophos   139
Endolaimus   145
Enoplacea   115, 217, 222, 245, 275, **296**, 301
Enoplata   120, 127–30
Enoplea   123
Enoplia   42, 46–7, 55–6, 62–65, 74, 108, 121, 123–**125**–6, 131, 133, 192–3, 195, 216–18, **220**, 222, 275–6, **286**, **294**
Enoplida   14, 22, 30, 34, 42–3, 45, 70, 74, 75–6, 78–9, 81, 83–6, 88–9, 91–2, 95–9, 101–03, 105–06, 108–09, 112, 114–15, 117, 123, 130, 133–4, 166–7, 185, 192, 195, 197, 205, 217–**220**, 222, 224, 233, 236, 238, 242, 244–6, 256, 259, 261–2, 267–71, 274–6, 278, 280, **294**, 301
Enoplidae   30, 37–8, 55, 57, 60, 73, 75–8, 93, 102–03, 107–09, 115–18, 194, 217, 221–224–226, 230–31, 268, **296**, 305, 322, 341, 354
Enoplimorpha   121
Enoplina   112, 116, 217–18, 221–3, 226, 246, 259, 280, **294**, **296**
Enoplinae   73, 78, 108–09, 322, 342
Enoplocheilus   278
Enoploidea   102, 112, 217, 221, 223, 226, 228, 231, 235, 240, 246, 267, 272, **296**, 300
Enoploides   230
   E. cirrhatus   103
   E. labrostriatus   38, 73, 116, 342
Enoploidinae   73, 109, 228, 230, 302, 305, 342
Enoplolaiminae   73, 109, 223, 228–30, **296**, 302, 305, 322, 342
Enoplolaimus   230
   E. connexus   73, 116, 342
   E. propinquus   73, 109, 117, 342
Enoplonema   230
Enoplostoma   225
Enoplus   68, 74, 221, 225, 304
   E. bracyuris   225
   E. brevis   37, 73, 75, 77, 117, 224, 342
   E. communis   35

E. michaelsensi   117, 342
E. schulzi   224, 304
Epacanthion   230
   E. buetschlii   38, 73, 75, 109, 116, 342
   E. multipapillatum   103, 223
epidermal glands   116, **117**–18, 224–6, 300
Epsilonella   154
Epsilonema   74, 153–4
   E. byssicola   37, 335
   E. dentatum   335
Epsilonematidae   37, 42, 53, 60, 66, 86, 90, 92–3, 95, 127, 134, 149–50, 153–5, 275, **288**, 319, 335, 351
Epsilonematina   154
Epsilonematinae   154, **288**
Epsilonoides   154
Ereptonema   173
Etamphidelus   279
Ethmolaimidae   136–7, 140, 142–3, **288**, 301
Ethmolaiminae   138, 140
Ethmolaimus   74, 86, 93, 95, 101, 137, 140, 352
   E. pratensis   336
Eubostrichus   103, 151
   E. cobbi   319
Eucephalobus paracornutus   344
Euchromadora   139, 320, 336
Euchromadorinae   33, 136, 138, 139, **288**, 304, 320, 336
Eudesmoscolex   163
Eudorylaimus   74, 356
   E. carteri   343
   E. doryuris   343
Eulinhomoeus   206
Eumorpholaimus   206
Eurystoma   [see *Eurystomina*]   258
Eurystomina   55, 247, 254–8
   E. assimilis   343
Eurystominidae   248
Eurystomininae   57, 64, 66, 73, 245, 247–8, 254, 256–7, **296**, 303, 343
Eusynonchus   241
Eutelolaimus   168
Euteratocephalus   92, 103, 173–5, 303, 347
   E. crassidens   107, 111, 123
   E. palustris   **110**, 326
Euthoracostomopsis   202
Eutricoma   163
Eutylenchus   322
excretory duct, of caudal gland   108
excretory gland   see cervical gland
excretory pore   see cervical pore

INDEX

Festuca 25
Fenestrolaimus 73, 116, 226, 230, 304, 322, 341
  F. antarcticus 230
Fiacra 241
Filipjeva 201
Filipjevia 230
Filipjeviella 269
Filipjevinema 269
Filoncholaimus 254
Fimbriella 269
Fimbrilla 259
Fotolaimus 253
Frostia 280
Frostina 280, 306
Fusonema 138
fusus, of amphids 48, 189, 191

Gairleanema 260–61, 305
  G. angremilae 55, **260**
Galeonema 234
Gammanema 148
  G. cancellatum 320
  G. conicauda 44
Gammarinema sp. 74, 199
  G. cardisoma 331
  G. gammari 26, 37, 331
  G. ligiae 25, 37, 331
  G. paratelphusae 199
Gammarus 26
ganglia, lumbar 111
gastrotrichs
  sensory organs of 60, 63
  gonads of 96–7
Gerlachiinae 165, **290**
Gerlachius 161, 164–5
Gerlachystomina [see *Eurystomina*] 258
Glochinema 127, 154, 275
Glochinematinae 154, **288**
Gnomoxyala 55, 201
Gomphionema 142, 320
gonads 79
  position relative to the intestine 15, 47, 65, 80, 89, **90**–92, 96, 98, 100–02, 124, 126, 132, 135–7, 141–4, 146, 148, 157, 159, 167, 172, 174, 176, 178, 181–2, 185, 188, 191, 196, 199–200, 202, 204–05, 208, 210, 215, 223, 231, 235, 237, 239–40, 242, 245–6, 261, 263, 266–9, 271–4, 282–3, 300, 324
  phylogenetic assessment of the number 94

369

Gonionchus 74, 201
  G. inaequalis 200
Goodeyus 357
Gradylaimus 151
Grahamius 213
Graphonema 139, 320
Greeffiella 74, 89, 163, 328
  G. dasyura 358
Greeffiellinae 86, 95, 111, 161, 164, 290, 302, 328, 349
Greeffiellopsis 164
Greenenema 196
Greenia 196
Gullmarnia 234
Gymnolaimus 178, 278

Haconnus 169
Halalaiminae 49, 55, 57, 67–8, 72, 109, 221, 242, 244, 275, **296**, 305, 321, 340, 354
Halalaimoides 244
Halalaimus 243, 304, 340
  H. borealis 244, 304
  H. aff. *longicaudatus* 72
  H. longistriatus 275
  H. gracilis 72, 76, 358
  H. similis [see also *H. borealis*] 244, 304
  H. terrestris 340
Halanonchinae 71, 270–71, 272, **294**, 340, 353
Halanonchus 71, 105, 129, 131, 271–2, 340
Halaphanolaimus 102, 104, 169, 348
Halaphanolaiminae 168
Halichoanolaimus 74, 146–8, 320
  H. longicaudatus 338
  H. robustus 338
Halicylindrolaimus 207
Halinema 197, 205–06
  H. varicans 44
Haliplectidae 53, 81, 92, 133, 167, 176–7, **290**, 318, 326, 347
Haliplectus 177
  H. dorsalis 326
  H. schulzi 326
Hapalominae 163, 302
Hapalomus 164
  H. terrestris 328
Haptotricoma 74, 165
Harpagonchidae 301
Harpagonchinae 127, 139, **288**
Harpagonchoides 139
Harpagonchus 139

Harveyjohnstonia 145
Haustrifera 145
Helalaimus 199
Helicotylenchus 74
H. varicaudatus 345
Helmabia 323
Hemicycliophora 74
Heterocephalus 233
Heterocyatholaimus 146
Heteroderidae 357
Heterodesmodora 150
Heterodesmoscolex 163
Hexamermis albicans 111
Hofmaenneria 101, 200–01
H. niddensis 200, 332
holophyletic species groups 19, 21, 40
holophyl 21
holophyletic taxa 23, 45, 299
holophyly 16, 19, 119, 299
  definition of 19
holapomorphy 16, 21, 299
  definition of 22
homoiology 102
homologous feature 21
homology 19
  definitions of 19
Hoplolaimidae 357
Hopperia 212
H. massiliensis 358
Hyalacanthion 230
Hypodontolaiminae 138–139, **288**, 304, 320, 337
Hypodontolaimus 139
Hyptiolaimus 230

Ichthyosdesmodora 150
Illium 259
Ingenia 57, 260–61
Innocunema 140
  I. tentabunda 337
Ionema 167, 170
Iotadorus 139
Iotalaimus 280
Ironella 30, 239, 321
  I. prismatolaima 38, 88
Ironidae 30, 38, 42, 56, 60, 71, 75–6, 78–9, 88, 93, 109, 112, 115, 195, 216–17, 221–2, 236–8, 264, 275, 278, **296**, 300, 305, 321, 339, 353
Ironinae 217, 237, 238, **296**, 305
Ironoidea 223, 236, **296**
Ironus 68, 77, 79, 109, 114, 236–8, 264, 305, 321, 339
  I. ignavus 71, 236, 339

I. tenuicaudatus 236
Isolaimida 277
Isolaimiidae 51, 53, 81, 93, 102, 108, 177, 277, 279, 318, 326, 347
Isolaimium 279
Isonemella [see *Symplocostoma*] 258
Ixonema 74, 108, 156, 158, 333

Jaegerskioeldia 241
Judonchulus 74

Keratonema 154
Keratonematinae 154
Kinonchulus 29, 105, 192, 274
kinorhynchs, gonads of 96
Klugea 234, 305
Kosswigonema 148
Krampia 251
  K. opaca 260
Krampiinae 73, 246, 248, 251, **296**, 355
Kraspedonema 144
Kreisia 181, 280, 306
Kreisoncholaimus 253, 359
Kreisonema 143
Kreisonematinae 93, 143, 168, 178, **288**, 304, 318, 348
Kurikania 264
Kurikaniinae 261, **294**, 303

Laimella 210–212, 320
  L. longicauda 336
Lamuania 264, 304
Lasiomitus [see *Symplocostoma*] 258
lateral line 56, 57, 59, 63–6, 74
Latilaimus 272
Latronema 74, 146–8, 320
Lauratonema 79, 274–5
  L. hospitum 72, 341
  L. reductum 359
  L. aff. *spiculifer* 37, 72, **87**, 341
Lauratonematidae 37, 56, 60, 72, 76, 87–8, 93, 96, 101, 109, 115, 219, 270, 274–5, **294**, 322, 341, 354
Lauratonemoides 274–5
Laxonema 153
Laxus 151
Ledovitia 256, 258, 343
Leoberginema 188, 303, 305
Leptodasynemellinae 181, **290**, 302, 319
Leptodasynemella 183, 303
Leptogastrella 201

Leptolaimidae 51, 53, 65, 68, 81, 84, 88–9, 91–3, 97, 100, 103–04, 107–08, 128, 132–3, 143, 166, 167–8, 171–2, 175–6, 178, 213, **290**, 304–5, 318, 327, 347
Leptolaimina 99, 124, **125**–6, 129, 133, 149, 157, 166, 170–1, 174–7, 181–2, 185, 186–9, 194–5, 197, 204–05, 213, 219, 238, 263, 266, 274, **286**, **290**, 301
Leptolaiminae 93, 103, 167, 168, 170, 176, **290**, 302, 305, 318, 327, 347
Leptolaimoides 51, 68, 167, 169, 348
L. thermastris 88, 108, 168, 327
Leptolaimus 51, 70, 167–8, 303, 305, 327, 348
L. acicula 108, 327
L. ampullacea 358
L. cupulatus 103–04
L. elegans 327
L. exilis 168
L. leptaleus 103–04
L. limicola 88, 327
L. minutus 168
L. mixtus 327
L. papilliger 88, 103–04, 327
L. puccinelliae 327
L. pumicosus 104
L. venustus 327
Leptonchidae 356
Leptonchus 356
Leptonemella 74, 151
L. aphanothecae 334
L. gorgo 334
Leptosomatidae 30, 44, 57, 60, 72, 75–7, 105, 108–09, 115, 235–6, 239–41, 245, 267, **296**, 322, 341, 354
Leptosomatides 240
Leptosomatina 241
Leptosomatinae 72, 240, **296**, 341
Leptosomatum 239–40
Leptosomella 239–40
L. phaustra 239
Ligia oceanica 25
Limonchulus 274
Lineola 280
Lineolia 280
Linhomoeidae 34–5, 37, 41, 44, 53, 60, 81, 84, 92, 126, 178, 185, 188, 197–8, 203, 204–05, 207, 215, 270, **292**, 305–6, 328, 350
Linhomoeinae 206, **292**, 331, 350
Linhomoella 206
Linhomoeus 206

L. filiaris 205, 323, 331
L. hirsutus 44, 102
L. timi 205
Linhystera 115, 127, 222
Linolaimus 281, 349
Litinium 243, 272, 304, 321
L. bananum 243, 340
L. simplex 243, 306
Litonema 281
Litotes 281
Longibulbophora 279
Longicyatholaiminae 144–5, 302
Longicyatholaimus 34, 145
L. complexus 338
Longidoridae 356
Longidorus 74, 343
Longilaimus 281
Longitubopharynx 281
Lumbricillus lineatus 188
Lynhystera 200–201, 271
Lyranema 258

Macfadyenia 281
Macramphis 201
Macronchus 241
Maldivea 245
mandibles 225–6, 228–30
Manunema 89, 99, 106, 175–6
M. annulata 88, 327
M. proboscidis 88, 327
Manunematinae 175–6, **290**, 302
Margonema 209, 349
Marinoplectus 173
Marionella see *Eurystomina*
Marylynnia 145
Mastodes 150
Megadesmolaimus 206
Megalamphis 201
Megalolaimus 201
Megeurystomina 55, 254–6, 258
Megodontolaimus 136, 140
Mehdilaimus 177–8, 305
Melepturus 153
Meloidogyne 89
Mermithida 123
Mermithidae 111, 113, 120
Mermithina 217, 279
Mermithoidea 55, 74, 95–6, 108, 113–14, 216
Meroviscosia [see also *Viscosia longicaudata*] 251, 304
Mesacanthion 230
M. arabium 102
M. diplechma 73, 75, 116, 342

M. monhystera 226, 354
Mesacanthoides 230
Mesodorus 150
Mesonchium 100, 212, 319, 358
Mesotheristus 201
Metachoniolaimus 145
Metachromadora 49, 74, 150–51
  M. remanei 334
Metachromadorinae 150
Metachromadoroides 151
Metacomesoma 210–11
Metacyatholaimus 145
Metacylicolaimus 241
Metadasynemella 181, 183, 303
  M. cassidiniensis 183
  M. elegans 183
  M. macrophallus 183
  M. picrocephala 183
Metadasynemellinae 181, **290**, 302, 319
Metadasynemoides 181, 183
  M. cristatus 183
  M. latus 183
  M. longicollis 183
Metadesmodora 150
Metadesmolaimus 201
Metalaimus 206
Metalinhomoeus 204, 206
  M. gracilior 358
  M. aff. *typicus* 328
  M. variabilis 350, 358
metanemes 15, 68, **69**–71, 74–9, 107, 130, 185, 219, 220, 223–6, 231–2, 234–5, 237–9, 242, 245, 259, 261–4, 266, 268–71, 273–6, 300
  caudal filaments of, 68–70, 77–8, 224, 226, 231–2, 239, 242, 245, 264, 268–9
Metaparoncholaimus 246, 253
  M. campylocercus 359
Metapelagonema 269
Metaraeolaimoides 215
Metasabatieria 212
Metateratocephalinae 166, 173, <u>175</u>, **290**
Metateratocephalus 175, 303
Metenoploides 230
  M. aff. *alatus* 117, 342, 354
Metepsilonema 153–4
  M. hagmeieri 335
  M. laterale 53
Metoncholaimoides 253
Metoncholaimus 224, 253
  M. pristiurus 359
Metonyx 151

Meyersia 246, 252–3
Meylia 74, 127, 161–2, 164
  M. alata 162, 164
  M. spinosa **160**, 162, 165, 328
Meyliidae 53, 99, 126, 161–3, <u>164</u>–5, 176, **290**, 305, 328, 349
Meyliinae 81, 99, 160–1, <u>165</u>, **290**
Meylonema 356
Micoletzkyia 234, 305
Micranthonchus 146
Microlaimella 159
Microlaimidae 82, 126–7, 131, 133–6, 141, 149, 151, 156–9, 187–8, 197, **288**, 302, 304
Microlaiminae 53, 81, 86, 92–3, 98–9, 108, 319, 333–4, 351
Microlaimoidae 159
Microlaimoidea 149, <u>156</u>, **288**
Microlaimoides 158
Microlaimus 74, 152, 158
  M. arcticus 151
  M. cervoides 358
  M. citrus 333
  M. conspicuous 333
  M. cyatholaimoides 358
  M. dentatus 333
  M. globiceps 157, 333
  M. aff. *honestus* 333
  M. pecticauda 358
  M. spirifer 358
  M. tenuispiculum 152, 158, 304, 333
  M. torosus 333, 351
  M. turgofrons 333
Micromicron 150
microscopy
  electron 47, 51, 69, 284
  light 47, 68, 111, 267
  phase contrast 33
Mikinema 188, 205, 303, 305
Minolaimus 144
Mirolaimus 278
Mitrephoros 175
Molgolaimidae 152, 157–8, 301, 304
Molgolaiminae 126, <u>152</u>, 157, **288**
Molgolaimus 92, 95, 149, 152, 157–8, 351
  M. demani [see also *Microlaimus tenuispiculum*] 152, 304
  M. tenuis 152
  M. tenuispiculum **82**, 152, 334
Monhystera 74, 199, 332
  M. anophthalma 331
  M. cuspidospiculum 331
  M. disjuncta 331
  M. multisetosa 331

INDEX 373

M. paludicola 331
M. paramacramphis 331
M. parasimplex 331
M. parva 331
M. pusilla 331
M. stagnalis 332
M. venusta 332
Monhysterata 98, 120, 127–9, 131, 210
Monhysterida 31, 33–5, 37, 42–3, 45, 51, 74, 81, 83–6, 92, 95, 96, 98, 101–2, 105–6, 124, **125**–6, 128, 131–4, 157, 160, 166, 176, 185, 188, 196–7, 199, 209–10, 219, 262, 271, **286**, **292**, 301, 328
Monhysteridae 86, 93–6, 101, 105, 108, 126, 128–29, 131, 198–9, 208, **292**, 331–2, 350
Monhysteriella 206
Monhysterina 127, 129, 131, 162, 210
Monhysteroidea 15, 53, 60, 96, 128, 197–200, 202, **292**, 317
Monhysteroides 206
Monhystrella 199
  M. inaequispiculum 332
  M. paramacrura 332
Monhystrium 106, 199–200
Monochoides 74
Monochromadora 194, 196, 353
  M. monohystera 358
Monochromadorinae 71, 75–6, 93, 133, 194–5, 219, 266, **290**, 302, 305, 321, 353
Mononchida 123, 250
Mononchidae 217–18, 247, 250, 263, 276, 280, 303, 305, 355
Mononchina 74, 92, 217, 276–7, 280, 343
Mononchinae 217, 247
Mononchoidea 217–18
Mononcholaimidae 248
Mononcholaiminae 249, 251
Mononcholaimus [see also *Viscosia*] 249, 251–2, 304, 359
  M. elegans see *Viscosia viscosa*
  M. rusticus 359
Mononchulidae 276–7, 278
Mononchulinae 355
Mononchulus 278
Mononchus 74, 280, 343
monophyletic taxa 23
monophyly 16, 19, 299
Monoposthia 74, 159, 358
  M. costata 335, 358
  M. mielcki 358
  M. mirabilis 335

Monoposthiidae 51, 53, 65, 92–3, 127, 149, 156, 159, **288**, 319, 335, 351
Monoposthioides 159
Monotrichodorus 279
Multidens 266
Mumtazium 356
Myctolaimus 357
Mylonchulus 74, 343, 355
Myolaimidae 57, 284, 357

Nanidorus 279
Nannolaimoides 144
Nannolaimus 140, 144
  N. volutus 144, 320
Nannonchus 261
Nanonema 238
Nasinema 232
neck gland see cervical gland
Necolaimus 170
Necticonema 145
Nema 281
Nemacoma [see also *Halalaimus*] 243–4, 304–5
Nemanema 243, 321
  N. filiforme 354, 358
  N. rotundicaudatum 109, 340, 354, 358
Nemanemella 243
Nemathalminthes 97, 111
Nematoda 123, **286**
Nemella 167, 170
Neochromadora 74, 140
  N. poecilosoma 337
Neochromadorina 138
Neonchus 192
Neonyx 151
Neoquadricoma 166
Neotheristus 201
Neotonchidae 136, 141, **288**, 301
Neotonchinae 46, 140, 142, 144, 320, 337, 352
Neotonchus 34, 142, 337
  N. corcundus 337
  N. meeki 337
Neotylenchidae 357
Neotylocephalus 173
nerve ring 38–9, 55, 57–8, 65
Neurella 170
Nicasolaimus 211
Nijhoffia 207, 306
Noffsingeria 165
Notocamacolaimus 281
Notochaetosoma 156
Notochaetosomatinae 156, 302

Notosabatieria 213
Notosouthernia 281
Nuada 244
Nuadella 281
Nualaimus [see also *Halalaimus*] 244, 304
Nudolaimus 241
Nudora 159
  N. besnardi 319
  N. crepidata 335
  N. omercooperi 319
Nummocephalus 145
Nunema 148
Nygellus 355
Nygmatonchus 138–9, 304, 320
Nygolaimellus 355
Nygolaimidae 355

ocelli 162, 196, 199, 213, 239
Octonchinae 73, 248, 254, **296**, 355
Octonchus 254
Odontanticoma 236
odontia 208
Odontobius 199–200
Odontocrius 140
Odontolaimidae 167, 191, 192–3, **290**, 301
Odontolaiminae 56, 71, 89, 219, 321, 339, 353
Odontolaimus 192–3
  O. aquaticus 71, **193**, 339, 353
  O. chlorurus 353
Odontonema 139
Odontopharynginae 356
Odontophora 208–09
  O. armata 329
  O. fatisca 358
  O. ornata 329
  O. peritricha 329
  O. phalarata 329
  O. villoti 349
  O. wieseri 349
Odontophoroides sp. 60, 63, 86, 208–09, 349
  O. monhystera 37, **61**, 329
oesophageal glands see pharyngeal glands
Ohridiidae 149, 157, 167, 187–8, 205, 212, **289**, 301, 305
Ohridius 157, 188, 303, 351
  O. bathybius 334
Oionchus 278
Oistolaimus 151
Oligomonohystera 209
Oligoplectus 173

Omicronema 201
onchium/-a 36, 224–6, 228–30, 233, 246–**255**, **257**
Onchium 167, 170
Oncholaimacea 217, 222–3, 245, **296**, 301
Oncholaimellinae 73, 247–9, 251–2, **296**, 305, 355
Oncholaimelloides 105, 246–7, 251–2, 355
  O. vonhaffneri 73, 342
Oncholaimellus 246, 249, 251–2
  O. labiatus 359
Oncholaimidae 30, 36–7, 57, 60, 73, 75–6, 78, 88, 93, 102, 105, 108–09, 115, 217, 222, 231, 246–249, 280, **296**, 305–6, 322, 342, 355
Oncholaimina 217, 246, 249
Oncholaiminae 73, 247–9, 252, 253, **296**, 305, 342, 355
Oncholaimium see *Oncholaimus*
Oncholaimoidea 78, 102, 116, 230, 245–249, 259, **296**
Oncholaimoides 246, 252
Oncholaimus 247, 253
  O. brachycercus 73, 342
  O. demani see *Ditlevsenella danica*
  O. keiensis 246, 253, 359
  O. leptos 246, 253
  O. longidentatus see *Viscosia longidentata*
  O. skawensis 342
  O. vesicularis 48
Onchulella 170
Onchulidae 192–3, 272–4, 277, **294**
Onchulinae 30, 34, 39, 44, 56, 71, 98, 219, 321, 339, 353
Onchulus 270, 273–4
  O. longicauda 105
  O. nolli 38, 71, 273, 339
Onyx 74, 103, 151
  O. perfectus 358
  O. setempapillatus 334
Opailaimidae 356
ovaries
  number of 14, 79, 84–5, 94–96, 133, 144, 158–9, 166–7, 170, 173, 177, 182, 185, 192, 194, 196, 198, 204, 207–08, 210, 221, 246, 250–53, 255, 265, 272, 274–6, 324
  outstretched 80, 82–4, 86, 91–93, 98–100, 106, 124, 127, 129–32, 152, 157–61, 165–6, 175–6, 181, 194, 196–8, 204–05, 207, 210, 213, 215–16, 272, 282, 324

INDEX

orientation of 97
position of 84–5, 90–1, 139, 142, 150, 153–4, 171, 177, 189, 191, 194, 251, 324
reflexed 80, 84, 100, 129–30, 152
antidromously 80–4, 86, 91–3, 97, 99, 100, 102, 106, 127, 131–133, 135–6, 142–4, 146, 149, 152–3, 157, 159–61, 165–7, 171–3, 176–80, 185, 187–9, 191–2, 194–5, 213, 216, 221, 224, 226, 231, 233, 235, 237, 239, 242, 260, 263, 265, 267–269, 272–6, 282, 324
homodromously 80–83, 91, 93, 98, 106, 131–2, 174, 282, 324
structure of 14, 79, 80–2, 86, 91–2, 129–32, 300
Oxyonchinae 73, 109, 228, 230, 302, 305, 322, 341
Oxyonchus 230
O. dentatus 73, 75, 109, **110**, 116, 341
Oxystoma 241, 243
Oxystomatina 243
Oxystomella 243
Oxystomina 242–3, 304, 321
O. alpha 38
O. astridae 243
O. brevicaudata 358
O. elongata 72, 109, 340, 354
O. novozemeliae 358
O. oxycaudata 358
O. tenuis 243
O. unguiculata 243
Oxystominidae 30, 32, 38, 40, 42, 44, 56, 72, 75–6, 78, 84, 105, 109, 115, 167, 177, 221–2, 236, 241, 272, 275–6, 279, **296**, 300, 305–6, 321, 340, 354
Oxystomininae 56, 72, 96, 109, 226, 242–4, 276, **296**, 305, 321, 340, 354

Pachydora 244, 304
Pakira 167–9, 172
Panagrolaimidae 93, 121, 284, 344, 357
Panagrolaimus 74, 344
Pandolaimidae 205, 231, 259, 267–8, 269, **294**, 301, 305
Pandolaimus 34, 108–09, 115, 205, 231, 269–70, 305, 322–3, 355
P. latilaimus 73, 342
Pandurinema 178
Panduripharynx 140

papillae [see also copulatory organs] 272
labial 184
pre-cloacal 141–2, 146, 159, 172, 174, 176–7
pre-anal 187–8, 191–2, 196, 207, 239, 263
Parabarbonema [see also *Platycoma*] 235, 240
Parabarbonematidae 240
Parabarbonematinae 235, 240, 303
Parabathylaimus 261
Paracanthonchinae **288**, 320, 338, 352
Paracanthonchus 34, 74, 103, 145, 338
P. caecus 338
P. falklandiae 358
P. macrodon 105, 338
P. maior 358
Parachromadora 141, 151
Parachromadorita 140
Parachromagaster 214
Parachromagasteriella 214–15, 304, 330, 349
P. annelifer 209, 305
Paracomesoma 210–11, 319
P. hexasetosum 43
P. inequale 43
P. sipho 43
Paraconthonolaimus 158
Paractinolaimus 74, 343
Paracyatholaimoides 145
Paracyatholaimus 34, 74, 145, 338
P. intermedius 34, 338
P. occultus 338
P. pentodon 37, 338
P. truncatus 144
Paracylicolaimus 241
Paradesmodora 150
P. immersa 334
Paradesmolaimus 206
Paradontophora 53
Paradoxolaimus 142–3
Paradraconema 156
Paraegialoalaimus 207, 306
Paraeolaimus nudus 53, 108
Paraleptosomatides 240
Paralinhomoeus 204, 206
P. fuscacephalum 350
P. lepturus 205
P. tenuicaudatus 358
P. uniovarium 350
Parallelocoilas 151
Paralongicyatholaimus 34, 145
P. minutus 338
Paramesacanthion 230

Paramesonchium 211–12, 319
Paramicrolaimidae 157, 167, 186, 187, 290, 301, 305
Paramicrolaimus 157, 187, 305, 319
P. primus 334, 351
P. spirulifera **186**
Paramonohystera 201
P. elliptica 51, 161
Paramphidelus 279
Paranticoma 235–6, 276, 354
P. caledoniensis 235
P. tubuliphora 72, 77, 235
Paraphanoderma 234, 305
Paraphanolaimus 167, 169, 348
P. anisitsi 167
Paraphelenchidae 357
paraphyletic taxa 22
paraphyly 22
  definition of 23
Parapinnanema 139
Paraplectonema 167, 169, 348
Paraplectus 169
Parapomponema 145, 303
Pararaeolaimus 215
P. nudus 214, 318, 330
Parasabatieria 212
Parasaveljevia 231
Parascolaimus 209
Paraseuratiella 145
Parasphaerolaiminae 202, **292**
Parasphaerolaimus 74, 203, 350
P. paradoxus 37, 332
Parasymplocostoma [see *Symplocostoma*] 258
Paratarvaia 213, 215–16, 304–5, 318, 349
Parateschellingia 181, 204, 305, 350
P. brevicaudatus 358
P. fusiforme 358
Parathalassoalaimus 153
Paratrichodorus 279
Paratricoma 165
Paratrilobus 264
Paratripyla 264–6, 353
Paratripyloides 261
Paratuerkiana 241
Parenoplus 231
Pareuchromadora 139
Pareudesmoscolex 163
P. laciniosus 328
Pareurystomina 55, 108, 247, 254–6, 258
Pareurystomininae 254, 257, **296**
Parironus 93, 109, 237, 239, 321
P. bicuspis 38, 71, 75, 109, 236–7, 339, 353
Parodontophora 209
P. quadristicha 329
Paroncholaimus [see also *Pontonema*] 254
Paronchulus 274
Paroxystomina 244
Paroxystomininae 72, 105, 109, 242, 245, **296**, 321, 354
parthenogenetic reproduction 142, 153, 172, 174, 191, 194
peduncles **160**–162
Pelagonema 250, 342, 359
P. omalum 359
Pelagonematidae 248
Pelagonematinae 73, 246, 248, 250, **296**, 303, 342, 355
Pelagonematoidea 248–9
Pelagonemella 250
Pendulumia 269
Penetrantia 122
Penzancia see *Theristus*
Pepsonema 212
Perepsilonema 153–4
P. crassum 335
P. papulosum 37
Peresiana 175–6
Peresianidae 133, 166–7, 169, 175–177, 197, **290**, 301–02
Peresianinae 81, 93, 96, 98–9, 103, 168, 318, 327, 347
Perilinhomoeus 206
Perioplectus 173
Periplectus 173
Perspiria 151
Phaenoncholaimus 250
Phanoderma 232, 234, 305
P. campbelli 72, 77, 117, 233, 341
P. (P.) etha 359
P. necta 233
Phanodermatidae 38, 57, 60, 72, 75–8, 93, 102–03, 107–09, 115–18, 223, 225–6, 232–3, 235, 239–40, **296**, 301, 305, 322, 341, 354
Phanodermatina 233
Phanodermatinae 233–4, **296**, 305
Phanodermella 234, 305
P. longicaudata 359
Phanodermina 233
Phanodermopsis 233–4, 305
P. necta 38, 72, 116, 305, 341
Phanoglene 281
pharyngeal bulb 141–4, 149, 153, 155–8, 161, 166–7, 169, 171–3, 176, 180, 185, 192, 194–5, 216, 237, 256

pharyngeal glands 14, 30, 40, 161, 182, 185, 191–2, 194, 203, 218–19, 221, 224, 228, 230–33, 235–7, 239, 242, 246, 264, 266–7, 269–71, 273, 275–6, 279–80
pharynx, structure of 99, 121, 127, 133, 136, 140, 146, 151, 165, 170, 172, 174–9, 181–2, 187–9, 191–4, 198, 202–04, 207, 210, 213, 215, 216, 222–6, 235, 237, 239, 246, 248, 250, 255–6, 259, 266, 272, 274–5, 277, 279, 283
phasmata 111, 113, 161, 300
Phasmidia 113, 120
phasmids
 occurence of 14, 110, 114, 124–5, 174, 300
 phylogenetic significance of 113–14, 123, 281
 structure of 14, 40, 110–11
Pheronus 239, 321
Phyllolaimus 146
phylogenesis 21, 23
phylogenetic
 relationships 13, 15, 23, 299
 systematics 19, 289
 systems 23, 119, 289, 299
 hierarchical structure of 22
 definition of 23
phylogeny 23
pigment particles see ocelli
Pierrickia 210, 212, 320
Pilosinema 240
Platycoma 57, 239–40
 P. sudafricana 72, 239, 341
Platycominae 235, 240, **296**, 303
Platycompsis 239–40
 P. dimorphus 57
Plectida 123
Plectidae 29, 37–8, 42, 51, 65, 81, 93, 103–04, 114, 120–1, 127, 132–3, 166–8, 170, 171–2, 208, 279, **290**, 318, 326, 347
Plectinae 38, 103, 105, 173, **290**, 326, 347
Plectoides 173
Plectolaimus 168–9, 348
Plectus 60–62, 70, 169, 172–3, 204, 326
 P. acuminatus 347
 P. cirratus 37, 39, 326
 P. parietinus 121, **122**
 P. parvus 121, **122**
plesiomorphous characters 20
Polydontus [see also *Octonchus*] 254
Polygastrophora 254–6, 258

P. omercooperi 256
P. quinquebulba 256
Polylaimium [see also *Leptolaimus*] 168–9, 213, 303, 305, 349
polyphyletic taxa 22
polyphyly 22–3
Polysigma 103, 151
Pomponema 34, 44, 88, 144–5, 303
 P. ammophilum 337
 P. astrodes 337
 P. compactum 337, 352
 P. debile 337
 P. elegans 337
 P. hastatum 145
 P. litorium 87, 352
 P. macrospirale 145
 P. multipapillatum **87**, 133, 135, 144, 337, 352
 P. syltense 144, 352
Pomponematinae 144–5, **288**, 302, 320, 337, 352
Pontonema 250, 304
 P. ardens 30, 246, 322
 P. balticum 359
 P. ditlevsensi **36**, 37, 60, 73, 253, 343
 P. filicaudatum 306
 P. parocellatum 37
 P. teissieri 306
Pontonematinae 73, 246, 248, 250, 253–4, **296**, 343, 355
Porocoma 75, 84, 96–7, 115, 243, 275–6, 305, 321
 P. striata 354
Potamonema 281
Praeacanthonchus 103, 146
 P. aff. *kreisi* 338
Pratylenchus 74
 P. macrophallus 344
 P. pratensis 344
pre-cloacal organs 168
Prionchulus 74, 343
Prismatolaimidae 71, 76, 101, 105, 108–09, 114–15, 127, 167, 191–193, 273–4, **290**, 321, 339, 353
Prismatolaiminae 34, 55, 57, 63–6, 71, 93, 101, 133, 219, 321, 339, 353
Prismatolaimus 65–6, 191, 339
 P. dolichurus **54**, 71, 339
 P. intermedius 105, 191
 P. aff. *verrucosus* 101, 339
Pristionema 182, 184, 306
Procamacolaiminae 93, 103, 169–70, 302, 318, 330, 348
Procamacolaimus 70, 170, 330

P. tubifer 104
Prochaetosoma 154, 156
Prochaetosomatidae 155, 301
Prochaetosomatinae 155, 156, **288**, 302
Prochromadora 138
Prochromadorella 138
  P. attenatea 336
  P. paramucrodonta 336
Prodesmodora 92, 95, 149, 152–3, 157, 304, 351
  P. circulata 334
Prodesmodorinae 152, 153, 157, **288**, 301, 304
Prodesmoscolex 163
Progreeffiella 164
Proleptonchus 356
Prolinhomoeus 206
Promonhystera 201
Promononchus 266
Prooncholaimus 253
Proplatycoma 240
Propomponema 145, 303
  P. foeticola 145
Prosphaerolaimus 206
Proteroplectus 173
Protochromadora 139
Protodesmoscolex 163
Prototricoma 166
Protorhabditis 344
Protricoma 165
Protricomoides 163
Pselionema 179, 181, 183, 303
  P. annulatum 183
  P. beauforti 183
  P. deconincki 183
  P. detriticola 183
  P. dissimile 182–3
  P. longiseta 183
  P. longissimum 183
  P. minutum 183
  P. parasimplex 183
  P. richardi 183
  P. rigidum 183
  P. simile 183
  P. simplex 183
Pselionematinae 181, **290**, 302, 319
Pselionemoides 183, 303
Pseudaraeolaimus 215, 349
Pseudoaulolaimus 177–8
  P. anchilocaudatus 177
Pseudobathylaimus 169
Pseudocella 241
Pseudochromadora 150
Pseudodesmodora 150

Pseudodilaimus 281, 306
Pseudodiplogasteroididae 284, 357
Pseudolella 53, 209
Pseudolelloides 209
Pseudometachromadora 151
Pseudomicrolaimus 57
Pseudonchinae 93, 151, **288**, 319, 334, 351
Pseudoncholaimus 253
Pseudonchulus 274
Pseudonchus 52, 151, 334
Pseudoparoncholaimus [see also *Filoncholaimus*] 254
Pseudopelagonema 250, 359
Pseudorhabdolaimus 280–1, 306
Pseudosteineria 201
Pseudotheristus 201
Pteronium 148
Pterygonema 181, 184
  P. alatum 184
  P. cambriense 184
  P. ornatum 184
Ptycholaimellus 140
Puccinellia 25
Pulchranemella 202
Punctodora 138
  P. exochopora 276
  P. ratzeburgensis 336
Pungentus 356
Pyenolaimus 173

Quadricoma 166
Quadricomoides 166
Quasibrilinae 262, **294**, 303
Quasibrilus 264, 304
  Q. kurikania 263

Rahmium 278
Raptothyreidae 109
Raracanthonchinae 145
rectal glands 40
Retrotheristus 55, 201
Rhadinematidae 167
Rhabditea 123
Rhabditia 123
Rhabditida 35, 38, 74, 81–3, 85–6, 91–2, 98, 102, 171, 174, 211, 280–3, 284, **286**, 322, 356
Rhabditidae 38, 42, 65, 82, 93, 120–1, 283–4, 343, 357
Rhabditis 42, 46, 48, 60, 62, 74, 121, 283–4
  R. elegans 35, 38, 40

R. marina **82**, 344
Rhabditoidea 81, 98, 172, 174, 208, 282–4, **286**, 322
Rhabdocoma **31**, 34, 38, 55, 71, 271–2, 321, 340, 353
R. americana 38, 55, 271, 340, 353
Rhabdodemania 55, 268–9, 304
R. brevicaudata 268, 306
R. minor 38, **56**, 73, 109, 267, 342, 359
Rhabdodemaniidae 38, 55, 67, 73, 75–6, 108–9, 115, 259, 267–70, **294**, 306, 322, 342, 354
Rhabdogaster 154
Rhabdolaimidae 51, 53, 81, 84, 86, 92, 98–100, 126, 133, 166–7, 194–197, 237, 238, 263, 266, **290**, 302, 305, 318, 326, 347
Rhabdolaimus 92, 100, 194–6
R. minor 326, 347
R. nannus 326, 347
R. terrestris 194, 326, 347
Rhabdotoderma 281
Rhadinema 52, 92, 168, 178, 305, 348
R. flexile 104, 327
Rhadinematidae 126, 178, **290**, 301, 305
Rhaptothyreidae 55, 73, 76, 84, 108–09, 115, 322, 355
Rhinema 159, 351
Rhinonema 281
Rhinoplostoma [see *Calyptronema*] 258
Rhips 139, 320
Rhynchonema 200–01, 332
Rhynchonematinae 200, **292**
Richtersia 44, 53, 74, 146–8
R. inaequalis 335
Richtersiella 148
Richtersiidae 30, 33, 44, 53, 86, 90, 92, 136, 146–7, **288**, 302, 319, 335, 351
Richtersiinae 134
Ritenbenkia 241
Robbea 151, 180
R. tenax 104
Rogerinae 194, **290**, 302
Rogerus 194, 196, 347
R. orientalis 318
R. rajasthanensis 195
Rondonia 274
rotifers, gonads of 96–7
Ruamowhitia 304
Ruamowhitia halophila [see also *Enoplus schulzi*] 225, 304

Ruamowhitia orae [see also *Enoplus schulzi*] 225, 304, 354

Sabatieria 74, 210, 212
S. celtica 37, 74, 209, 336
S. hilarula 210, 336
S. supplicans 320
Sabatieriinae 212, **292**, 319, 336
Sadkonavis 241
Sarsonia 206, 323, 350
Saveljevia 73, 116, 231, 341
Scaptrella 201
Scaptrellidae 200, **292**
scapulus, of a metaneme 68, 226
Schistodera 243
Scholpaniella 212
Scolelepsis squamata 200
Secernentea 29, 33, 35, 40, 42–3, 47–8, 57, 59, 60, 62–5, 80–1, 83, 85–6, 89, 91, 95–6, 98, 105–07, 111, 113–14, 120–1, 124, **125**–6, 133, 172, 174, 216, 218, 273, 281–4, **286**, 317
Secernentia 122
Selachinema 148
Selachinematidae 30, 34, 44, 47, 92, 103, 134–6, 146–8, **288**, 302, 320, 338, 352
sense organs
  lateral see dereids
  papilliform see sensilla, cephalic
  setiform see sensilla, labial
sensilla
  cephalic 14, 29, 30–1, 33, 35, 37, 39, 41–7, 54–7, 59, 62–4, 98–9, 100, 126, 129, 133, 135–8, 140–44, 147, 149, 157–9, 161–2, 166–7, 169, 172–3, 176–8, 180–82, 185, 187, 194, 196, 199, 205, 207–08, 210, 216, 221, 236–9, 242–3, 246, 260, 263, 265, 270, 276, 284, 300, 317–323
  jointed 34, 42, 146, 221, 260, 265, 273, 300, 318–323
  labial 14, 29, 30–1, 33–5, 39, 41, 43–5, 98–9, 100, 129, 135–136, 144, 147, 149, 157, 159, 161, 166, 173, 176–7, 181, 186, 189, 191–2, 196, 198, 205, 207, 210, 221–2, 224–6, 231–2, 239, 242, 244, 246, 260, 262, 265, 268, 271, 273–6, 300, 317–323
  postembryonic development 35, 37
  papilliform 43, 45, 189, 191, 194, 198, 207, 221, 224, 226, 232, 236, 239, 243, 246, 262, 265, 268, 274,

276, 318–323
setiform 32–3, 43, 135, 167, 173,
   194, 198, 221, 225–6, 239, 246, 265,
   270–71, 273, 276, 318–323
sensilla circles 30–6, 39, 41–7, 54–6,
   62–3, 124, 127, 133, 135–136, 140,
   142–4, 146, 149, 157–9, 161, 172,
   179–81, 185, 187, 191, 194, 196,
   198, 200, 202, 204–05, 207–08, 210,
   213, 215–16, 232, 236, 239, 242,
   246, 263, 265, 268–75, 318–323
setae 14, 164
   adhesive 155, 163
   cephalic 30, 32, 34–5, 41, 56, 99,
      124, 161, 171, 175, 179, 185, 187,
      189, 191–3, 200, 202, 204, 210, 213,
      215, 232, 244, 262, **265**–6, 268–76,
      282
   lateral 34, 38, 40, 271
   rigid 153
   somatic 32, 39, 42, 163, 175
   stilt 155
Setoplectus 177, 318
Setosabatieria 211
Seuratiella see Acanthoncus
Sigmophora 151
Sigmophoranema 103, 151
   S. rufum 334
Simpliconema 271
Simpliconematidae 271, **294**
Sinanema 194, 196, 321, 353
Siphonolaimidae 37, 43, 53, 81, 84, 96,
   181, 197, 198, 203, 204, **290**, 305,
   328, 350
Siphonolaimoidea 128, 161, 185, 197,
   203–04, 219, **292**
Siphonolaimus sp. 74, 204, 350
   S. cobbi 37, 44, 74, 328
Sitadevinema 200
Smalsundia 148
socles, associated with setae 175–6
Solenolaimus 204
Southernia see Cyartonema
Southerniella 214–15, 304
   S. arctica 214
   S. conicauda 214
   S. cylindricauda 214
   S. zosterae 214
Spartina 25
Sphaerocephalum 205–06
   S. chabaudi 35, 37, 60
   S. crassicauda 350
Sphaerolaimidae 37, 46, 92, 96, 108,
   126, 197–8, 202–03, **292**, 332, 350
Sphaerolaiminae 202, **292**

Sphaerolaimus 74, 202
   S. gracilis 37, 332
   S. hadalis 350
   S. peruanus 350
Sphaerotheristus 201
Sphagnum 27
spicular apparatus of cloaca 14, 79,
   128, 141, 170, 188, **190**, 274, 277
Spiliphera 137–8
Spilipherinae 136–138, 143, **288**, 302,
   320, 336
Spilophora 138
Spilophorella papillata 320, 337
   S. paradoxa 337
Spilophorium 138
Spira 151
Spiramphinema 201
Spirina 151
Spirinia 74, 151
   S. laevis 74, 334
Spiriniinae 150–51, **288**, 304, 319, 334,
   351
Spirophorella 140
Spirotheristus 201
Squanema 151
Statenia 138
Steineria 183, 200–01, 252
   S. ericia 37, 332
   S. pilosa 37, 101, 200, 332
Steineridora 139, 320
Steineriella [see also Viscosia] 252
Steiner's organs 57–8, 232
Stenolaimus 236
Stenonchulus 273–4
   S. troglodytes 39, 71, 339
Stephanium 278
Stephanolaimus 51, 70, 102, 166–7,
   169, 318, 348
   S. elegans 70, 104, 327
   S. flevensis 330
   S. paraflevensis 330
   S. spartinae 330
Stilbonema 151
Stilbonematinae 134, 151, 187, **288**,
   319, 334, 351
stomatostyle 282
stretch receptors see metanemes
Striatodora 215, 349
Strongylida 74
stylet 99
Stylotheristus 201
Subsphaerolaimus 202
   S. litoralis 37, 332
supplements, pre-anal 256, 262, 267,
   272–3, 275

post-anal 272
symplesiomorphy 20-2, 299
  definition of 20
Symplocostoma 255, 258, 268
  S. tenuicolle 355, 359
Symplocostomatinae 254
Symplocostomella 258
synapomorphy 16, 19-22, 299
  definition of 20
Synodontium 86, 208-09, 349
  S. monhystera see *Odontophoroides monhystera*
Synodontoides see *Apodontium*
  S. procerus 329
Synonchiella 74, 146, 148
  S. major 358
  S. minor 358
  S. riemanni 338
Synonchinae 72, 241, 296
Synonchium 148
Synonchoides 241
Synonchus 239, 241
  S. longisetosus 72, 239, 359
Synonema 74, 159
  S. ochrum 58
Synonemoides 159
Syringolaimus 76, 92, 133, 195, 216, 236-9, 305, 318, 347
  S. aff. *brevicaudatus* 358
  S. striatocaudatus 111, 326
Syringonomus 239-40
systematics, theory of 13, 299
sytematic criteria 14, 80, 123, 130, 287

Tachyhodites 199
tail, shape of 14, 141, 167-8, 178, 185, 198, 202-03, 208, 210, 213-14, 250, 264
Takakia 191
Tarvaia sp. **52**, 67, 179, 213, 240, 349
  T. donsi 179
Tarvaiidae 167, 179, 213, **290**, 301
teeth 47, 127, 133-4, 136-7, 140-44, 146-8, 153, 155-7, 159, 166, 170-71, 173, 185-6, 191-2, 194-6, 207, 210, 233, 236-7, 239, 242, 254, 260, 262-4, **265**-7, 270, 273, 277
Teratocephalus 93, 170, 173-5, 326, 347
Teratocephalidae 81, 86, 92-3, 103, 106-07, 133, 166-7, 170-71, 173-4, **290**, 318, 326, 347
Teratocephalinae 175, **290**
Terschellingia 206

T. distalamphida 328
T. longicaudata 330, 358
T. monhystera 350
T. parva 358
Terschellingiinae 206, 302
Terschellingioides 206
  T. filiformis 350
testes (testis) 65, 80, 86-90, 94-7, 99, 104, 127, 133, 135-7, 139-42, 144, 146, 148, 151-3, 156-7, 159, 161, 163-7, 171-3, 175-80, 182, 185, 187, 189, 191, 196, 198, 200, 202, 204-05, 207-08, 210, 213, 215, 221, 224, 226, 231-2, 235, 237, 239, 242, 246, 259, 263, 265, 267-9, 271-6, 300, 324
Thalassironidae 238
Thalassironinae 238-9, **296**, 303, 305
Thalassironus 109, 236-7, 239, 321
  T. jungi 236
Thalassoalaimus 242-3, 321
  T. aquaedulcis 178, 243, 305
  T. egregius 340
  T. oxycauda 358
  T. pirum **32**, 38-9, 109, 340
  T. septentrionalis 72, 340
  T. tardus 38, 72, 109, 340
Thalassogenus 250, 280, 305
Theristus 74, 201
  T. aculeatus 44
  T. polychaetophilus 127, 200
  T. wegelinae 332
Thoonchinae 57, 64, 66, 73, 254, 257, **296**, 303, 343
Thooncus 255, 257-8
Thoracostoma 239, 241
  T. trachygaster 244
Thoracostomatinae 72, 241, **296**, 341
Thoracostomopsidae 225, 226-7, 229, 233, 268, **296**, 305
Thoracostomopsinae 73, 109, 228, **296**, 305, 322, 341
Thoracostomopsis 225, 228
  T. barbata 73, 109, **117**, **227**, 341
Thornenema 356
Timmia 138
Tobrilia 75, 79, 133, 194-6, 219, 263, 321
  T. longicaudata 353
Tobrilidae 195, 259, 261-4, 266, 278, **294**, 303
Tobriliinae 194-5, **290**, 302, 305
Tobrilinae 34, 37, 60, 71, 75, 79, 109, 133, 195, 219, 266, 321, 339, 353
Tobriloides 263-4, 321

Tobriloidinae 262, **294**, 303
Tobrilus 14, **48**, 128, 264
T. gracilis 71, 339
T. grandipapillatus **48**, **262**, 339
T. longus 37, 41
Torquentia 122
Trefusia 190, 271–2, 340
T. helgolandica 38, 71
T. litoralis 38
T. cf. *longicauda* **32**, 38, 71, 340
T. varians 57
Trefusialaimus 270–72, 321, 353
Trefusiida 149, 182, 216–17, 219, **220**, 223, 243, 270–71, 276, **294**, 301
Trefusiidae 30–2, 34, 38, 40, 42, 44, 49, 55–7, 64, 66, 71, 76, 81, 84, 93, 98–9, 105, 108, 115, 216, 219, 243, 270, 271–2, **294**, 300, 305, 321, 340, 353
Trefusiinae 71, 271, 272, **294**, 340, 353
Triaulolaimus 241
Triceratonema 241
Trichenoplus see *Fenestrolaimus*
Trichethmolaimus 140
Trichocephalida 123
Trichoderma 164
Trichodorida 105
Trichodoridae 274, 277, 279
Trichodorus 105, 279
T. monhystera 356
Trichotheristus 201
Trichromadora 138
Trichromadorita 138
Trichuroidea 95, 108, 113–14, 216
Tricoma 74, 111, 164, 166, 328
T. similis 162
T. spuria 358
Tricominae 33, 99, 111, 133, 161–165, **290**, 305, 328, 349
Tridentella 138
Tridentellia 138
Tridontolaimus 141
Trigonolaimus 209
Trileptiinae 73, 109, 228, **296**, 305, 322, 341
Trileptium 73, 109, 116, **117**, 228, **229**, 341
T. salvadoriense 103
Trilobidae 217
Trilobinae 217, 261
Trilobus 261, 264
Triodontolaimidae 57, 72, 75–6, 79, 89, 93, 109, 115, 259, 262, 266–7, **294**, 322, 341, 354
Triodontolaimus 267

T. acutus 72, 267, 341
Triplonchium 279
Tripyla 68, 105, 266, 353
T. glomerans 71, 339
T. setifera **265**
Tripylidae 56, 71, 75–6, 78, 89, 93, 105, 114–15, 194–5, 217, 238, 259, 263, 264, **265**–6, 274, **294**, 305, 321, 339, 353
Tripylina 217, 266
Tripylinae 34, 71, 75, 79, 109, 266, 321, 339, 353
Tripylium 105, 199–200
T. carcinicola 37, 332
Tripyloidae 109
Tripyloidea 130, 217, 250
Tripyloides 260–61
T. acherusius 260–61
T. amazonicus 260
T. marinus 91, 339
Tripyloididae 14, 30, 34, 44, 49, 55–7, 64, 71, 75–6, 78, 86, 89, 93, 95, 115, 127, 129–30, 134, 136, 216–17, 259–**260**–261, **294**, 305, 321, 339, 353
Tripyloidina 217, 221–223, 231, 259, 262, 264, 266, 268, **294**
Tripyloidinae 217
Trischistoma 265–6, 321, 353
Trissonchulus 236–7, 238, 321
T. acutus 71, 236, 339
T. oceanus 236–37, 353
T. raskii 237, 353
T. reversus 236
Tristicochaeta 155
Trochamus 138–9, 304
Trogolaimus 148
Tubolaimella 185, 235, 240
Tubolaimoides 74, 81, **184**–186, 205, 219, 305, 331, 350
T. bullatus 185
Tubolaimoididae 126, 161, 167, 182, 184–186, 205, 213, 219, **290**, 301, 305
Tubulaimus 169, 201
tubules, copulatory see copulatory organs
  pre-cloacal/pre-anal 99, 103, 133, 137, 144, 146, 149, 166–7, 169–172, 174, 178, 196, 223–4, 232–3, 235, 239, 300
Tubuligula 181, 204
Tuerkiana 241
Tycnodora 244
Tylenchida 74, 81, 83, 85, 89, 92, 94,

96, 98–9, 101, 161, 281–4, <u>285</u>, **286**,
  322, 344–5, 357
Tylenchidae 281, 357
Tylenchoidea 92, 285, 344, 357
Tylencholaimellus 356
Tylencholaimidae 105
Tylencholaiminae 217
Tylencholaimus 356
  T. zeelandicus 356
Tylenchorhynchus 74, 344
Tylenchulidae 357
Tyleptus 356
Tylocephalus 173
Tylolaimophorus 279
Tylopharynginae 356

Udonchus 194, 196, 353
Ungulilaimella 158
Ungulilaimus 213
Usarpnema 164–6

Valvaelaimus 201
valvular apparatus, of the pharynx 121,
  122, 166–7, 171–4, 176, 283
Vanderlinida 105
Valvaelaimus 74
  V. maior 37
Vasculonema 250
Vasostoma 212
ventral gland see cervical gland
vestibulum, structure of 14, 99, 124,
  129, 134–5, 143, 156–7, 159, 174,
  178, 182, 188–9, 194–5
Viscosia 247, 249, 251–2, 304
  V. bandaensilis 251, 301
  V. bandaensis [see also
    V. bandaensilis] 251, 301
  V. brevidendata 251
  V. cobbi 342
  V. conicaudata 251
  V. diodon 251
  V. elegans 251
  V. filiformis 251
  V. gabriolae 251
  V. glaberoides 251
  V. keiensilis 252, 301
  V. keiensis [see also *V. keiensilis*]
    252, 301
  V. klatti 252
  V. longicaudata 252
  V. longidentata 252
  V. norvegica 252
  V. papillata 252, 301

V. papillatula 252, 301
V. parasetosa 252
V. profunda 252
V. rustica 73, 252, 342
V. separabilis 252
V. setosa 252
V. tasmaniensis 252
V. viscosa [see also *V. viscosula*]
  251
V. viscosula 252, 301
vulva 79, 83–4, 96–7, 124, 150, 274–5

Walcherenia 171
Wieseria 243, 321, 358
  W. hispida 44
  W. inaequalis 44
  W. pica 44
Wieserius 201
Wiesoncholaimus 246, 253
Wilsereptus 173
Wilsonema 173, 326
  W. otophorum 347
Wilsonematidae 38
Wilsonematinae 86, 103, 172, <u>173</u>,
  290, 326, 347
Wilsotylus 173

Xanthodora 150
Xenella 74, 133, 275–6
  X. suecica 335
Xenellidae 51, 86, 93, 102, 149, 182,
  243, <u>275</u>–6, **294**, 305, 319, 335, 351
Xenocyatholaiminae <u>145</u>, **288**, 320,
  352
Xenocyatholaimus 145
Xenodesmodora 150
Xenolaimus 201
Xenonema 153
Xinema 212
Xiphinema 356
Xyala 74, 201
Xyalidae 31, 33–4, 37, 41, 44–6, 51,
  55, 91, 93, 96, 101, 108, 114, 127,
  161–2, 197–<u>200</u>, 202, 205, 210, 222,
  262, 271, **292**, 332–3, 350
Xyzzors 146

Ypsilon 170

Zalonema 150
Zanema 206
Zygonemella 201